CONTINUUM MECHANICS

CONTINUUM MECHANICS

The Birthplace of Mathematical Models

Myron B. Allen
Department of Mathematics
University of Wyoming
Laramie, WY

Published by John Wiley & Sons, Inc., Hoboken, New Jersey
Published simultaneously in Canada

Library of Congress Cataloging-in-Publication Data:

Allen, Myron B., 1954-
  Continuum mechanics : the birthplace of mathematical models / Myron B. Allen.
    pages cm
  Includes bibliographical references and index.
  ISBN 978-1-118-90937-9 (cloth)
  1. Continuum mechanics–Textbooks. I. Title.
  QA808.2.A45 2015
  531.076–dc23

2015016165

Printed in the United States of America

10 9 8 7 6 5 4 3 2 1

To Adele

PREFACE

American universities often relegate continuum mechanics to advanced graduate courses, mainly in engineering. In this setting, the subject remains obscure to other important audiences. I hope that this semester-length book will help to introduce the rudiments of continuum mechanics to first-year graduate students in engineering, physics, and geophysics, as well as to advanced mathematics undergraduates. In all of these fields, continuum mechanics lies at the foundation of prominent mathematical models. And in my experience the concepts are accessible to these students.

The title deserves a comment. This is a textbook on continuum mechanics. It includes derivations of commonly used mathematical models based on continuum mechanical concepts, to illustrate the philosophy that good mathematical models come from rigorous foundations. I distinguish this approach from that employed in many mathematical modeling textbooks, where the range of models may also include molecular, biological, financial, and stochastic systems. In those settings, methodologies beyond those normally associated with continuum physics often appear; see [23], [24], and [42] for examples. While I applaud trend toward teaching mathematical modeling in this broader sense, this book focuses on continua.

Anyone who teaches a first-level course in continuum mechanics must choose among several stylistic and conceptual approaches. I've tried to adopt an approach that meshes with undergraduate courses on vector calculus and linear algebra in the following respects:

- Wherever possible, I use direct notation instead of index notation. From a pedagogic point of view, I think this choice is easier on newcomers. It also minimizes reliance on particular coordinate systems. That said, index notation appears in several places, and I don't shy away from using Cartesian coordinates for examples and problems.

- I use infinitesimals sparingly. When they appear, it is largely for heuristic purposes. Most undergraduate mathematics courses emphasize parametrizations, Taylor approximations, and limits in lieu of infinitesimals.

- Students increasingly encounter partial differential equations (PDEs)—Nature's mother tongue—early in their college careers. I try to show how many of the most commonly encountered PDEs arise from continuum mechanical principles. In places, the book introduces qualitative attributes of these equations that I believe promote physical understanding but that students typically encounter in other courses.

The order of topics in continuum mechanics varies widely according to authors' tastes. To me it seems natural to start with the geometric, algebraic, and analytic foundations—especially the facts about vectors and tensors that the standard undergraduate curriculum does not reliably cover. Chapter 1 reviews this material. With this foundation in place, Chapters 2 and 3 cover topics in kinematics, the descriptive language for continuum motions. Chapter 4 covers balance laws, from which arise many of the governing equations of interest in applied mathematics and physical science. Chapter 5 introduces constitutive laws through a short tour of common mathematical models encountered in heat flow, fluid dynamics, and solid mechanics. This chapter includes an optional section on potential theory. Chapter 6 views constitutive laws from a more theoretical vantage point, outlining an elementary treatment of constitutive axioms, in response to the laudable trend toward including this material in introductory texts. This discussion of constitutive theory really amounts to a first glimpse, illustrating a handful of concepts and techniques for a particular type of material, thermoviscous fluids.

In Chapter 7, the book attempts something unusual for texts at this level, introducing multiconstituent continua. I base the treatment on mixture theory, which has limitations, unsettled issues, and much more depth than a single chapter can probe. Nevertheless, Chapter 7 provides a streamlined way to introduce diffusion, a topic of immense importance in environmental science and engineering, and it enables rational derivations of the field equations most commonly used to model flows in porous media.

Even a book this slender contains more material than many instructors will try to cover in a semester. It may help to know that one can skip the following sections without missing essential material used later:

- Section 2.3: Pathlines, Streamlines, and Streaklines.
- Section 3.4: Vorticity and Circulation (if one skips Section 5.3).
- Section 4.6: Jump Conditions.
- Any sections in Chapter 5; they are largely independent of each other.
- Chapter 7; however, later sections in this chapter depend on earlier ones.

One can also consider skipping Chapter 1, if the audience is unusually well prepared.

The book has at least three quirks.

1. Although more general settings exist, for a course at this level I restrict attention to the three-dimensional Euclidean space of geometric vectors. I refer to this space as $\mathbb{E}$, reserving the notation $\mathbb{R}^3$ for the space of ordered triples of real numbers. The difference may seem fussy to some, but I want to distinguish 3-vectors from the coordinates used to represent them. In examples and exercises, where concrete numerical representations of vectors are indispensible, I typically fix a basis and then use $\mathbb{R}^3$ freely.

2. For examples and exercises, I stick to Cartesian coordinates. Curvilinear and even nonorthogonal coordinate systems arise frequently in applications. But from a modeler's perspective continuum mechanics serves primarily as a framework for deriving governing equations. At this level direct notation, with occasional reference to Cartesian coordinates, seems pedagogically sufficient.

3. The treatment of PDEs is far from systematic. That subject deserves courses of its own. But I have selected from the theory of PDEs a small number of largely qualitative topics, in an effort to promote intuition about the physics. I hope these topics will either refresh material learned in previous courses on PDEs or prompt students to take such courses.

Interspersed throughout the text are over 200 exercises. Most ask the reader to fill in logical connections that the narrative elides. Some provide glimpses of related topics. I don't intend for these exercises to be difficult, but in case my judgment is flawed I've included an appendix with hints and solutions.

I owe thanks to several people at John Wiley & Sons and their affiliates for guidance during the production of the book. In particular, I'm grateful to Susanne Steitz-Filler, Sari Friedman, Katrina Maceda, and Lincy Priya for their expertise and for the time they spent on the manuscript after it was under contract.

Figure 3.9, adapted from [50, Figure 111, p. 341], appears with kind permission of Springer Science+Business Media.

Anyone who studies continuum mechanics owes a tremendous debt to the giants. I learned a great deal from the work of Ray Bowen, Bernard Coleman, Jerald Ericksen, A. Cemal Eringen, A.E. Green, Morton Gurtin, Paul Naghdi, Walter Noll, Ronald Rivlin, and Clifford Truesdell, among many other masters of the field.

I also owe to my teachers George F. Pinder and William G. Gray the conviction that these concepts lie at the proper foundation of any efforts at mathematical modeling. I'm indebted to Aleksey Telyakovskiy, Li Wu, and Stefan Heinz for reviewing the manuscript. I owe many thanks to several students at the University of Wyoming who endured preliminary notes and early drafts of this material. They asked many questions that helped me correct errors, and they challenged me to develop clearer explanations. Special thanks go to Stephen Bagley, Yi-Hung Kuo, Kevin Lenth, and Xiaoban Wu, who, in addition, developed many hints for the exercises.

Finally, I owe an enormous debt of gratitude to my colleagues at the University of Wyoming, who helped shape my views of this subject over many years and convinced me that it deserves more attention. Special among them is my friend and colleague, Andy Hansen, for more improvements, insight, and encouragement than he knows.

MYRON B. ALLEN

Laramie, Wyoming
May 2015

CONTENTS

CHAPTER 1

GEOMETRIC SETTING

Physicists recognize two branches of mechanics: classical and quantum. For the past century, the quantum view, emphasizing the corpuscular nature of matter at atomic and finer scales, has played a dominant role in most universities' physics curricula. In those settings, classical mechanics enjoys a distinguished mathematical pedigree, being based on ideas developed by Galileo Galilei, Johannes Kepler, Isaac Newton, Leonhard Euler, Joseph Louis Lagrange, William Rowan Hamilton, and others. Nevertheless, many academic physics departments treat classical mechanics as a mathematical training ground for undergraduates preparing to study the principles of subatomic particles and quantum fields, subjects commonly regarded as more fundamental.

Applied mathematicians and engineers tend to view classical mechanics from a different perspective. While the most elementary formulations of Newton's laws and the Lagrangian and Hamiltonian formalisms focus on idealized particles with mass, many natural phenomena appear to macroscopic observers—those whose scales of observation are significantly larger than 10^{-10} meters—as continuous in space and

Continuum Mechanics: The Birthplace of Mathematical Models, First Edition. M.B. Allen.

time. For these phenomena, fruitful mathematical descriptions typically arise from extensions of classical mechanics pioneered, most notably, by Leonhard Euler and Augustin-Louis Cauchy and refined during the last half of the twentieth century by a large community of scientists, some of whom are mentioned in the preface. In these extensions, matter appears to be continuous, in a sense to be made more precise in the next chapter.

Continuum mechanics embodies these extensions, furnishing useful mathematical models of fluids, elastic solids, and viscoelastic materials. These models describe phenomena that we see and feel in our everyday interactions with the world: rocks in the Earth's crust, water on and beneath its surface, weather, the structures that humans build, and the biological tissues that we occupy. The models typically take the form of partial differential equations describing rates of change with respect to spatial position and time. Advances in our ability to understand and solve these types of equations—especially using high-performance computers—have made continuum mechanics one of the most powerful tools in applied mathematics and engineering. For this reason, in developing the elements of the subject, this book frequently draws connections between its core concepts and the qualitative theory of partial differential equations.

1.1 VECTORS AND EUCLIDEAN POINT SPACE

From a mathematical perspective, continuum mechanics has roots in geometry. In the most natural geometric setting, basic principles do not depend on any observer's particular frame of reference or choice of coordinate systems. One aim of this book is to develop the rudiments of continuum mechanics in a manner that minimizes reliance on particular coordinates, recognizing that using these concepts in specific problems often requires the adoption of a well-chosen coordinate system.

The geometric setting here is relatively simple, relying on ideas familiar to anyone who has studied multivariable calculus and linear algebra. For a more sophisticated approach, refer to [38].

1.1.1 Vectors

Fundamental to continuum mechanics is the three-dimensional Euclidean vector space over the field $\mathbb{R}$ of real numbers. This space, which we denote as $\mathbb{E}$, has features that do not depend on any system for assigning numbers to the vectors in it. In particular, $\mathbb{E}$ has three attributes beyond those common to all vector spaces.

Although the attributes are elementary, it is useful to review them in coordinate-free language and to show how coordinates arise.

• **Inner product.** $\mathbb{E}$ possesses an **inner product**, that is, a binary operation that maps each pair of vectors $\mathbf{x}, \mathbf{y} \in \mathbb{E}$ onto a real number $\mathbf{x} \cdot \mathbf{y}$ with the following properties:

$\mathbf{x} \cdot \mathbf{y} = \mathbf{y} \cdot \mathbf{x}$	symmetry
$(\mathbf{x} + \mathbf{y}) \cdot \mathbf{z} = \mathbf{x} \cdot \mathbf{z} + \mathbf{y} \cdot \mathbf{z}$	additivity
$(a\,\mathbf{x}) \cdot \mathbf{y} = a(\mathbf{x} \cdot \mathbf{y})$ for all $a \in \mathbb{R}$	homogeneity
$\mathbf{x} \cdot \mathbf{x} \geqslant 0$; $\mathbf{x} \cdot \mathbf{x} = 0$ if and only if $\mathbf{x} = \mathbf{0}$.	positive definiteness

Geometry in $\mathbb{E}$ arises from the inner product. It allows us to associate with each vector in $\mathbb{E}$ a **length**,

$$\|\mathbf{x}\| = \sqrt{\mathbf{x} \cdot \mathbf{x}},$$

and, when $\mathbf{x} \neq \mathbf{0}$, a **direction** $\mathbf{x}/\|\mathbf{x}\|$. Two nonzero vectors $\mathbf{x}$ and $\mathbf{y}$ have **angle** θ given by

$$\cos\theta = \frac{\mathbf{x} \cdot \mathbf{y}}{\|\mathbf{x}\|\,\|\mathbf{y}\|}.$$

Two vectors $\mathbf{x}$ and $\mathbf{y}$ are **orthogonal** if $\mathbf{x} \cdot \mathbf{y} = 0$. For two arbitrary vectors $\mathbf{x}, \mathbf{y} \in \mathbb{E}$ with $\mathbf{x} \neq 0$, the **orthogonal projection** of $\mathbf{y}$ onto $\mathbf{x}$ is

$$\mathrm{Proj}_{\mathbf{x}}(\mathbf{y}) = \frac{\mathbf{x} \cdot \mathbf{y}}{\|\mathbf{x}\|^2}\mathbf{x}. \tag{1.1.1}$$

See Figure 1.1.

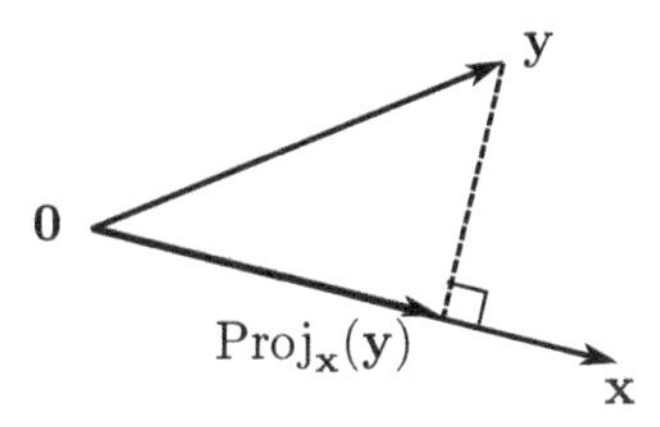

Figure 1.1. The orthogonal projection of $\mathbf{y}$ onto $\mathbf{x}$.

• **Orthonormal basis.** There is an **orthonormal basis** $\{\mathbf{e}_1, \mathbf{e}_2, \mathbf{e}_3\}$ for $\mathbb{E}$, that is, a basis such that $\mathbf{e}_i \cdot \mathbf{e}_i = 1$ for $i = 1, 2, 3$ and $\mathbf{e}_i \cdot \mathbf{e}_j = 0$ when $i \neq j$.

The basis $\{\mathbf{e}_1, \mathbf{e}_2, \mathbf{e}_3\}$, shown in Figure 1.2, establishes a Cartesian coordinate system on $\mathbb{E}$. Using this system, we represent any vector $\mathbf{x} \in \mathbb{E}$ as a point in the vector space $\mathbb{R}^3$ of ordered triples of real numbers: if $\mathbf{x} = x_1\mathbf{e}_1 + x_2\mathbf{e}_2 + x_3\mathbf{e}_3$, then

$$\begin{bmatrix} x_1 \\ x_2 \\ x_3 \end{bmatrix}$$

denotes its representation in $\mathbb{R}^3$ with respect to the basis. A different choice of basis vectors for $\mathbb{E}$—even a different choice of orthonormal basis—yields a different representation in $\mathbb{R}^3$, but the vector in $\mathbb{E}$ remains fixed in magnitude and direction. For this reason we distinguish $\mathbb{E}$ from $\mathbb{R}^3$.

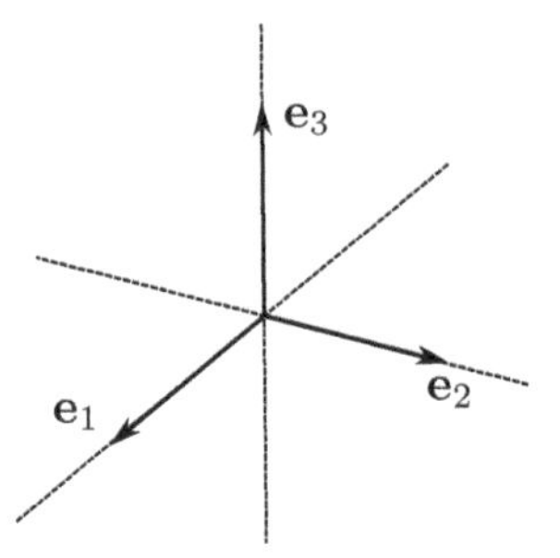

Figure 1.2. Standard orthonormal basis vectors defining a Cartesian coordinate system.

EXERCISE 1 *For a given orthonormal basis* $\{\mathbf{e}_1, \mathbf{e}_2, \mathbf{e}_3\}$, *it is possible to determine the coefficients* x_i *knowing the vector* $\mathbf{x}$. *Show that* $x_i = \mathbf{x} \cdot \mathbf{e}_i$.

For consistency with the conventions of matrix multiplication, discussed later, we write representations of vectors in $\mathbb{R}^3$ as column arrays. Under this convention, when the basis for $\mathbb{E}$ is understood, we sometimes abuse notation by writing as if the vector equals its representation:

$$\mathbf{e}_1 = \begin{bmatrix} 1 \\ 0 \\ 0 \end{bmatrix}, \quad \mathbf{e}_2 = \begin{bmatrix} 0 \\ 1 \\ 0 \end{bmatrix}, \quad \mathbf{e}_3 = \begin{bmatrix} 0 \\ 0 \\ 1 \end{bmatrix}.$$

For typesetting convenience we sometimes denote column vectors as formal transposes of row vectors, for example,

$$\begin{bmatrix} x_1 \\ x_2 \\ x_3 \end{bmatrix} = (x_1, x_2, x_3)^\top.$$

With respect to the orthonormal basis $\{\mathbf{e}_1, \mathbf{e}_2, \mathbf{e}_3\}$, the inner product of two vectors $\mathbf{x}, \mathbf{y} \in \mathbb{E}$ has the value

$$\mathbf{x} \cdot \mathbf{y} = \sum_{i=1}^{3} x_i y_i.$$

The following exercise gives a coordinate-free expression for the inner product in terms of lengths.

EXERCISE 2 *Prove the* **polarization identity***:* $\mathbf{x} \cdot \mathbf{y} = \frac{1}{4}(\|\mathbf{x} + \mathbf{y}\|^2 - \|\mathbf{x} - \mathbf{y}\|^2)$.

• **Cross product.** $\mathbb{E}$ admits a second form of vector multiplication. If $\mathbf{x}, \mathbf{y} \in \mathbb{E}$ have angle θ, then their **cross product** is the vector $\mathbf{x} \times \mathbf{y} \in \mathbb{E}$ that has length $\|\mathbf{x}\| \, \|\mathbf{y}\| \, |\sin\theta|$ and direction orthogonal to both $\mathbf{x}$ and $\mathbf{y}$, with sense given by the right-hand rule, as illustrated in Figure 1.3.

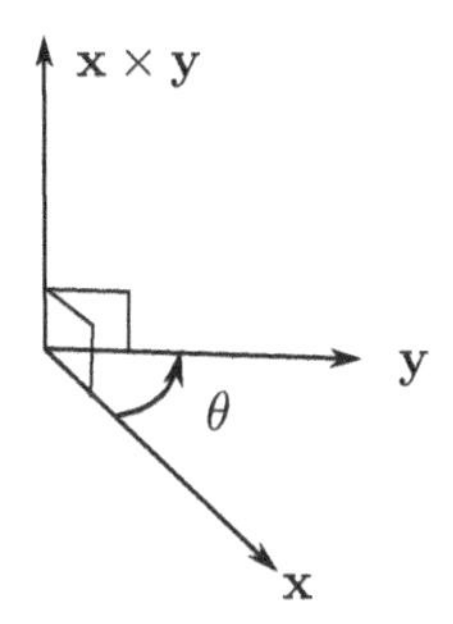

Figure 1.3. The cross product $\mathbf{x} \times \mathbf{y}$, showing the right-hand rule.

As a binary operation on vectors, the cross product is endemic to three space dimensions, as discussed further in Section 3.3. In all that follows, we assume that the orthonormal basis $\{\mathbf{e}_1, \mathbf{e}_2, \mathbf{e}_3\}$ for $\mathbb{E}$ has **positive orientation**, meaning that $\mathbf{e}_i \times \mathbf{e}_j = \mathbf{e}_k$ whenever $(i, j, k) = (1, 2, 3)$ or $(2, 3, 1)$ or $(3, 1, 2)$, that is, whenever (i, j, k) is an even permutation of $(1, 2, 3)$. Under this convention, with respect to the basis $\{\mathbf{e}_1, \mathbf{e}_2, \mathbf{e}_3\}$, the cross product has the value

$$\mathbf{x} \times \mathbf{y} = (x_2 y_3 - x_3 y_2)\mathbf{e}_1 + (x_3 y_1 - x_1 y_3)\mathbf{e}_2 + (x_1 y_2 - x_2 y_1)\mathbf{e}_3. \qquad (1.1.2)$$

Later sections explore additional algebraic and geometric interpretations of the cross product.

Representations with respect to the basis $\{\mathbf{e}_1, \mathbf{e}_2, \mathbf{e}_3\}$ can be useful for calculations, but two caveats are in order. First, they are not the only numerical representations available for vectors $\mathbf{x} \in \mathbb{E}$. Infinitely many orthonormal bases exist for $\mathbb{E}$, and it is possible to construct infinitely many nonorthonormal, non-Cartesian bases. Second, the principles of continuum mechanics do not require *any* choice of basis or associated coordinate system. Nevertheless, this book frequently uses $\{\mathbf{e}_1, \mathbf{e}_2, \mathbf{e}_3\}$ and the Cartesian coordinate system it defines to discuss examples, since this basis furnishes a computationally familiar setting. The remainder of this chapter reviews further aspects of the algebra and geometry of $\mathbb{E}$, with an attempt to minimize unnecessary references to coordinates.

Subsequent chapters refer to vectors that have a variety of physical dimensions. For example, the dimension of position vectors is length, denoted by [L], while velocity vectors have dimension length/time, or [LT^{-1}]. Fastidious readers may anticipate some apparent anomalies associated with algebraic operations involving pairs of vectors having different physical dimensions. Section 1.2 proposes a resolution.

1.1.2 Euclidean Point Space

As fundamental as the Euclidean vector space $\mathbb{E}$ may be to the mathematics of continuum mechanics, the objects of interest—sets of material points defined in Chapter 2—do not reside there. Instead, consistent with experience, these objects occupy points P in a type of space $\mathbb{X}$ that has no intrinsic algebraic structure. To apply the tools of algebra and calculus, we attach the vector space $\mathbb{E}$ to $\mathbb{X}$, choosing a point $O \in \mathbb{X}$ to serve as the origin and assigning to every point $P \in \mathbb{X}$ a vector in $\mathbb{E}$ that translates O into P. This approach allows us to refer to points $P \in \mathbb{X}$ in a way that facilitates the mathematical analysis available for vectors $\mathbf{x} \in \mathbb{E}$. This subsection furnishes details of the association between $\mathbb{X}$ and the space $\mathbb{E}$.

Recognizing the distinction between the set $\mathbb{X}$ of points and the algebraic structure $\mathbb{E}$ may seem pedantic. But without doing so, we cannot correctly account for disparate descriptions made by different observers, who assign vectors to points in different ways. Sections 3.5 and 6.3 examine the effects of such differences.

DEFINITION. *A set $\mathbb{X}$ of points is a* **Euclidean point space** *over $\mathbb{E}$ if there is a* **translation mapping** $\mathbf{d}\colon \mathbb{X} \times \mathbb{X} \to \mathbb{E}$ *having the following properties:*

1. $\mathbf{d}(P, P) = \mathbf{0} \in \mathbb{E}$ *for every* $P \in \mathbb{X}$.
2. $\mathbf{d}(P, R) = \mathbf{d}(P, Q) + \mathbf{d}(Q, R)$ *for all* $P, Q, R \in \mathbb{X}$.
3. *For every fixed $P \in \mathbb{X}$ the mapping $\mathbf{d}(P, \cdot)$ is one-to-one and onto. In other words, for every point $P \in \mathbb{X}$ and every vector $\mathbf{t} \in \mathbb{E}$ there exists a unique point $Q \in \mathbb{X}$ such that $\mathbf{d}(P, Q) = \mathbf{t}$.*

EXERCISE 3 *From these properties show that* $\mathbf{d}(P, Q) = -\mathbf{d}(Q, P)$.

We interpret this definition as follows: to each pair (P, Q) of points in $\mathbb{X}$, the mapping $\boldsymbol{d}$ assigns the vector $\mathbf{t} \in \mathbb{E}$ that translates P to Q. With the set $\mathbb{X}$ of points and the vector space $\mathbb{E}$ related in this way, we call $\mathbb{E}$ the **translation space** for $\mathbb{X}$.

Using $\boldsymbol{d}$ it is possible to assign to each point in $\mathbb{X}$ a unique vector $\mathbf{x} \in \mathbb{E}$. To construct such an assignment $\boldsymbol{f}\colon \mathbb{X} \to \mathbb{E}$, first pick a point $O \in \mathbb{X}$ and define $\boldsymbol{f}(O) = \mathbf{0}$. The point O is the **origin** of $\mathbb{X}$ under $\boldsymbol{f}$. Then for any other point $P \in \mathbb{X}$, define $\boldsymbol{f}(P) = \boldsymbol{d}(O, P)$, that is, the vector that translates the origin to P. As shown in Figure 1.4, the translation vector connecting two points $P, Q \in \mathbb{X}$ is

$$\mathbf{t} = \boldsymbol{d}(P, Q) = \boldsymbol{d}(P, O) + \boldsymbol{d}(O, Q) = \boldsymbol{f}(Q) - \boldsymbol{f}(P).$$

In this way, $\boldsymbol{f}$ furnishes a distance function on $\mathbb{X}$:

$$\text{distance}\,(P, Q) = \|\boldsymbol{f}(P) - \boldsymbol{f}(Q)\|.$$

The mapping $\boldsymbol{f}$ also provides access to the algebraic tools available in the Euclidean vector space $\mathbb{E}$. In this way, $\mathbb{E}$ constitutes a space of **position vectors** for $\mathbb{X}$ under $\boldsymbol{f}$.

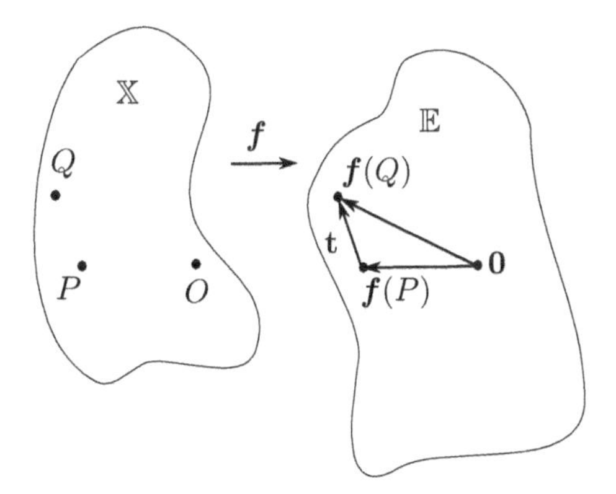

Figure 1.4. Points in $\mathbb{X}$ and vectors in $\mathbb{E}$, showing the translation vector $\mathbf{t} = \boldsymbol{d}(P, Q)$.

Through the mapping $\boldsymbol{f}\colon \mathbb{X} \to \mathbb{E}$, any coordinate system on $\mathbb{E}$ induces a coordinate system on the Euclidean point space $\mathbb{X}$. In particular, the orthonormal basis $\{\mathbf{e}_1, \mathbf{e}_2, \mathbf{e}_3\}$ introduced earlier induces a Cartesian coordinate system on $\mathbb{X}$.

The mapping $\boldsymbol{f} = \boldsymbol{d}(O, \cdot)$ is not unique. Infinitely many translation maps $\boldsymbol{d}$ exist, and for each there are infinitely many possible choices of the origin $O \in \mathbb{X}$. In addition, the analysis of continua involves time t as a variable, so we may as well regard the objects of interest as residing in **space-time**, $\mathbb{X} \times \mathbb{T}$, where $\mathbb{T}$ is the time interval of interest. There are infinitely many ways to assign a temporal instant $t_0 \in \mathbb{T}$ to the origin of the time axis $\mathbb{R}$.

DEFINITION. *A* **frame of reference** *is a choice* $(\boldsymbol{f}, t_0)$ *of one-to-one correspondence* $\boldsymbol{f}: \mathbb{X} \to \mathbb{E}$, *having the properties listed above, together with a choice* t_0 *of instant to be assigned to the origin of the time axis* $\mathbb{R}$.

Because any frame of reference amounts to a choice made by an observer, we cannot treat it as an essential aspect of the physics. For this reason, it will be necessary, eventually, to examine (1) which quantities associated with the physics remain invariant under changes in frame of reference and (2) how noninvariant quantities vary under such changes. We return to these questions in Section 3.5.

1.1.3 Summary

Vectors in this text belong to a three-dimensional Euclidean vector space $\mathbb{E}$. This space possesses an inner product, which gives rise to such geometric concepts as length, angle, and orthogonality. It also possesses a standard orthonormal basis $\{\mathbf{e}_1, \mathbf{e}_2, \mathbf{e}_3\}$ and a cross product. The orthonormal basis enables us to represent vectors in $\mathbb{E}$ as elements of $\mathbb{R}^3$, which facilitates calculations in specific problems.

The objects of interest in continuum mechanics reside in a Euclidean point space. The vector space $\mathbb{E}$ serves as a translation space for this space, through a mapping $\boldsymbol{f}$ that assigns to a unique point O the vector $\mathbf{0}$ and assigns to every other point P the vector $\mathbf{x}$ that translates O to P. Any choice of (1) a mapping $\boldsymbol{f}$ and (2) a time instant t_0 to assign to the origin 0 of the time axis $\mathbb{R}$ constitutes a frame of reference, which different observers may choose in different ways.

1.2 TENSORS

The analysis of motion involves linear mappings that act on vectors as geometric objects. Coordinate systems play an important role in calculations, but they are ancillary in the sense that the mappings' geometric actions remain invariant under a change of coordinates. Linear mappings having this property are tensors. In this text we encounter two orders of tensors.

1.2.1 First-Order Tensors and Vectors

Many types of vectors have representations as elements of $\mathbb{R}^3$. Examples include position vectors in $\mathbb{E}$, velocities, forces, and other entities that have both direction and magnitude. Logically, these different types of vectors belong to different vector spaces, all of which are inner-product spaces having the same algebraic structure as $\mathbb{R}^3$.

DEFINITION. *A* **Euclidean 3-space** *is a three-dimensional vector space* $\mathbb{V}$ *for which there is a one-to-one, onto mapping* $\Phi\colon \mathbb{V} \to \mathbb{R}^3$ *such that*

$$\begin{aligned}\Phi(\mathbf{x}+\mathbf{y}) &= \Phi(\mathbf{x}) + \Phi(\mathbf{y}),\\ \Phi(s\mathbf{x}) &= s\Phi(\mathbf{x}),\\ \Phi(\mathbf{x})\cdot\Phi(\mathbf{y}) &= \mathbf{x}\cdot\mathbf{y},\end{aligned}$$

for all $\mathbf{x}, \mathbf{y} \in \mathbb{V}$ *and for all* $s \in \mathbb{R}$.

In words, the algebra of $\mathbb{V}$ mirrors that of $\mathbb{R}^3$. $\mathbb{E}$ is such a space. Because this definition admits spaces associated with different physical dimensions, it gives rise to potential anomalies in vector algebra alluded to in Section 1.1. For example, if $\mathbf{v}$ is a velocity vector with dimension [LT^{-1}], computing its component in the direction of a unit vector $\mathbf{n}$ having dimension [L] requires calculating $\mathbf{v}\cdot\mathbf{n}$. Yet because they have different physical dimensions, $\mathbf{v}$ and $\mathbf{n}$ belong to different Euclidean 3-spaces, and in a fussy sense their inner product *per se* appears to make little sense.

In Euclidean spaces, we can resolve the matter by viewing vectors not only as geometric objects but also as mappings acting on other vectors.

DEFINITION. *A* **first-order tensor** *on a Euclidean 3-space* $\mathbb{V}$ *is a linear functional on* $\mathbb{V}$*, that is, a mapping* $\ell\colon \mathbb{V} \to \mathbb{R}$ *that satisfies the following properties:*

$$\begin{aligned}\ell(\mathbf{x}+\mathbf{y}) &= \ell(\mathbf{x}) + \ell(\mathbf{y}),\\ \ell(a\mathbf{x}) &= a\ell(\mathbf{x}) \quad \text{for any } a \in \mathbb{R}.\end{aligned}$$

It is possible to represent any first-order tensor on an inner-product space $\mathbb{V}$ as a vector in $\mathbb{R}^3$. In particular, if $\mathbf{x} \in \mathbb{V}$ has representation $x_1\mathbf{w}_1 + x_2\mathbf{w}_2 + x_3\mathbf{w}_3$ with respect to a basis $\{\mathbf{w}_1, \mathbf{w}_2, \mathbf{w}_3\}$ for $\mathbb{V}$, then any linear functional ℓ defined on $\mathbb{V}$ has an action that decomposes as follows:

$$\ell(\mathbf{x}) = \sum_{i=1}^{3} x_i \ell(\mathbf{w}_i) = \begin{bmatrix} c_1 \\ c_2 \\ c_3 \end{bmatrix} \cdot \begin{bmatrix} x_1 \\ x_2 \\ x_3 \end{bmatrix} = \mathbf{c}\cdot\mathbf{x},$$

where each $c_i = \ell(\mathbf{w}_i)$. In this sense, first-order tensors are vectors.

EXERCISE 4 *Let* $\mathbf{n} \in \mathbb{E}$ *be a fixed vector, and let* $\mathbb{V}$ *be the Euclidean 3-space of velocities. Define* $\ell_{\mathbf{n}}\colon \mathbb{V} \to \mathbb{R}$ *by* $\ell_{\mathbf{n}}(\mathbf{v}) = \mathbf{v}\cdot\mathbf{n}$. *Show that* $\ell_{\mathbf{n}}$ *is a first-order tensor on* $\mathbb{V}$.

The apparent anomaly in the expression $\mathbf{v} \cdot \mathbf{n}$ now disappears. The inner product simply furnishes a computational device for representing the linear functional defined by $\ell_{\mathbf{n}}(\mathbf{v}) = \mathbf{v} \cdot \mathbf{n}$. This first-order tensor maps $\mathbf{v}$ onto the quantity that results when we formally compute the inner product $\mathbf{v} \cdot \mathbf{n}$ and assign the appropriate physical dimension, $[\mathrm{LT}^{-1}][\mathrm{L}] = [\mathrm{L}^2\mathrm{T}^{-1}]$. With this observation in mind, we henceforth denote linear functionals on Euclidean 3-spaces as vectors in $\mathbb{E}$.

Representing linear functionals as vectors in $\mathbb{E}$ involves no serious risks of confusion in this book. In contrast, representing vectors as elements of $\mathbb{R}^3$ requires a strong caveat. Any numerical representation of a first-order tensor depends on the choice of basis. When we change to a new basis—for example, by changing Cartesian coordinate systems—we leave the geometric action of the linear functional unchanged. But its numerical representation as an ordered triple in $\mathbb{R}^3$ changes.

EXERCISE 5 *In terms of the orthonormal basis $\{\mathbf{e}_1, \mathbf{e}_2, \mathbf{e}_3\}$, a typical vector $\mathbf{x} \in \mathbb{E}$ has an expansion of the form $\mathbf{x} = x_1\mathbf{e}_1 + x_2\mathbf{e}_2 + x_3\mathbf{e}_3$. Change to a coordinate system having the same origin but a new orthonormal basis $\{\hat{\mathbf{e}}_1, \hat{\mathbf{e}}_2, \hat{\mathbf{e}}_3\}$, where $\hat{\mathbf{e}}_1 = (\mathbf{e}_1 + \mathbf{e}_3)/\sqrt{2}$, $\hat{\mathbf{e}}_2 = \mathbf{e}_2$, $\hat{\mathbf{e}}_3 = (-\mathbf{e}_1 + \mathbf{e}_3)/\sqrt{2}$. We now have two expansions for a typical vector $\mathbf{x}$:*

$$\mathbf{x} = \sum_{j=1}^{3} x_j \mathbf{e}_j = \sum_{j=1}^{3} \hat{x}_j \hat{\mathbf{e}}_j.$$

Find expressions for the coefficients $\hat{x}_j$ in terms of the coefficients x_i.

Exercises 6 and 7 establish the transformation rules that representations of vectors in $\mathbb{R}^3$ must obey to keep their lengths and directions—and their actions as first-order tensors—invariant under coordinate changes.

EXERCISE 6 *For a more general change of orthonormal basis from $\{\mathbf{e}_1, \mathbf{e}_2, \mathbf{e}_3\}$ to $\{\hat{\mathbf{e}}_1, \hat{\mathbf{e}}_2, \hat{\mathbf{e}}_3\}$, leaving the origin fixed, show that the following relationship holds between the three coefficients x_j and the three coefficients $\hat{x}_i$:*

$$\begin{bmatrix} \hat{x}_1 \\ \hat{x}_2 \\ \hat{x}_3 \end{bmatrix} = \begin{bmatrix} \hat{\mathbf{e}}_1 \cdot \mathbf{e}_1 & \hat{\mathbf{e}}_1 \cdot \mathbf{e}_2 & \hat{\mathbf{e}}_1 \cdot \mathbf{e}_3 \\ \hat{\mathbf{e}}_2 \cdot \mathbf{e}_1 & \hat{\mathbf{e}}_2 \cdot \mathbf{e}_2 & \hat{\mathbf{e}}_2 \cdot \mathbf{e}_3 \\ \hat{\mathbf{e}}_3 \cdot \mathbf{e}_1 & \hat{\mathbf{e}}_3 \cdot \mathbf{e}_2 & \hat{\mathbf{e}}_3 \cdot \mathbf{e}_3 \end{bmatrix} \begin{bmatrix} x_1 \\ x_2 \\ x_3 \end{bmatrix}. \tag{1.2.1}$$

Thus we can account for this type of change in coordinates through matrix multiplication. We call the entries $Q_{ij} = \hat{\mathbf{e}}_i \cdot \mathbf{e}_j$ of the transformation matrix the **direction cosines** *of the transformation. Why?*

EXERCISE 7 *Write the matrix multiplication in Equation (1.2.1) as follows:*

$$\hat{x}_m = \sum_{i=1}^{3} Q_{mi} x_i.$$

Show that

$$x_j = \sum_{l=1}^{3} Q_{lj} \hat{x}_l = \sum_{l=1}^{3} Q_{jl}^{\top} \hat{x}_l,$$

where $Q_{jl}^{\top}$ *denotes the* 3×3 *matrix obtained by reflecting the entries of the matrix* Q_{jl} *in Equation (1.2.1) across the diagonal. Thus the expression giving each* x_j *in terms of the coordinates* $\hat{x}_l$ *also takes the form of matrix multiplication by direction cosines.*

1.2.2 Second-Order Tensors

The idea of mappings on Euclidean 3-spaces extends further. Let $\mathbb{V}$ and $\mathbb{W}$ be two such spaces.

DEFINITION. *A* **second-order tensor** *from* $\mathbb{V}$ *to* $\mathbb{W}$ *is a linear transformation* $\mathsf{A}: \mathbb{V} \to \mathbb{W}$. *We denote the set of all second-order tensors from* $\mathbb{V}$ *to* $\mathbb{W}$ *as* $\mathrm{L}(\mathbb{V}, \mathbb{W})$ *or, when* $\mathbb{W} = \mathbb{V}$, *as* $\mathrm{L}(\mathbb{V})$.

The second-order tensors of chief interest in this text are those in $\mathrm{L}(\mathbb{E})$, where we focus attention from now on. These entities need not be exotic. For example, the identity tensor I, defined by the action $\mathsf{I}\mathbf{x} = \mathbf{x}$ for all $\mathbf{x} \in \mathbb{E}$, is a second-order tensor. Less trivially, for fixed $\mathbf{x} \in \mathbb{E}$, let $\mathrm{Proj}_{\mathbf{x}}: \mathbb{E} \to \mathbb{E}$ be the standard orthogonal projection operator onto $\mathbf{x}$, described in Equation (1.1.1).

EXERCISE 8 *Show that* $\mathrm{Proj}_{\mathbf{x}}$ *is a second-order tensor.*

The following example furnishes many more possibilities:

DEFINITION. *The* **dyadic product** *of two vectors* $\mathbf{a}, \mathbf{b} \in \mathbb{E}$ *is the mapping* $\mathbf{a} \otimes \mathbf{b}: \mathbb{E} \to \mathbb{E}$ *defined by*

$$(\mathbf{a} \otimes \mathbf{b})(\mathbf{x}) = \mathbf{a}(\mathbf{b} \cdot \mathbf{x}).$$

EXERCISE 9 *Show that* $\mathbf{a} \otimes \mathbf{b}$ *is a second-order tensor. Find a representation for* $\mathrm{Proj}_{\mathbf{x}}$ *as a dyadic product.*

With respect to a fixed choice of basis for $\mathbb{E}$, any second-order tensor $\mathsf{A} \in \mathrm{L}(\mathbb{E})$ has a numerical representation. If $\{\mathbf{v}_1, \mathbf{v}_2, \mathbf{v}_3\}$ and $\{\mathbf{w}_1, \mathbf{w}_2, \mathbf{w}_3\}$ are the bases for the domain and target space of $\mathsf{A} \in \mathrm{L}(\mathbb{E})$, then the corresponding **matrix representation** for A is the 3×3 array of real numbers in which the entry in row i, column j is the coefficient A_{ij} multiplying $\mathbf{w}_i$ in the expansion

$$\mathsf{A}\mathbf{v}_j = \sum_{i=1}^{3} A_{ij}\mathbf{w}_i \in \mathbb{E}.$$

EXERCISE 10 *With respect to the basis* $\{\mathbf{e}_1, \mathbf{e}_2, \mathbf{e}_3\}$ *for* $\mathbb{E}$ *defined in Section 1.1, show that the matrix representation of* $\mathsf{A} \in \mathrm{L}(\mathbb{E})$ *has entries* $A_{ij} = \mathbf{e}_i \cdot \mathsf{A}\mathbf{e}_j$. *Hence, the matrix representation of* $\mathbf{e}_1 \otimes \mathbf{e}_3$ *is*

$$\begin{bmatrix} 0 & 0 & 1 \\ 0 & 0 & 0 \\ 0 & 0 & 0 \end{bmatrix}.$$

With respect to this same basis, show that $\mathbf{a} \otimes \mathbf{b}$ *has matrix representation*

$$\begin{bmatrix} a_1b_1 & a_1b_2 & a_1b_3 \\ a_2b_1 & a_2b_2 & a_2b_3 \\ a_3b_1 & a_3b_2 & a_3b_3 \end{bmatrix}.$$

Thus any second-order tensor $\mathsf{A} \in \mathrm{L}(\mathbb{E})$ has an expansion in terms of the elementary dyadic products $\mathbf{e}_i \otimes \mathbf{e}_j$:

$$\mathsf{A} = \sum_{i=1}^{3}\sum_{j=1}^{3} A_{ij}\mathbf{e}_i \otimes \mathbf{e}_j.$$

Sums over coordinate indices appear so often in continuum mechanics that we frequently use a common shorthand for them. Consider the expressions

$$\mathbf{x} \cdot \mathbf{y} = \sum_{i=1}^{3} x_i y_i, \qquad \mathsf{A} = \sum_{i=1}^{3}\sum_{j=1}^{3} A_{ij}\mathbf{e}_i \otimes \mathbf{e}_j. \tag{1.2.2}$$

In these equations the notations $\sum_{i=1}^{3}$ and $\sum_{i=1}^{3}\sum_{j=1}^{3}$ become superfluous if we adopt the **Einstein summation convention**:

1. If a letter index such as i or j appears exactly once in a term, it stands for any of the values $1, 2, 3$.

2. If a letter index appears exactly twice in a term, we treat the index as a dummy index and sum the terms over the index values $1, 2, 3$.

3. If a term requires a letter index to appear more than twice, we agree to write any required summation signs explicitly.

Under this convention, repetition of the dummy indices i (in the term $x_i y_i$) and i, j (in the term $A_{ij}\mathbf{e}_i \otimes \mathbf{e}_j$) implies summation from 1 to 3 over the repeated indices. Thus we write $x_i y_i$ and $A_{ij}\mathbf{e}_i \otimes \mathbf{e}_j$ as shorthand for the more explicit summation notations in Equation 1.2.2.

Matrix representations and the Einstein summation convention yield a sleek way to represent compositions of second-order tensors. If $\mathsf{A}, \mathsf{B} \in \mathrm{L}(\mathbb{E})$, we denote their composition as a product: $(\mathsf{A} \circ \mathsf{B})(\mathbf{x}) = \mathsf{A}(\mathsf{B}(\mathbf{x})) = (\mathsf{AB})\mathbf{x}$. The notation is suggestive.

EXERCISE 11 *Show that* $\mathsf{AB} \in \mathrm{L}(\mathbb{E})$. *Show that*

$$\mathsf{AB} \quad = \quad \sum_{j=1}^{3}\sum_{l=1}^{3}\sum_{k=1}^{3} A_{jk}B_{kl}\mathbf{e}_j \otimes \mathbf{e}_l \quad = \quad A_{jk}B_{kl}\mathbf{e}_j \otimes \mathbf{e}_l$$

Standard notation Summation convention

Thus the matrix representation of the composition AB is the ordinary matrix product of the 3×3 representations for A and B. In shorthand, the (j, l) entry in the matrix representation for AB is $A_{jk}B_{kl}$.

As with first-order tensors, the representation of a second-order tensor as an array of numbers depends on the choice of bases for the Euclidean 3-spaces involved. When we change from one orthonormal basis to another, we leave the geometric action of the linear transformation unchanged but change its numerical representation as a matrix. To see how, consider the vector $\mathbf{y} = \mathsf{A}\mathbf{x}$ resulting from the action of a second-order tensor A on a vector $\mathbf{x}$. The coordinates of $\mathbf{y}$ with respect to the orthonormal basis $\{\mathbf{e}_1, \mathbf{e}_2, \mathbf{e}_3\}$ are $y_i = \mathbf{y} \cdot \mathbf{e}_i$. Substituting for $\mathbf{y}$ in terms of a new orthonormal basis $\{\hat{\mathbf{e}}_1, \hat{\mathbf{e}}_2, \hat{\mathbf{e}}_3\}$ and for $\hat{x}_m$ using the results of Exercise 6 yields (using the summation convention):

$$\begin{aligned} y_i = \hat{y}_l \hat{\mathbf{e}}_l \cdot \mathbf{e}_i &= \hat{A}_{lm}\hat{x}_m \hat{\mathbf{e}}_l \cdot \mathbf{e}_i \\ &= \hat{A}_{lm}\hat{x}_m Q_{li} \\ &= \hat{A}_{lm}Q_{mj}x_j Q_{li} = (Q_{il}^{\top}\hat{A}_{lm}Q_{mj})x_j. \end{aligned} \tag{1.2.3}$$

Here, Q_{il} denotes entries of the direction cosine matrix introduced in Exercise 6. The expression (1.2.3) for $y_i = A_{ij}x_j$ holds for all possible vectors $\mathbf{x}$. It follows that the entries of A_{ij} transform according to the following matrix multiplication:

$$A_{ij} = Q_{il}^{\top}\hat{A}_{lm}Q_{mj}.$$

EXERCISE 12 *Show that* $\hat{A}_{lm} = Q_{li} A_{ij} Q^{\top}_{jm}$.

Caution is in order here. We do not regard the matrix Q_{il} of direction cosines as a representation of a second-order tensor. Second-order tensors act on vectors in $\mathbb{E}$ to produce (possibly new) vectors in $\mathbb{E}$. The matrix Q_{il} acts on *numerical representations* of vectors in $\mathbb{E}$ and tensors in $\mathrm{L}(\mathbb{E})$ to produce (possibly new) representations of these same entities.

The concept of a tensor generalizes to orders higher than one and two. Although higher-order tensors do not figure prominently in the mechanics covered in this book, there is utility in the approach. Recall that a first-order tensor (which we regard as a vector in $\mathbb{E}$) is a linear functional $\ell\colon \mathbb{E} \to \mathbb{R}$. By extension, we regard a second-order tensor $\mathsf{A} \in \mathrm{L}(\mathbb{E})$ as a **bilinear functional** $\mathsf{A}\colon \mathbb{E} \times \mathbb{E} \to \mathbb{R}$, that is, a mapping $\mathsf{A}\colon (\mathbf{a}_1, \mathbf{a}_2) \mapsto \mathbf{a}_1 \cdot \mathsf{A}\mathbf{a}_2$ that is linear in both of the vector arguments $\mathbf{a}_1, \mathbf{a}_2$. To generalize this idea, define an **nth-order tensor** as an n-linear mapping, that is, a function $\mathsf{A}\colon \mathbb{E}^n \to \mathbb{R}$ that is linear in each of its n arguments.

For example, recall that any second-order tensor is a linear combination of dyadic products $\mathbf{a}_1 \otimes \mathbf{a}_2$, and observe that the actions of these products as bilinear functionals take the following form:

$$\mathbf{a}_1 \otimes \mathbf{a}_2 \colon (\mathbf{b}_1, \mathbf{b}_2) \mapsto \mathbf{b}_1 \cdot (\mathbf{a}_1 \otimes \mathbf{a}_2)\mathbf{b}_2 = (\mathbf{a}_1 \cdot \mathbf{b}_1)(\mathbf{a}_2 \cdot \mathbf{b}_2).$$

More generally, the n-linear functional $\mathbf{a}_1 \otimes \mathbf{a}_2 \otimes \cdots \otimes \mathbf{a}_n$ acts on an n-tuple $(\mathbf{b}_1, \mathbf{b}_2, \ldots, \mathbf{b}_n) \in \mathbb{E}^n$ as follows:

$$(\mathbf{a}_1 \otimes \mathbf{a}_2 \otimes \cdots \otimes \mathbf{a}_n)(\mathbf{b}_1, \mathbf{b}_2, \ldots, \mathbf{b}_n) = \prod_{i=1}^{n} \mathbf{a}_i \cdot \mathbf{b}_i.$$

As in the case $n = 2$, any nth-order tensor is a linear combination of products of this form.

This view of second-order tensors as bilinear functionals gives rise to two additional concepts from elementary linear algebra.

DEFINITION. *If* $\mathsf{A} \in \mathrm{L}(\mathbb{E})$, *then its* **transpose** *is the second-order tensor* $\mathsf{A}^{\top} \in \mathrm{L}(\mathbb{E})$ *defined by the equation* $\mathbf{a} \cdot \mathsf{A}^{\top}\mathbf{b} = \mathsf{A}\mathbf{a} \cdot \mathbf{b}$ *for all* $\mathbf{a}, \mathbf{b} \in \mathbb{E}$.

One easily verifies that $\mathbf{b} \otimes \mathbf{a} = (\mathbf{a} \otimes \mathbf{b})^{\top}$. Corresponding to this observation is the fact that one obtains the matrix representation of $\mathsf{A}^{\top}$ with respect to any orthonormal basis simply by reflecting the matrix for A across its main diagonal: $A^{\top}_{ij} = A_{ji}$.

DEFINITION. $\mathsf{A} \in \mathrm{L}(\mathbb{E})$ *is* **invertible** *if there exists a tensor* $\mathsf{A}^{-1} \in \mathrm{L}(\mathbb{E})$ *such that* $\mathsf{A}\mathsf{A}^{-1} = \mathsf{A}^{-1}\mathsf{A} = \mathsf{I}$.

Each of the following conditions is necessary and sufficient for A to be invertible:

1. $\mathsf{A}\colon \mathbb{E} \to \mathbb{E}$ is one-to-one, that is, $\mathsf{A}\mathbf{x} = \mathsf{A}\mathbf{y}$ only if $\mathbf{x} = \mathbf{y}$.

2. $\mathsf{A}\mathbf{x} = \mathbf{0}$ only if $\mathbf{x} = \mathbf{0}$.

EXERCISE 13 *Show that each of these conditions implies the other.*

1.2.3 Cross Products, Triple Products, and Determinants

Three concepts—cross products, scalar triple products, and determinants—connect the algebra of $\mathbb{E}$ and $\mathrm{L}(\mathbb{E})$ to the geometry of volumes. As introduced in Section 1.1, the cross product $\mathbf{x} \times \mathbf{y}$ is orthogonal to the plane spanned by $\mathbf{x}$ and $\mathbf{y}$, as illustrated in Figure 1.3. It has length $\|\mathbf{x}\|\,\|\mathbf{y}\|\,|\sin\theta|$, where θ is the angle between $\mathbf{x}$ and $\mathbf{y}$. Equivalently, the length of $\mathbf{x} \times \mathbf{y}$ is the area of the parallelogram with sides $\mathbf{x}$ and $\mathbf{y}$. Figure 1.5 also shows this relationship.

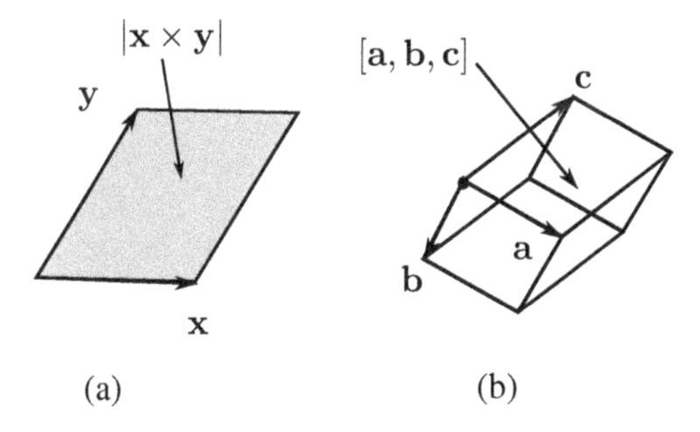

Figure 1.5. (a) Relationship between $\mathbf{x}$, $\mathbf{y}$, and $\mathbf{x} \times \mathbf{y}$. (b) The parallelepiped having adjacent edges $\mathbf{a}, \mathbf{b}, \mathbf{c}$.

Equation (1.1.2) gives the components of the cross product $\mathbf{x} \times \mathbf{y}$ with respect to the orthonormal basis $\{\mathbf{e}_1, \mathbf{e}_2, \mathbf{e}_3\}$. Occasionally an indicial form for $\mathbf{x} \times \mathbf{y}$ is useful:

$$\mathbf{x} \times \mathbf{y} = \varepsilon_{ijk} x_j y_k \mathbf{e}_i.$$

Here, ε_{ijk} is the **Levi-Civita symbol**, defined as follows:

$$\varepsilon_{ijk} = \begin{cases} 1, & \text{if} \quad (i,j,k) = (1,2,3), (3,1,2), (2,3,1); \\ -1, & \text{if} \quad (i,j,k) = (1,3,2), (2,1,3), (3,2,1); \\ 0, & \text{otherwise.} \end{cases}$$

EXERCISE 14 *Verify that*

$$\varepsilon_{ijk}\varepsilon_{lmk} = \begin{cases} 1, & \text{if } \; i = l \text{ and } j = m; \\ -1, & \text{if } \; i = m \text{ and } j = l; \\ 0, & \text{otherwise.} \end{cases}$$

The connection between the cross product and volumes hinges on the following algebraic operation.

DEFINITION. *The* **scalar triple product** *of* $\mathbf{a}, \mathbf{b}, \mathbf{c} \in \mathbb{E}$ *is* $[\mathbf{a}, \mathbf{b}, \mathbf{c}] = \mathbf{a} \cdot (\mathbf{b} \times \mathbf{c})$.

Geometrically, $[\mathbf{a}, \mathbf{b}, \mathbf{c}]$ gives the signed volume of the parallelepiped having $\mathbf{a}, \mathbf{b}, \mathbf{c}$ as adjacent edges, as shown in Figure 1.5.

From an algebraic perspective, the scalar triple product is an **alternating 3-linear form**, that is, the mapping $[\cdot, \cdot, \cdot] \colon \mathbb{E}^3 \to \mathbb{R}$ has the following properties:

1. $[\cdot, \cdot, \cdot]$ is linear in each argument.
2. For all ordered triples $(\mathbf{v}_1, \mathbf{v}_2, \mathbf{v}_3) \in \mathbb{E}^3$, $[\mathbf{v}_i, \mathbf{v}_j, \mathbf{v}_k] = \varepsilon_{ijk}[\mathbf{v}_1, \mathbf{v}_2, \mathbf{v}_3]$.

EXERCISE 15 *Show that* $[\mathbf{v}_1, \mathbf{v}_2, \mathbf{v}_3] = 0$ *whenever* $\{\mathbf{v}_1, \mathbf{v}_2, \mathbf{v}_3\} \subset \mathbb{E}$ *is a linearly dependent set. Show that* $[\mathbf{v}_1, \mathbf{v}_2, \mathbf{v}_3] = \pm 1$ *whenever* $\{\mathbf{v}_1, \mathbf{v}_2, \mathbf{v}_3\}$ *is an orthonormal basis for* $\mathbb{E}$.

When the numbering of an orthonormal basis $\{\mathbf{v}_1, \mathbf{v}_2, \mathbf{v}_3\}$ satisfies the condition $[\mathbf{v}_1, \mathbf{v}_2, \mathbf{v}_3] = 1$, we say that the basis has **positive orientation**.

Alternating 3-linear forms stand in an interesting relationship to each other:

THEOREM 1.2.1 (PROPORTIONALITY OF ALTERNATING 3-LINEAR FORMS). *If* $f \colon \mathbb{E}^3 \to \mathbb{R}$ *is a nonzero alternating 3-linear form and* g *is any other alternating 3-linear form, then there exists a scalar* $D \in \mathbb{R}$ *such that*

$$g(\mathbf{v}_1, \mathbf{v}_2, \mathbf{v}_3) = Df(\mathbf{v}_1, \mathbf{v}_2, \mathbf{v}_3) \tag{1.2.4}$$

for all ordered triples $(\mathbf{v}_1, \mathbf{v}_2, \mathbf{v}_3) \in \mathbb{E}^3$.

Crucially, the scalar D is independent of the vectors $\mathbf{v}_1, \mathbf{v}_2, \mathbf{v}_3$.

PROOF: Since f is nonzero, it is possible to pick $(\mathbf{u}_1, \mathbf{u}_2, \mathbf{u}_3) \in \mathbb{E}^3$ such that $f(\mathbf{u}_1, \mathbf{u}_2, \mathbf{u}_3) \neq 0$. By Exercise 15, $\{\mathbf{u}_1, \mathbf{u}_2, \mathbf{u}_3\}$ cannot be linearly dependent, so it constitutes a basis for $\mathbb{E}$. Define

$$D = \frac{g(\mathbf{u}_1, \mathbf{u}_2, \mathbf{u}_3)}{f(\mathbf{u}_1, \mathbf{u}_2, \mathbf{u}_3)}.$$

We must show that Equation (1.2.4) holds using this choice of D, for all ordered triples $(\mathbf{v}_1, \mathbf{v}_2, \mathbf{v}_3) \in \mathbb{E}^3$. Each vector in such an ordered triple has an expansion of the form $\mathbf{v}_i = c_{ij}\mathbf{u}_j$. By 3-linearity,

$$\begin{aligned} f(\mathbf{v}_1, \mathbf{v}_2, \mathbf{v}_3) &= f(c_{1j}\mathbf{u}_j, c_{2k}\mathbf{u}_k, c_{3l}\mathbf{u}_l) \\ &= c_{1j}c_{2k}c_{3l}f(\mathbf{u}_j, \mathbf{u}_k, \mathbf{u}_l) \\ &= \varepsilon_{jkl}c_{1j}c_{2k}c_{3l}f(\mathbf{u}_1, \mathbf{u}_2, \mathbf{u}_3) \\ &= C\, f(\mathbf{u}_1, \mathbf{u}_2, \mathbf{u}_3) \neq 0, \end{aligned}$$

where $C = \varepsilon_{jkl}c_{1j}c_{2k}c_{3l}$. Similarly, $g(\mathbf{v}_1, \mathbf{v}_2, \mathbf{v}_3) = C\, g(\mathbf{u}_1, \mathbf{u}_2, \mathbf{u}_3)$, for the same constant C. The identity (1.2.4) follows. ■

Of special interest are alternating 3-linear forms generated when second-order tensors act on ordered triples of vectors. If $\mathsf{A} \in \mathrm{L}(\mathbb{E})$ and f is any nonzero alternating 3-linear form, then the mapping defined by

$$(\mathbf{v}_1, \mathbf{v}_2, \mathbf{v}_3) \mapsto f(\mathsf{A}\mathbf{v}_1, \mathsf{A}\mathbf{v}_2, \mathsf{A}\mathbf{v}_3) \tag{1.2.5}$$

is also an alternating 3-linear form. Hence, by Theorem 1.2.1, there exists a scalar $D_A \in \mathbb{R}$ such that

$$f(\mathsf{A}\mathbf{v}_1, \mathsf{A}\mathbf{v}_2, \mathsf{A}\mathbf{v}_3) = D_A f(\mathbf{v}_1, \mathbf{v}_2, \mathbf{v}_3), \tag{1.2.6}$$

for all ordered triples $(\mathbf{v}_1, \mathbf{v}_2, \mathbf{v}_3) \in \mathbb{E}^3$. More remarkable is the following fact:

THEOREM 1.2.2 (A SCALAR INVARIANT). *The scalar factor D_A, defined by Equation (1.2.6) for the tensor $\mathsf{A} \in \mathrm{L}(\mathbb{E})$ and the alternating 3-linear form f, is independent of f.*

PROOF: If g is any other nonzero alternating 3-linear form, we must show that $g(\mathsf{A}\mathbf{v}_1, \mathsf{A}\mathbf{v}_2, \mathsf{A}\mathbf{v}_3) = D_A g(\mathbf{v}_1, \mathbf{v}_2, \mathbf{v}_3)$ for the scalar D_A defined by Equation (1.2.6). By Theorem 1.2.1, $g = Df$ for some nonzero scalar D. For the alternating 3-linear form defined by the mapping

$$(\mathbf{v}_1, \mathbf{v}_2, \mathbf{v}_3) \mapsto g(\mathsf{A}\mathbf{v}_1, \mathsf{A}\mathbf{v}_2, \mathsf{A}\mathbf{v}_3),$$

we have

$$\begin{aligned} g(\mathsf{A}\mathbf{v}_1, \mathsf{A}\mathbf{v}_2, \mathsf{A}\mathbf{v}_3) &= D\, f(\mathsf{A}\mathbf{v}_1, \mathsf{A}\mathbf{v}_2, \mathsf{A}\mathbf{v}_3) \\ &= D D_A f(\mathbf{v}_1, \mathbf{v}_2, \mathbf{v}_3) \\ &= D_A D\, f(\mathbf{v}_1, \mathbf{v}_2, \mathbf{v}_3) = D_A\, g(\mathbf{v}_1, \mathbf{v}_2, \mathbf{v}_3). \end{aligned}$$

Thus the factor D_A is a characteristic of A, independent of the alternating 3-linear form f. ■

This fact justifies a special name and notation for the invariant factor D_A:

DEFINITION. *The* **determinant** *of* $\mathsf{A} \in \mathrm{L}(\mathbb{E})$ *is the number* $\det \mathsf{A} \in \mathbb{R}$ *such that*

$$(\det \mathsf{A}) f(\mathbf{v}_1, \mathbf{v}_2, \mathbf{v}_3) = f(\mathsf{A}\mathbf{v}_1, \mathsf{A}\mathbf{v}_2, \mathsf{A}\mathbf{v}_3),$$

for every ordered triple $(\mathbf{v}_1, \mathbf{v}_2, \mathbf{v}_3) \in \mathbb{E}^3$, *for every alternating 3-linear form* f.

An immediate consequence is the following observation about the effect of a second-order tensor $\mathsf{A} \in \mathrm{L}(\mathbb{E})$ on volumes. Given three vectors $\mathbf{a}, \mathbf{b}, \mathbf{c} \in \mathbb{E}$, A maps their scalar triple product as follows:

$$[\mathbf{a}, \mathbf{b}, \mathbf{c}] \longmapsto [\mathsf{A}\mathbf{a}, \mathsf{A}\mathbf{b}, \mathsf{A}\mathbf{c}].$$

This mapping carries the parallelepiped corresponding to $\mathbf{a}, \mathbf{b}, \mathbf{c}$ into a new parallelepiped, as shown in Figure 1.6. The scalar quantity $\det \mathsf{A}$ gives the ratio of the transformed volume to the original volume: for all $(\mathbf{a}, \mathbf{b}, \mathbf{c}) \in \mathbb{E}^3$,

$$[\mathbf{a}, \mathbf{b}, \mathbf{c}] \det \mathsf{A} = [\mathsf{A}\mathbf{a}, \mathsf{A}\mathbf{b}, \mathsf{A}\mathbf{c}]. \tag{1.2.7}$$

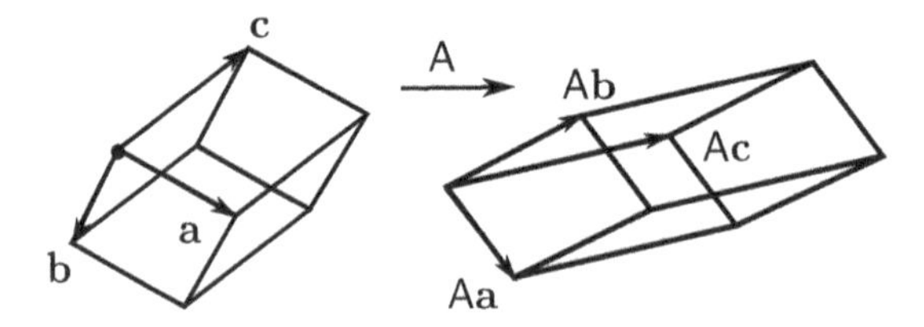

Figure 1.6. Effect of the linear transformation A on the parallelepiped having edges $\mathbf{a}, \mathbf{b}, \mathbf{c}$.

The definition of the determinant adopted here makes no reference to a basis for $\mathbb{E}$, and as a consequence the numerical value of $\det \mathsf{A}$ is independent of any matrix representation used for A. However, a calculation similar to that used in Theorem 1.2.2 yields a familiar formula for $\det \mathsf{A}$ in terms of its matrix representation A_{ij}

with respect to a basis $\{\mathbf{u}_1, \mathbf{u}_2, \mathbf{u}_3\}$ for $\mathbb{E}$:

$$\begin{aligned}(\det \mathsf{A})\left[\mathbf{u}_1, \mathbf{u}_2, \mathbf{u}_3\right] &= \left[\mathsf{A}\mathbf{u}_1, \mathsf{A}\mathbf{u}_2, \mathsf{A}\mathbf{u}_3\right] \\ &= \left[A_{i1}\mathbf{u}_i, A_{j2}\mathbf{u}_j, A_{k3}\mathbf{u}_k\right] \\ &= A_{i1}A_{j2}A_{k3}\left[\mathbf{u}_i, \mathbf{u}_j, \mathbf{u}_k\right] \\ &= \varepsilon_{ijk}A_{i1}A_{j2}A_{k3}\left[\mathbf{u}_1, \mathbf{u}_2, \mathbf{u}_3\right].\end{aligned}$$

It follows that

$$\det \mathsf{A} = \varepsilon_{ijk}A_{i1}A_{j2}A_{k3}. \tag{1.2.8}$$

The following theorem summarizes additional facts about determinants.

THEOREM 1.2.3 (PROPERTIES OF DETERMINANTS). *Let* $\mathbf{a}, \mathbf{b} \in \mathbb{E}$ *and* $\mathsf{A}, \mathsf{B} \in \mathrm{L}(\mathbb{E})$.

1. $\det \mathbf{a} \otimes \mathbf{b} = 0$.
2. $\det \mathsf{A} \neq 0$ *if and only if* A *is invertible.*
3. $\det(\mathsf{AB}) = \det \mathsf{A} \det \mathsf{B}$.
4. $\det \mathsf{A}^\top = \det \mathsf{A}$.
5. $\det(c\mathsf{A}) = c^3 \det \mathsf{A}$.

EXERCISE 16 *Prove Theorem 1.2.3.*

The connection between determinants and invertibility merits a brief digression. Let

$$\mathrm{GL}(\mathbb{E}) = \left\{\mathsf{A} \in \mathrm{L}(\mathbb{E}) \;\middle|\; \det \mathsf{A} \neq 0\right\}.$$

EXERCISE 17 *Show that* $\mathrm{GL}(\mathbb{E})$ *is a* **group** *under composition. That is:*

1. $\mathrm{GL}(\mathbb{E})$ *is closed under composition: if* $\mathsf{A}, \mathsf{B} \in \mathrm{GL}(\mathbb{E})$*, then* $\mathsf{AB} \in \mathrm{GL}(\mathbb{E})$.
2. *Composition is associative:* $(\mathsf{AB})\mathsf{C} = \mathsf{A}(\mathsf{BC})$.
3. $\mathrm{GL}(\mathbb{E})$ *possesses an identity* I.
4. *Every* $\mathsf{A} \in \mathrm{GL}(\mathbb{E})$ *possesses an inverse* $\mathsf{A}^{-1} \in \mathrm{GL}(\mathbb{E})$ *satisfying the equations* $\mathsf{AA}^{-1} = \mathsf{A}^{-1}\mathsf{A} = \mathsf{I}$.

$\mathrm{GL}(\mathbb{E})$ *is the* **general linear group** *over* $\mathbb{E}$.

EXERCISE 18 *Show that, with respect to any orthonormal basis for* $\mathbb{E}$, *the identity tensor* $\mathsf{I} \in \mathrm{GL}(\mathbb{E})$ *has a matrix representation in which* $I_{ij} = \delta_{ij}$, *where* δ_{ij} *is the* **Kronecker symbol**,

$$\delta_{ij} = \begin{cases} 1, & \text{if } i = j, \\ 0, & \text{otherwise.} \end{cases}$$

EXERCISE 19 *Verify the* **$\varepsilon\delta$ identity**,

$$\varepsilon_{ijk}\varepsilon_{lmk} = \delta_{il}\delta_{jm} - \delta_{im}\delta_{jl}. \tag{1.2.9}$$

EXERCISE 20 *Using index notation and the orthonormal basis* $\{\mathbf{e}_1, \mathbf{e}_2, \mathbf{e}_3\}$, *prove the following identity for the* **vector triple product**:

$$\mathbf{a} \times (\mathbf{b} \times \mathbf{c}) = \mathbf{b}(\mathbf{a} \cdot \mathbf{c}) - \mathbf{c}(\mathbf{a} \cdot \mathbf{b}). \tag{1.2.10}$$

EXERCISE 21 *Prove that*

$$(\mathbf{a} \times \mathbf{b}) \cdot (\mathbf{a} \times \mathbf{b}) = \|\mathbf{a}\|^2\|\mathbf{b}\|^2 - (\mathbf{a} \cdot \mathbf{b})^2. \tag{1.2.11}$$

1.2.4 Orthogonal Tensors

The following definition identifies a special class of invertible tensors encountered repeatedly in mechanics.

DEFINITION. *A second-order tensor* $\mathsf{Q} \in \mathrm{L}(\mathbb{E})$ *is* **orthogonal** *if* Q *is invertible and* $\mathsf{Q}^{-1} = \mathsf{Q}^\top$.

Clearly the identity tensor I is orthogonal—a fact that leads to a more refined algebraic observation:

EXERCISE 22 *Denote*

$$\mathrm{O}(\mathbb{E}) = \left\{ \mathsf{Q} \in \mathrm{L}(\mathbb{E}) \;\middle|\; \mathsf{Q} \text{ is orthogonal} \right\}.$$

Show that $\mathrm{O}(\mathbb{E})$ *is a* **subgroup** *of* $\mathrm{GL}(\mathbb{E})$, *that is,* $\mathrm{O}(\mathbb{E}) \subset \mathrm{GL}(\mathbb{E})$ *and is a group. Show that* $\det \mathsf{Q} = \det \mathsf{Q}^\top$ *whenever* $\mathsf{Q} \in \mathrm{O}(\mathbb{E})$, *and hence* $\det \mathsf{Q} = \pm 1$.

EXERCISE 23 *Show that the set*

$$\mathrm{O}^+(\mathbb{E}) = \left\{\mathsf{Q} \in \mathrm{O}(\mathbb{E}) \;\middle|\; \det \mathsf{Q} = 1\right\}$$

of **proper orthogonal transformations** *is a subgroup of* $\mathrm{O}(\mathbb{E})$.

Orthogonal tensors share with I an important geometric property:

EXERCISE 24 *Show that, if* $\mathsf{Q} \in \mathrm{O}(\mathbb{E})$, *then* $\mathsf{Q}\mathbf{u} \cdot \mathsf{Q}\mathbf{v} = \mathbf{u} \cdot \mathbf{v}$ *for all* $\mathbf{u}, \mathbf{v} \in \mathbb{E}$. *In particular,* $\|\mathsf{Q}\mathbf{u}\| = \|\mathbf{u}\|$ *for every* $\mathbf{u} \in \mathbb{E}$. *We call any length-preserving tensor an* **isometry**.

From the geometric point of view, proper orthogonal tensors act by rotating vectors, leaving their lengths unchanged. Section 2.1 explores this notion in more detail. An orthogonal tensor having negative determinant effects, in addition, a reflection across an axis. Exercise 24 shows that all orthogonal tensors preserve inner products. The following exercise shows that cross products behave somewhat differently when the transformation involves a reflection.

EXERCISE 25 *Let* $\mathsf{Q} \in \mathrm{O}(\mathbb{E})$ *and* $\mathbf{a}, \mathbf{b}, \mathbf{c} \in \mathbb{E}$. *Show that* $(\mathsf{Q}\mathbf{a} \times \mathsf{Q}\mathbf{b}) \cdot \mathsf{Q}\mathbf{c} = (\det \mathsf{Q})\,(\mathbf{a} \times \mathbf{b}) \cdot \mathbf{c}$. *Show also that* $\mathsf{Q}(\mathbf{a} \times \mathbf{b}) \cdot \mathsf{Q}\mathbf{c} = (\mathbf{a} \times \mathbf{b}) \cdot \mathbf{c}$. *Conclude that* $\mathsf{Q}\mathbf{a} \times \mathsf{Q}\mathbf{b} = (\det \mathsf{Q})\,\mathsf{Q}(\mathbf{a} \times \mathbf{b})$. *Thus, only proper orthogonal tensors preserve cross products, even after accounting for the rotation.*

EXERCISE 26 *Let* $\mathsf{Q} \in \mathrm{O}(\mathbb{E})$, *and suppose that* $\{\mathbf{p}_1, \mathbf{p}_2, \mathbf{p}_3\}$ *is an orthonormal basis for* $\mathbb{E}$. *Show that* $\{\mathsf{Q}\mathbf{p}_1, \mathsf{Q}\mathbf{p}_2, \mathsf{Q}\mathbf{p}_3\}$ *is also an orthonormal basis for* $\mathbb{E}$.

EXERCISE 27 *The converse of Exercise 26 also holds: suppose* $\{\mathsf{Q}\mathbf{p}_1, \mathsf{Q}\mathbf{p}_2, \mathsf{Q}\mathbf{p}_3\}$ *is an orthonormal basis for* $\mathbb{E}$ *whenever* $\{\mathbf{p}_1, \mathbf{p}_2, \mathbf{p}_3\}$ *is an orthonormal basis. Prove that* $\mathsf{Q} \in \mathrm{O}(\mathbb{E})$.

In light of Exercises 26 and 27, to define an orthogonal tensor Q, it suffices to define $\mathsf{Q}\mathbf{p}_1, \mathsf{Q}\mathbf{p}_2, \mathsf{Q}\mathbf{p}_3$ for some orthonormal basis $\{\mathbf{p}_1, \mathbf{p}_2, \mathbf{p}_3\}$, in such a way that $\{\mathsf{Q}\mathbf{p}_1, \mathsf{Q}\mathbf{p}_2, \mathsf{Q}\mathbf{p}_3\}$ also constitutes an orthonormal basis.

1.2.5 Invariants of a Tensor

The determinant of a tensor $\mathsf{A} \in \mathrm{L}(\mathbb{E})$ is just one of many scalars associated with A that do not depend on the choice of bases or coordinate systems. We call such scalars **invariants** of A.

As we establish shortly, the following quantity associated with A is also invariant:

DEFINITION. *The* **trace** *of a dyadic product* $\mathbf{a} \otimes \mathbf{b}$ *is* $\operatorname{tr}(\mathbf{a} \otimes \mathbf{b}) = \mathbf{a} \cdot \mathbf{b}$.

Since every second-order tensor A is a linear combination of dyadic products, the definition of the trace extends by linearity to all of $\mathrm{L}(\mathbb{E})$:

$$\operatorname{tr} \sum_i c_i \, \mathbf{a}_i \otimes \mathbf{b}_i = \sum_i c_i \operatorname{tr}(\mathbf{a}_i \otimes \mathbf{b}_i).$$

EXERCISE 28 *Show that, with respect to any orthonormal basis,*

$$\operatorname{tr} \mathsf{A} = A_{11} + A_{22} + A_{33} = A_{ii}.$$

EXERCISE 29 *Show that* $\operatorname{tr}(\mathsf{AB}) = \operatorname{tr}(\mathsf{BA})$. *Conclude that* $\operatorname{tr}(\mathsf{BAB}^{-1}) = \operatorname{tr}(\mathsf{A})$ *whenever* B *is invertible.*

Using the trace we define, for later chapters, a scalar-valued product of second-order tensors as follows:

DEFINITION. *The* **double-dot product** *of* $\mathsf{A}, \mathsf{B} \in \mathrm{L}(\mathbb{E})$ *is*

$$\mathsf{A} : \mathsf{B} = \operatorname{tr}(\mathsf{A}^\top \mathsf{B}). \tag{1.2.12}$$

Clearly $\operatorname{tr}(\mathsf{A}) = \mathsf{I} : \mathsf{A}$. Additional properties follow almost as easily:

EXERCISE 30 *(a) Show that, with respect to any orthonormal basis,* $\mathsf{A} : \mathsf{B} = A_{ij} B_{ij}$. *(b) Show that the operation* $:$ *possesses the properties of an inner product, as listed in Section 1.1.*

To show that $\operatorname{tr} \mathsf{A}$ is an invariant and to identify other invariants require some basic facts about characteristic polynomials and eigenvalues.

DEFINITION. *If* $\mathsf{A} \in \mathrm{L}(\mathbb{E})$, *its* **characteristic polynomial** *is the cubic expression*

$$p_A(\lambda) = \det(\lambda \mathsf{I} - \mathsf{A}). \tag{1.2.13}$$

The **eigenvalues** *of* A *are the roots* $\lambda_1, \lambda_2, \lambda_3$ *of* p_A.

The condition $p_A(\lambda) = 0$ for a root implies the existence of at least one nonzero vector $\mathbf{p}$ such that $(\lambda \mathsf{I} - \mathsf{A})\mathbf{p} = \mathbf{0}$, that is, $\mathsf{A}\mathbf{p} = \lambda \mathbf{p}$. Geometrically, the action of A in the direction of $\mathbf{p}$ reduces to scalar multiplication. Since A is linear, whenever $\mathbf{p}$ is an eigenvector, so is any scalar multiple of $\mathbf{p}$. Hence, we can choose eigenvectors that have unit length whenever doing so is convenient.

DEFINITION. *If* $\mathsf{A} \in \mathrm{L}(\mathbb{E})$, *a nonzero vector* $\mathbf{p}$ *for which* $\mathsf{A}\mathbf{p} = \lambda\mathbf{p}$ *is an* **eigenvector** *of* A *associated with* λ.

A priori, a root λ of the characteristic polynomial p_A can be either real or complex with nonzero imaginary part. The fact that $p_A(\lambda)$ is cubic with real coefficients guarantees that at least one eigenvalue is real. The other two eigenvalues may be either (1) both real or (2) complex conjugates of each other. If $\lambda = d + iw$ is a complex eigenvalue with nonzero imaginary part w, then any eigenvector that corresponds to λ may have the complex form $\mathbf{p}+i\mathbf{q}$, where $\mathbf{p}, \mathbf{q} \in \mathbb{E}$. We call an eigenvector $\mathbf{p}+i\mathbf{q}$ of A for which $\mathbf{q} = \mathbf{0}$ a **real eigenvector**.

The eigenvalues of A are arguably its most important invariants. But computing them requires finding the roots of a cubic polynomial, which can be inconvenient. To identify another set of invariants for A, observe that $p_A(\lambda) = \lambda^3 + \cdots$ is the only cubic polynomial having leading coefficient 1 that has the roots $\lambda_1, \lambda_2, \lambda_3$. Therefore, A uniquely determines the remaining three coefficients in $p_A(\lambda)$. It follows that these coefficients must be invariants for A.

DEFINITION. *The coefficients* $\mathrm{I}_A, \mathrm{II}_A, \mathrm{III}_A$ *in the expansion*

$$p_A(\lambda) = \lambda^3 - \mathrm{I}_A\lambda^2 + \mathrm{II}_A\lambda - \mathrm{III}_A \tag{1.2.14}$$

are the **principal invariants** *of* A.

EXERCISE 31 *Use the property (1.2.7) of* $\det \mathsf{A}$ *to show that* $p_A(\lambda) = 0$ *if and only if* λ *satisfies the equation*

$$\left[\lambda\mathbf{a} - \mathsf{A}\mathbf{a}, \lambda\mathbf{b} - \mathsf{A}\mathbf{b}, \lambda\mathbf{c} - \mathsf{A}\mathbf{c}\right] = 0 \tag{1.2.15}$$

for all $\mathbf{a}, \mathbf{b}, \mathbf{c} \in \mathbb{E}$.

Straightforward expansion of the vector triple product in Equation (1.2.15) and dividing through by the common factor $\left[\mathbf{a}, \mathbf{b}, \mathbf{c}\right]$ (which we can always arrange to be nonzero by choosing $\mathbf{a}, \mathbf{b}, \mathbf{c}$ appropriately) yield:

$$\begin{aligned} 0 = \lambda^3 &- \frac{\left(\left[\mathsf{A}\mathbf{a}, \mathbf{b}, \mathbf{c}\right] + \left[\mathbf{a}, \mathsf{A}\mathbf{b}, \mathbf{c}\right] + \left[\mathbf{a}, \mathbf{b}, \mathsf{A}\mathbf{c}\right]\right)}{\left[\mathbf{a}, \mathbf{b}, \mathbf{c}\right]}\lambda^2 \\ &+ \frac{\left(\left[\mathbf{a}, \mathsf{A}\mathbf{b}, \mathsf{A}\mathbf{c}\right] + \left[\mathsf{A}\mathbf{a}, \mathbf{b}, \mathsf{A}\mathbf{c}\right] + \left[\mathsf{A}\mathbf{a}, \mathsf{A}\mathbf{b}, \mathbf{c}\right]\right)}{\left[\mathbf{a}, \mathbf{b}, \mathbf{c}\right]}\lambda - \det \mathsf{A} = 0. \end{aligned} \tag{1.2.16}$$

Since it has leading coefficient 1, the right side of this equation is $p_A(\lambda)$. It follows from Equation (1.2.14) that

$$\begin{aligned}
\mathrm{I}_A &= \frac{[\mathsf{A}\mathbf{a}, \mathbf{b}, \mathbf{c}] + [\mathbf{a}, \mathsf{A}\mathbf{b}, \mathbf{c}] + [\mathbf{a}, \mathbf{b}, \mathsf{A}\mathbf{c}]}{[\mathbf{a}, \mathbf{b}, \mathbf{c}]}, \\
\mathrm{II}_A &= \frac{[\mathbf{a}, \mathsf{A}\mathbf{b}, \mathsf{A}\mathbf{c}] + [\mathsf{A}\mathbf{a}, \mathbf{b}, \mathsf{A}\mathbf{c}] + [\mathsf{A}\mathbf{a}, \mathsf{A}\mathbf{b}, \mathbf{c}]}{[\mathbf{a}, \mathbf{b}, \mathbf{c}]}, \\
\mathrm{III}_A &= \det \mathsf{A}, \qquad (1.2.17)
\end{aligned}$$

for any $\mathbf{a}, \mathbf{b}, \mathbf{c}$ having nonzero scalar triple product.

Equations (1.2.17) confirm, in particular, that $\det \mathsf{A}$ is an invariant for A. But directly reducing the more opaque expressions for I_A and II_A requires some effort. The fact that these quantities are invariant under changes in basis for $\mathbb{E}$ affords an alternative approach, as the following exercise illustrates.

EXERCISE 32 *Expand $p_A(\lambda)$ with respect to the orthonormal basis $\{\mathbf{e}_1, \mathbf{e}_2, \mathbf{e}_3\}$ to prove the following identities:*

$$\mathrm{I}_A = \operatorname{tr} \mathsf{A}, \qquad \mathrm{II}_A = \tfrac{1}{2}[(\operatorname{tr} \mathsf{A})^2 - \operatorname{tr}(\mathsf{A}^2)], \qquad \mathrm{III}_A = \det \mathsf{A}. \qquad (1.2.18)$$

From Equations (1.2.17) and (1.2.18) it follows that, for any $\mathbf{a}, \mathbf{b}, \mathbf{c} \in \mathbb{E}$,

$$[\mathbf{a}, \mathbf{b}, \mathbf{c}] \operatorname{tr} \mathsf{A} = [\mathsf{A}\mathbf{a}, \mathbf{b}, \mathbf{c}] + [\mathbf{a}, \mathsf{A}\mathbf{b}, \mathbf{c}] + [\mathbf{a}, \mathbf{b}, \mathsf{A}\mathbf{c}] \qquad (1.2.19)$$

whenever $\mathsf{A} \in \mathrm{L}(\mathbb{E})$.

1.2.6 Derivatives of Tensor-Valued Functions

The remainder of this chapter briefly reviews concepts from the differential calculus of tensor-valued functions. Section 2.2 presents further aspects of differential calculus specific to continuum mechanics, and concepts from integral calculus appear in Section 2.4 and Appendix B.

First, consider vector- and tensor-valued functions of time, regarded as a single parameter $t \in \mathbb{R}$. For such functions, the concept of differentiability extends the concept for scalar-valued functions in a straightforward manner:

DEFINITION. *A vector-valued function $\mathbf{a}\colon \mathbb{R} \to \mathbb{E}$ is* **differentiable** *at $t \in \mathbb{R}$ if the real-valued function $\mathbf{a}(t) \cdot \mathbf{e}$ is differentiable at t for every unit vector $\mathbf{e} \in \mathbb{E}$. A tensor-valued function $\mathsf{A}\colon \mathbb{R} \to \mathrm{L}(\mathbb{E})$ is differentiable at $t \in \mathbb{R}$ if the real-valued function $\mathbf{d} \cdot \mathsf{A}(t)\mathbf{e}$ is differentiable at t for all unit vectors $\mathbf{d}, \mathbf{e} \in \mathbb{E}$.*

The following theorem proves useful in the next chapter:

THEOREM 1.2.4 (DERIVATIVE OF A DETERMINANT). *If* $\mathsf{A}\colon \mathbb{R} \to \mathrm{GL}(\mathbb{E})$ *is differentiable at every* $t \in \mathbb{R}$, *then*

$$\frac{d}{dt}\det \mathsf{A} = \operatorname{tr}\left(\frac{d\mathsf{A}}{dt}\mathsf{A}^{-1}\right)\det \mathsf{A}. \tag{1.2.20}$$

PROOF: Applying the product rule to Equation (1.2.7) yields

$$\begin{aligned}
[\mathbf{a},\mathbf{b},\mathbf{c}]\frac{d}{dt}\det \mathsf{A} &= \left[\frac{d\mathsf{A}}{dt}\mathbf{a},\mathsf{A}\mathbf{b},\mathsf{A}\mathbf{c}\right] + \left[\mathsf{A}\mathbf{a},\frac{d\mathsf{A}}{dt}\mathbf{b},\mathsf{A}\mathbf{c}\right] + \left[\mathsf{A}\mathbf{a},\mathsf{A}\mathbf{b},\frac{d\mathsf{A}}{dt}\mathbf{c}\right]\\
&= \left[\frac{d\mathsf{A}}{dt}\mathsf{A}^{-1}\mathsf{A}\mathbf{a},\mathsf{A}\mathbf{b},\mathsf{A}\mathbf{c}\right] + \left[\mathsf{A}\mathbf{a},\frac{d\mathsf{A}}{dt}\mathsf{A}^{-1}\mathsf{A}\mathbf{b},\mathsf{A}\mathbf{c}\right] + \left[\mathsf{A}\mathbf{a},\mathsf{A}\mathbf{b},\frac{d\mathsf{A}}{dt}\mathsf{A}^{-1}\mathsf{A}\mathbf{c}\right]\\
&= \operatorname{tr}\left(\frac{d\mathsf{A}}{dt}\mathsf{A}^{-1}\right)[\mathsf{A}\mathbf{a},\mathsf{A}\mathbf{b},\mathsf{A}\mathbf{c}] = [\mathbf{a},\mathbf{b},\mathbf{c}]\operatorname{tr}\left(\frac{d\mathsf{A}}{dt}\mathsf{A}^{-1}\right)\det \mathsf{A},
\end{aligned}$$

the last two steps following from the identities (1.2.19) and (1.2.7). Since $\mathbf{a},\mathbf{b},\mathbf{c}$ are arbitrary, Equation (1.2.20) follows. ∎

Now consider vector- and tensor-valued functions of position $\mathbf{x} \in \mathbb{E}$. The concepts of differentiability and the derivatives of real-valued functions of real variables generalize to the following types of functions that arise throughout continuum mechanics:

1. **scalar fields**, that is, real-valued functions $f\colon \mathbb{E} \to \mathbb{R}$ of position;
2. **vector fields**, that is, vector-valued functions $\mathbf{a}\colon \mathbb{E} \to \mathbb{E}$ of position; and
3. **tensor fields**, that is, tensor-valued functions $\mathsf{A}\colon \mathbb{E} \to \mathrm{L}(\mathbb{E})$ of position.

DEFINITION. *A scalar field* $f\colon \mathbb{E} \to \mathbb{R}$ *is* **differentiable** *at* $\mathbf{x}$ *if there exists a unique vector* $\mathbf{g} \in \mathbb{E}$ *such that*

$$\mathbf{g}\cdot\mathbf{e} = \lim_{h\to 0}\frac{f(\mathbf{x}+h\mathbf{e}) - f(\mathbf{x})}{h} \tag{1.2.21}$$

for all unit vectors $\mathbf{e} \in \mathbb{E}$. *In this case, the vector* $\mathbf{g}$ *is the* **derivative** *of* f *at* $\mathbf{x}$, *denoted by* $\mathbf{g} = \operatorname{grad} f(\mathbf{x})$ *and often called the* **gradient** *of* f *at* $\mathbf{x}$.

The quantity $\mathbf{g}\cdot\mathbf{e} = \operatorname{grad} f(\mathbf{x})\cdot\mathbf{e}$ is the **directional derivative** of f at $\mathbf{x}$ in the direction $\mathbf{e}$. When f is differentiable at every point $\mathbf{x}$ in a region of $\mathbb{E}$, we treat $\operatorname{grad} f$ as a function of $\mathbf{x}$. In Equation (1.2.21), substituting for $\mathbf{e}$ the basis vectors

$\mathbf{e}_1, \mathbf{e}_2, \mathbf{e}_3$ gives the partial derivatives $\partial f / \partial x_i$ associated with a Cartesian coordinate system for $\mathbb{E}$ and yields the following representation for $\operatorname{grad} f$ in $\mathbb{R}^3$:

$$\begin{bmatrix} \partial f / \partial x_1 \\ \partial f / \partial x_2 \\ \partial f / \partial x_3 \end{bmatrix}.$$

In using coordinate representations of vector and tensor fields and their derivatives, the following index notation meshes well with the Einstein summation convention: for any differentiable scalar field f,

$$\frac{\partial f}{\partial x_i} = f_{,i}.$$

Thus, for example, $\operatorname{grad} f = f_{,i}\mathbf{e}_i$.

These concepts extend to vector and tensor fields as follows:

DEFINITION. *A vector field* $\mathbf{a}\colon \mathbb{E} \to \mathbb{E}$ *is* **differentiable** *at* $\mathbf{x}$ *if the scalar field* $\mathbf{a}(\mathbf{x})\cdot\mathbf{e}$ *is differentiable at* $\mathbf{x}$ *for every unit vector* $\mathbf{e} \in \mathbb{E}$. *A tensor field* $\mathsf{A}\colon \mathbb{E} \to \mathrm{L}(\mathbb{E})$ *is* **differentiable** *at* $\mathbf{x}$ *if the scalar field* $\mathbf{d} \cdot \mathsf{A}(\mathbf{x})\mathbf{e}$ *is differentiable at* $\mathbf{x}$ *for all unit vectors* $\mathbf{d}, \mathbf{e} \in \mathbb{E}$. *We denote the derivatives of* $\mathbf{a}$ *and* A *at* $\mathbf{x}$ *by* $\operatorname{grad}\mathbf{a}(\mathbf{x})$ *and* $\operatorname{grad}\mathsf{A}(\mathbf{x})$, *respectively.*

As a function of $\mathbf{x}$, the derivative $\operatorname{grad}\mathbf{a}$ is a tensor field, defined by its action: for any vector $\mathbf{b} \in \mathbb{E}$:

$$(\operatorname{grad}\mathbf{a})^{\top}\mathbf{b} = \operatorname{grad}(\mathbf{a}\cdot\mathbf{b}).$$

By extending the reasoning given for $\operatorname{grad} f$, substituting the basis vectors $\mathbf{e}_1, \mathbf{e}_2, \mathbf{e}_3$ for the general unit vector $\mathbf{e}$ yields a coordinate representation for $\operatorname{grad}\mathbf{a}(\mathbf{x})$:

$$\operatorname{grad}\mathbf{a} = \frac{\partial a_i}{\partial x_j}\mathbf{e}_i \otimes \mathbf{e}_j = a_{i,j}\mathbf{e}_i \otimes \mathbf{e}_j,$$

which has the following 3×3 matrix representation:

$$\begin{bmatrix} \partial a_1/\partial x_1 & \partial a_1/\partial x_2 & \partial a_1/\partial x_3 \\ \partial a_2/\partial x_1 & \partial a_2/\partial x_2 & \partial a_2/\partial x_3 \\ \partial a_3/\partial x_1 & \partial a_3/\partial x_2 & \partial a_3/\partial x_3 \end{bmatrix}.$$

Several related differential operators arise in mechanics:

DEFINITION. *The* **divergence** *of a differentiable vector field* $\mathbf{a}$ *is*

$$\operatorname{div}\mathbf{a} = \operatorname{tr}(\operatorname{grad}\mathbf{a}). \tag{1.2.22}$$

*The **curl** of $\mathbf{a}$ is a vector field having the following action: for any $\mathbf{b} \in \mathbb{E}$,*

$$(\operatorname{curl} \mathbf{a}) \cdot \mathbf{b} = \operatorname{div}(\mathbf{a} \times \mathbf{b}). \tag{1.2.23}$$

Although the direct use of grad A arises far more rarely, we have occasion to use the following quantity:

DEFINITION. *The **divergence**[1] of a differentiable tensor field A is a vector field having the following action: for all $\mathbf{b} \in \mathbb{E}$,*

$$(\operatorname{div} \mathsf{A}) \cdot \mathbf{b} = \operatorname{div}(\mathsf{A}^\top \mathbf{b}). \tag{1.2.24}$$

EXERCISE 33 *Suppose that $\mathbf{a}(\mathbf{x})$ and $\mathbf{b}(\mathbf{x})$ are differentiable vector fields. Show that*

$$\operatorname{div}(\mathbf{a} \otimes \mathbf{b}) = \begin{bmatrix} \operatorname{div}(a_1 \mathbf{b}) \\ \operatorname{div}(a_2 \mathbf{b}) \\ \operatorname{div}(a_3 \mathbf{b}) \end{bmatrix}.$$

EXERCISE 34 *Using Cartesian coordinates, show that if $\mathbf{a} \colon \mathbb{E} \to \mathbb{E}$ and $\mathsf{A} \colon \mathbb{E} \to \mathrm{L}(\mathbb{E})$ are differentiable,*

$$\operatorname{div} \mathbf{a} = \frac{\partial a_i}{\partial x_i} \quad \textit{and} \quad \operatorname{curl} \mathbf{a} = \varepsilon_{ijk} \frac{\partial a_i}{\partial x_j} \mathbf{e}_k = \varepsilon_{ijk} a_{i,j} \mathbf{e}_k. \tag{1.2.25}$$

Show that

$$\operatorname{div} \mathsf{A} = \frac{\partial A_{ij}}{\partial x_j} \mathbf{e}_i = A_{ij,j} \mathbf{e}_i.$$

1.2.7 Summary

Tensors arise in continuum mechanics as linear transformations of vectors in $\mathbb{E}$. We denote the space of all such transformations as $\mathrm{L}(\mathbb{E})$. Although matrix representations of tensors are useful for some purposes, their essential properties—including their actions on vectors and such scalar invariants as the trace and determinant—have geometric and algebraic significance independent of any choice of basis for $\mathbb{E}$. Differentiation of vector- and tensor-valued functions simply extends the concept of differentiation of scalar-valued functions.

[1] Some but not all authors adopt the transpose of this definition, along with corresponding changes in Theorem B.2.2 and its consequences. See, for example, [12, p. 40]. The notation adopted here is consistent with that used, for example, in [35, p. 278] and [9, p. 247].

CHAPTER 2

KINEMATICS I: THE CALCULUS OF MOTION

Kinematics describes motion, without reference to masses or forces. This chapter introduces the elements of kinematics, starting with the primitive concepts of particles and bodies and developing the basic analytic and algebraic notions required to describe the time-dependent behavior of these objects in space-time.

2.1 BODIES, MOTIONS, AND DEFORMATIONS

A **body** is a set $\mathcal{B}$ of **particles**, sometimes called **material points** X. To give the body mathematical structure, we postulate the existence of a mapping that associates the body $\mathcal{B}$ with an open, connected subset of $\mathbb{E}$. Specifically,

1. There exists a one-to-one, continuously differentiable function $\chi\colon \mathcal{B} \to \mathbb{E}$ that has continuously differentiable inverse $\chi^{-1}\colon \chi(\mathcal{B}) \to \mathcal{B}$. A function having these properties is a **diffeomorphism**.

Continuum Mechanics: The Birthplace of Mathematical Models, First Edition. M.B. Allen.

2. The image $\chi(\mathcal{B})$ of $\mathcal{B}$ under the function χ is a connected, open subset of $\mathbb{E}$.

We call any such mapping a **configuration** of $\mathcal{B}$.

Conceptually, a configuration assigns to each particle $X \in \mathcal{B}$ a position $\mathbf{x} \in \mathbb{E}$. It need not result in an arrangement of particles that the body ever actually occupies.

This foundation bears two technical remarks. First, to refer to a diffeomorphism presupposes that it makes sense to speak of open sets of particles in $\mathcal{B}$. All of the subsets of $\mathcal{B}$ that are of interest are obtainable from open subsets by at most countably many operations involving unions, intersections, and complements. This condition allows us to integrate over the subsets. We call these subsets **parts** of $\mathcal{B}$. We write $\mathcal{P} \subset \mathcal{B}$ to mean "$\mathcal{P}$ is a part of $\mathcal{B}$," even though in common usage the symbol $\subset$ strictly admits all subsets of $\mathcal{B}$, some far less tame.

Second, while these assumptions about parts suffice for the most rudimentary purposes, to reach all of the results presented in this text we frequently make further assumptions. In many instances we analyze integrals over images of parts under configurations, implicitly assuming that these subsets of $\mathbb{E}$ are regions for which classic theorems of vector calculus hold. This assumption, which we henceforth make without further notice, imposes additional requirements outlined in Appendix B.

Given a configuration χ, we call the image $\chi(X)$ of a particle $X \in \mathcal{B}$ the **position** of X in the configuration, sketched in Figure 2.1. The following definition accommodates the possibility that positions can change over time.

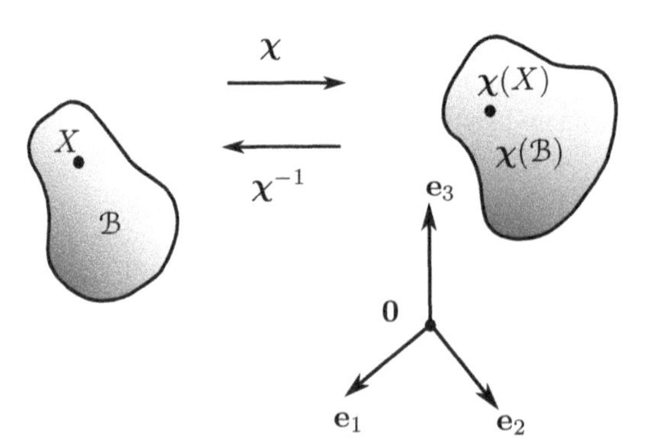

Figure 2.1. A configuration of $\mathcal{B}$, showing the position of a particle X.

DEFINITION. *A* **motion** *of $\mathcal{B}$ is a one-parameter family $\chi(\cdot, t)\colon \mathcal{B} \to \mathbb{E}$ of configurations, with the parameter t ranging over an interval in $\mathbb{R}$.*

The vector $\mathbf{x} = \chi(X, t)$ is the **position** of X at time t. Conversely, the relationship $X = \chi^{-1}(\mathbf{x}, t)$ tells which particle X occupies the spatial position $\mathbf{x}$ at t. Figure 2.2 illustrates the image of $\mathcal{B}$ under a motion at two distinct times. In what follows, we assume that $\chi(X, t)$ is differentiable with respect to t to whatever order the context requires, except for special cases noted explicitly.

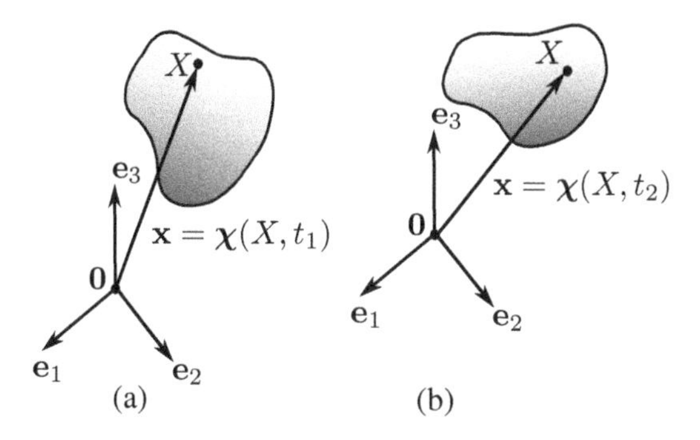

Figure 2.2. Image of a body $\mathcal{B}$ under a motion shown at two distinct times (a) t_1 and (b) t_2.

The program of axiomatizing continuum mechanics is not entirely settled. This book tries to identify axioms that seem to have attained broad acceptance. The first is basic to all that follows.

AXIOM (CONTINUITY). *Associated with any body is a motion.*

Labeling particles by abstract symbols such as X is similar to common practice in the classical mechanics of discrete particles, where one uses labels such as "1," "2," and so forth. But in many respects it is too abstract. Let us associate with $\mathcal{B}$ a **reference configuration** $\boldsymbol{\kappa}\colon \mathcal{B} \to \mathbb{E}$. It is useful in some applications to choose $\boldsymbol{\kappa}(\mathcal{B})$ to be the set of positions occupied by the particles of $\mathcal{B}$ at some particular time, but such a choice is not necessary. The reference configuration $\boldsymbol{\kappa}$ furnishes a more concrete labeling system $\mathbf{X} = \boldsymbol{\kappa}(X)$, which has the potential virtue of identifying the particle in terms of its position in some configuration, as shown in Figure 2.3. Think of $\boldsymbol{\kappa}(X)$ as the home address of the particle X.

We now have two addresses for any point $X \in \mathcal{B}$. The first is the vector of **referential coordinates** $\mathbf{X}$. With respect to the standard orthonormal basis $\{\mathbf{e}_1, \mathbf{e}_2, \mathbf{e}_3\}$,

$$\boldsymbol{\kappa}(X) = \mathbf{X} = \begin{bmatrix} X_1 \\ X_2 \\ X_3 \end{bmatrix}.$$

These coordinates refer to particles by their positions in the reference configuration. The second is the vector of **spatial coordinates** $\mathbf{x}$. With respect to the same

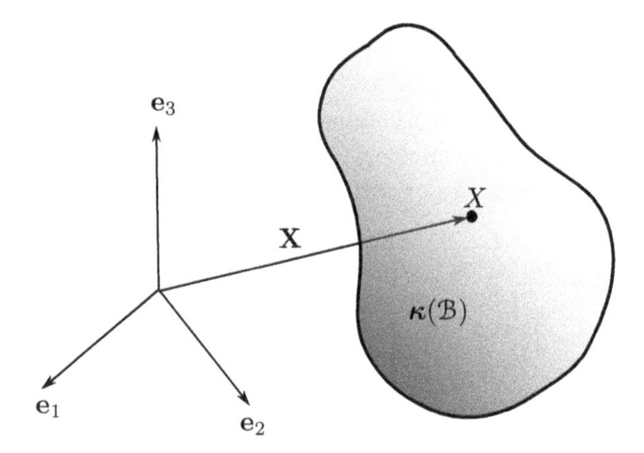

Figure 2.3. Image of a body $\mathcal{B}$ under a reference configuration $\boldsymbol{\kappa}$, showing the referential coordinates $\mathbf{X}$ of a particle X.

orthonormal basis $\{\mathbf{e}_1, \mathbf{e}_2, \mathbf{e}_3\}$,

$$\chi(X,t) = \mathbf{x} = \begin{bmatrix} x_1 \\ x_2 \\ x_3 \end{bmatrix}.$$

These coordinates give the position of particle X at time t under the motion χ.

2.1.1 Deformation

To gain access to the tools of calculus, it is useful to define a mapping that carries the reference configuration to the configuration at time t under the motion. Figure 2.4 suggests how to do this, and the following definition makes the notion precise.

DEFINITION. *The* **deformation** *is the mapping* $\chi_\kappa\colon \boldsymbol{\kappa}(\mathcal{B}) \times \mathbb{R} \to \mathbb{E}$ *defined by the rule*

$$\chi_\kappa(\mathbf{X},t) = \chi(\boldsymbol{\kappa}^{-1}(\mathbf{X}),t) = (\chi \circ \boldsymbol{\kappa}^{-1})(\mathbf{X},t). \tag{2.1.1}$$

EXERCISE 35 *Verify that* $(\chi \circ \boldsymbol{\kappa}^{-1})^{-1} = \boldsymbol{\kappa} \circ \chi^{-1}$.

Informally, we often write $\mathbf{x} = \mathbf{x}(\mathbf{X},t) = \chi_\kappa(\mathbf{X},t)$. The deformation has an inverse $\chi_\kappa^{-1}\colon \chi(\mathcal{B},t) \times \mathbb{R} \to \mathbb{E}$, defined as follows:

$$\chi_\kappa^{-1}(\mathbf{x},t) = \boldsymbol{\kappa}(\chi^{-1}(\mathbf{x},t)).$$

Informally, we write $\mathbf{X} = \mathbf{X}(\mathbf{x}, t) = \chi_\kappa^{-1}(\mathbf{x}, t)$. The deformation and its inverse play central roles in the remainder of the text.

Some applications require a gauge of how far the motion has displaced a particle $\mathbf{X}$ from its reference position.

DEFINITION. *The* **displacement** *of* $\mathbf{X}$ *is* $\mathbf{u}(\mathbf{X}, t) = \chi_\kappa(\mathbf{X}, t) - \mathbf{X}$.

We sometimes write this definition as $\mathbf{u} = \mathbf{x} - \mathbf{X}$. Figure 2.4 illustrates the displacement.

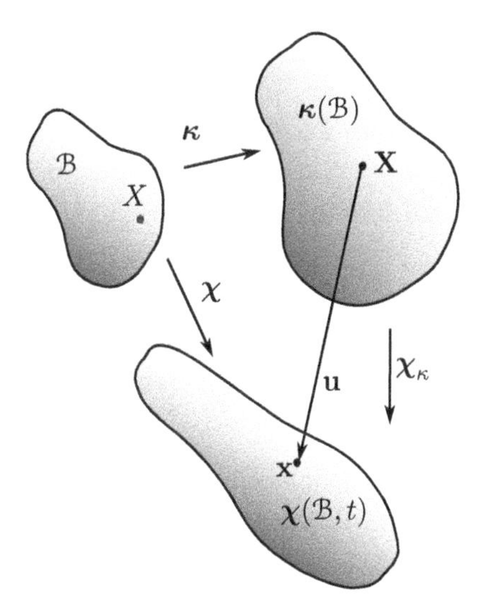

Figure 2.4. Mapping the reference configuration into the current configuration given by the motion.

The definition of χ_κ guarantees that it is a diffeomorphism at each time t, since both χ and κ are diffeomorphisms. The fact that χ_κ and χ_κ^{-1} are also one-to-one implies that the deformation maps distinct particles onto distinct particles. In the literature on continuum mechanics one sometimes sees this property referred to as the impenetrability of matter.

2.1.2 Examples of Motions

Three simple building blocks of motion help to illustrate the concepts. The first is **pure translation**,

$$\mathbf{x}(\mathbf{X}, t) = \mathbf{X} + \mathbf{c}(t),$$

drawn in Figure 2.5. This motion displaces each particle $\mathbf{X}$ from its reference position by a continuously differentiable, vector-valued function $\mathbf{c}(t)$. Since $\mathbf{c}(t)$ is independent of $\mathbf{X}$, the body retains its original shape and orientation throughout the motion.

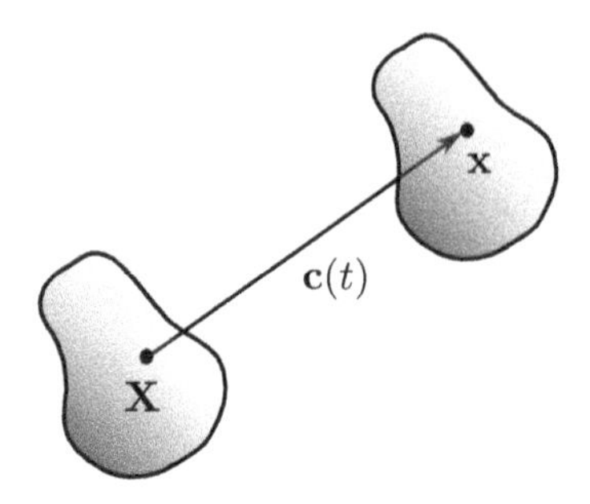

Figure 2.5. Pure translation.

The second example is **pure rotation**, illustrated in Figure 2.6. This motion has the form

$$\mathbf{x}(\mathbf{X}, t) = \mathsf{Q}(t)\mathbf{X}.$$

Here $\mathsf{Q}(t)$ is a continuously differentiable, one-parameter family $\mathsf{Q}\colon \mathbb{R} \to \mathrm{L}(\mathbb{E})$ of functions with each $\mathsf{Q}(t) \in \mathrm{O}^+(\mathbb{E})$. Thus $\mathsf{Q}(t)$ is an isometry at each instant t, as discussed in Exercise 24.

Examining the action of this motion on orthonormal basis vectors reveals more about the properties of Q. In particular, $\mathbf{x}(\mathbf{e}_j, t) = \mathsf{Q}(t)\mathbf{e}_j$, or in Cartesian coordinates,

$$x_i(\mathbf{e}_j, t) = \sum_{k=1}^{3} Q_{ik}(t)\delta_{kj} = Q_{ik}(t)\delta_{kj}, \qquad \text{(summation convention)}$$

where δ_{kj} is the Kronecker symbol defined in Exercise 18. It follows that $x_i(\mathbf{e}_j, t) = Q_{ij}(t)$. But by Exercise 1, $x_i(\mathbf{e}_j, t) = \mathbf{x}(\mathbf{e}_j, t) \cdot \mathbf{e}_i$. Because Q is an isometry, $\|\mathbf{x}(\mathbf{e}_j, t)\| = \|\mathbf{e}_j\| = 1$, and

$$x_i(\mathbf{e}_j, t) = \mathbf{x}(\mathbf{e}_j, t) \cdot \mathbf{e}_i = \cos\theta_{ij}(t),$$

where $\theta_{ij}(t)$ denotes the angle that $\mathbf{e}_i$ makes with the image of $\mathbf{e}_j$ at time t under the motion, as drawn in Figure 2.7. Therefore, the entry $Q_{ij}(t)$ with respect to the basis $\{\mathbf{e}_1, \mathbf{e}_2, \mathbf{e}_3\}$ is the direction cosine of the motion at time t.

Both pure translations and pure rotations are examples of **rigid** motions, that is, motions that preserve shapes. More specifically, a motion is rigid if all distances between particles and all angles between line segments connecting pairs of particles remain invariant. The most general rigid motion is a superposition of pure translation and pure rotation,

$$\mathbf{x}(\mathbf{X}, t) = \mathsf{Q}(t)\mathbf{X} + \mathbf{c}(t).$$

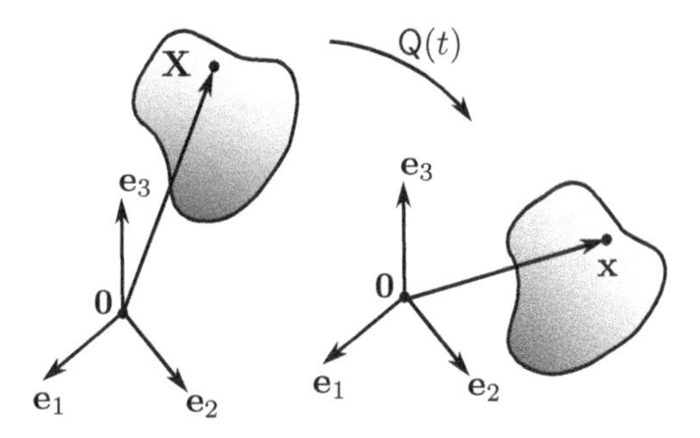

Figure 2.6. Pure rotation.

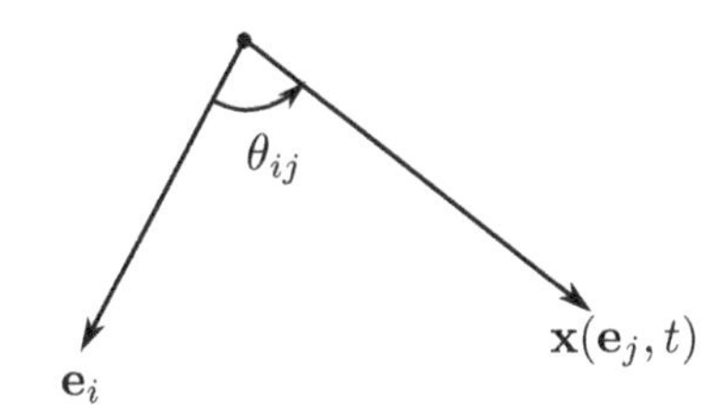

Figure 2.7. The angle $\theta_{ij}(t)$ produced by pure rotation.

The third example is **simple shear** parallel to $\mathbf{e}_1$, which has the form

$$\mathbf{x}(\mathbf{X}, t) = [\mathsf{I} + \alpha(t)\mathbf{e}_1 \otimes \mathbf{e}_2]\mathbf{X}. \tag{2.1.2}$$

With respect to the standard orthonormal basis, this motion has matrix representation

$$\mathbf{x}(\mathbf{X}, t) = \begin{bmatrix} 1 & \alpha(t) & 0 \\ 0 & 1 & 0 \\ 0 & 0 & 1 \end{bmatrix} \mathbf{X}.$$

It has the effect shown in Figure 2.8. The scalar function $\alpha(t)$ gives the tangent of the angle through which the motion displaces particles that are initially located along lines parallel to $\mathbf{e}_2$. In this motion, sets of material points that lie in a plane of the form $X_2 = \text{constant}$ in the reference configuration undergo uniform translation in the plane $x_2 = X_2$, so that at time t the translation vector is $\alpha(t)X_2\mathbf{e}_1$. We refer to planes in which the shearing reduces to uniform translation as **glide planes**.

EXERCISE 36 *Write a matrix representation of simple shear parallel to* $\mathbf{e}_2$. *Do the same for* $\mathbf{e}_3$. *Identify the glide planes for each motion.*

EXERCISE 37 *The general form of a simple shear parallel to one of the standard orthonormal basis vectors involves a tensor-valued function of the form* $\mathsf{S}(t) = \mathsf{I} +$

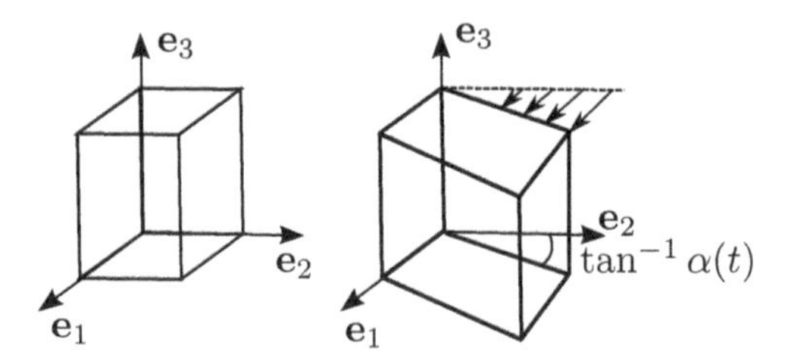

Figure 2.8. Simple shear for $\alpha(t)$.

$\alpha(t)\mathbf{e}_i \otimes \mathbf{e}_j$, *where* $i \neq j$. *Under what conditions do tensors of this form (and hence the associated simple shears) commute?*

2.1.3 Summary

A body $\mathcal{B}$ is a set of particles in Euclidean point space. A configuration of $\mathcal{B}$ maps $\mathcal{B}$ into the vector space $\mathbb{E}$, and a motion is a time-dependent family $\chi(\cdot, t)\colon \mathcal{B} \to \mathbb{E}$ of configurations that gives the spatial position of each particle at every time t.

Assigning a reference configuration $\boldsymbol{\kappa}\colon \mathcal{B} \to \mathbb{E}$ allows us to refer to particles by their positions $\mathbf{X}$ in that configuration. Once we have a reference configuration it is useful to define the deformation, that is, the mapping $\chi_\kappa = \chi \circ \boldsymbol{\kappa}^{-1}$. Knowing this mapping, we can then refer to a particle $\mathbf{X}$ by its current position $\mathbf{x} = \chi_\kappa(\mathbf{X}, t) \in \mathbb{E}$.

2.2 DERIVATIVES OF MOTION

As is typical in physics, it is often easier to get information about rates of change of the motion than to find the motion directly. This section examines the derivatives of the deformation $\chi_\kappa(\mathbf{X}, t)$ and related functions. In mechanics, at least as much as in other fields, it is crucial to keep in mind what the arguments of a function are and which ones are fixed during partial differentiation. We begin with time derivatives, then consider derivatives with respect to position. Throughout this section, we assume that $\chi_\kappa(\mathbf{X}, t)$ is differentiable to whatever order is required for the exposition to make sense.

2.2.1 Time Derivatives

DEFINITION. *The* **velocity** *of the particle* $\mathbf{X}$ *is*

$$\frac{\partial \mathbf{x}}{\partial t}(\mathbf{X}, t) = \frac{\partial \boldsymbol{\chi}_\kappa}{\partial t}(\mathbf{X}, t).$$

The **acceleration** *is*

$$\frac{\partial^2 \mathbf{x}}{\partial t^2}(\mathbf{X}, t) = \frac{\partial^2 \boldsymbol{\chi}_\kappa}{\partial t^2}(\mathbf{X}, t).$$

Since $\mathbf{X}$ is fixed during the differentiation, one can recast these definitions in terms of the displacement:

$$\frac{\partial \mathbf{x}}{\partial t}(\mathbf{X}, t) = \frac{\partial \mathbf{u}}{\partial t}(\mathbf{X}, t), \qquad \frac{\partial^2 \mathbf{x}}{\partial t^2}(\mathbf{X}, t) = \frac{\partial^2 \mathbf{u}}{\partial t^2}(\mathbf{X}, t).$$

These definitions serve as continuum analogs of those used in classical discrete mechanics, where, for example, the velocity of particle j is $d\mathbf{x}_j/dt$. In continuum mechanics the existence of several ways of referring to particles complicates the situation. Any quantity that varies with particle and time admits three descriptions:

- The **material description**, where we regard the quantity as a function of (X, t). We use this description sparingly.
- The **referential description**, where we regard the quantity as a function of $(\mathbf{X}, t) = (\boldsymbol{\kappa}(X), t)$. Formally, this description depends on the choice of reference configuration: $\boldsymbol{\kappa}$. In practice we often suppress reference to $\boldsymbol{\kappa}$. It is common to call this description the **Lagrangian description**.
- The **spatial description**, where we regard the quantity as a function of $(\mathbf{x}, t) = (\boldsymbol{\chi}_\kappa(\mathbf{X}, t), t) = (\boldsymbol{\chi}(X, t), t)$. Some people call this the **Eulerian description**.

The referential description tracks what happens to individual particles $\mathbf{X}$, which presumably are the real focus of the physics. The spatial description is useful in settings, such as certain fluid-mechanical applications, where it is easier in practice to observe what is happening at specified positions $\mathbf{x}$. For the velocity, especially, we use both descriptions frequently, distinguishing them via notation:

$$\underbrace{\frac{\partial \mathbf{x}}{\partial t}(\mathbf{X}, t) = \frac{\partial \boldsymbol{\chi}_\kappa}{\partial t}(\mathbf{X}, t)}_{\text{Referential velocity}}, \qquad \underbrace{\mathbf{v}(\mathbf{x}, t) = \frac{\partial \mathbf{x}}{\partial t}(\boldsymbol{\chi}_\kappa^{-1}(\mathbf{x}, t), t).}_{\text{Spatial velocity}}$$

EXERCISE 38 *Show that $\mathbf{v}(\mathbf{x}, t)$ is independent of the choice of reference configuration.*

Differentiation with respect to t, holding the particle $\mathbf{X}$ fixed, is such a useful operation that we adopt notation for it that we use in both the referential and spatial descriptions. First, consider the referential description.

DEFINITION. *The* **material derivative** *of a differentiable function $f(\mathbf{X}, t)$ is*

$$\frac{Df}{Dt}(\mathbf{X}, t) = \frac{\partial f}{\partial t}(\mathbf{X}, t).$$

The idea is to evaluate the temporal rate of change of f following a fixed particle $\mathbf{X}$—a process more natural in referential coordinates than in spatial coordinates.

Nevertheless, this rate of change is also important in spatial coordinates. We examine shortly how to evaluate Df/Dt when $f = f(\mathbf{x}, t)$.

EXERCISE 39 *Compute $\partial\mathbf{x}/\partial t$ for each of the three examples in Section 2.1.*

2.2.2 Derivatives With Respect to Position

The availability of more than one set of coordinates for particles makes it necessary to define derivatives with respect to position with some care.

Following the review in Section 1.2, for a real-valued differentiable function $f \colon \mathbb{E} \times \mathbb{R} \to \mathbb{R}$, we denote by $\operatorname{grad} f(\mathbf{x}, t)$ the derivative of f with respect to the spatial position $\mathbf{x}$. This derivative is the gradient of f, which has the following representation in Cartesian coordinates:

$$\operatorname{grad} f(\mathbf{x}, t) = \left(\frac{\partial f}{\partial x_1}(\mathbf{x}, t), \frac{\partial f}{\partial x_2}(\mathbf{x}, t), \frac{\partial f}{\partial x_3}(\mathbf{x}, t)\right)^{\top} = \frac{\partial f}{\partial x_j}(\mathbf{x}, t)\, \mathbf{e}_j, \qquad (2.2.1)$$

with the summation convention understood in the last expression. The compact notation for spatial partial derivatives introduced in Section 1.2 allows us to write Equation (2.2.1) more briefly as follows:

$$\operatorname{grad} f(\mathbf{x}, t) = f_{,j}(\mathbf{x}, t)\, \mathbf{e}_j.$$

For a function $f(\mathbf{X}, t)$ in the referential description, the appropriate gradient is

$$\operatorname{Grad} f(\mathbf{X}, t) = \frac{\partial f}{\partial X_j}(\mathbf{X}, t)\, \mathbf{e}_j$$

in Cartesian coordinates. To reduce the risk of confusion, we avoid using index notation to abbreviate $\partial/\partial X_j$.

THEOREM 2.2.1 (MATERIAL DERIVATIVE IN THE SPATIAL DESCRIPTION). *If $f(\mathbf{x}, t)$ is differentiable, then*

$$\frac{Df}{Dt} = \frac{\partial f}{\partial t} + \mathbf{v} \cdot \operatorname{grad} f. \tag{2.2.2}$$

PROOF: We work in Cartesian coordinates. The chain rule (with the summation convention) gives

$$\frac{Df}{Dt}(\mathbf{x}, t) = \frac{\partial f}{\partial x_j}(\mathbf{x}(\mathbf{X}, t), t)\frac{\partial x_j}{\partial t}(\mathbf{X}, t) + \partial_2 f(\mathbf{x}(\mathbf{X}, t), t).$$

The notation ∂_2 in the last term indicates differentiation of f with respect to its last argument, keeping other arguments fixed. But by definition,

$$\frac{\partial x_j}{\partial t}(\mathbf{X}(\mathbf{x}, t), t) = v_j(\mathbf{x}, t),$$

and

$$\partial_2 f(\mathbf{x}, t) = \frac{\partial f}{\partial t}(\mathbf{x}, t).$$

It follows that

$$\begin{aligned} \frac{Df}{Dt}(\mathbf{x}, t) &= \frac{\partial f}{\partial t}(\mathbf{x}, t) + \operatorname{grad} f(\mathbf{x}, t) \cdot \mathbf{v}(\mathbf{x}, t) \\ &= \frac{\partial f}{\partial t}(\mathbf{x}, t) + v_j(\mathbf{x}, t)\, f_{,j}(\mathbf{x}, t). \end{aligned}$$

This completes the proof. ■

Recall that for a differentiable vector field $\mathbf{f}(\mathbf{x}, t)$,

$$\operatorname{grad} \mathbf{f}(\mathbf{x}, t) = \frac{\partial f_j}{\partial x_k}\, \mathbf{e}_j \otimes \mathbf{e}_k = f_{j,k}\, \mathbf{e}_j \otimes \mathbf{e}_k,$$

with respect to the orthonormal basis $\{\mathbf{e}_1, \mathbf{e}_2, \mathbf{e}_3\}$. In terms of the coordinate functions $f_i(\mathbf{x}, t) = \mathbf{e}_i \cdot \mathbf{f}(\mathbf{x}, t)$, the identity (2.2.2) extends to vector fields as follows:

$$\frac{Df_i}{Dt} = \frac{\partial f_i}{\partial t} + v_j \frac{\partial f_i}{\partial x_j},$$

with the summation convention understood. To write this expression in direct notation, we extend the notation $v_j \partial/\partial x_j = \mathbf{v} \cdot \operatorname{grad}$ to vector-valued functions, yielding

$$\frac{D\mathbf{f}}{Dt} = \frac{\partial \mathbf{f}}{\partial t} + (\mathbf{v} \cdot \operatorname{grad})\mathbf{f} = \left(\frac{\partial f_i}{\partial t} + v_j \frac{\partial f_i}{\partial x_j}\right) \mathbf{e}_i. \tag{2.2.3}$$

The gradient $\operatorname{Grad}\mathbf{f}(\mathbf{X},t)$ with respect to referential coordinates has a representation analogous to that for $\mathbf{f}(\mathbf{x},t)$. With respect to this same basis, divergences of differentiable functions of $(\mathbf{x},t)$ and $(\mathbf{X},t)$ have the following representations:

$$\operatorname{div}\mathbf{f}(\mathbf{x},t) = \frac{\partial f_j}{\partial x_j}(\mathbf{x},t),$$

$$\operatorname{Div}\mathbf{f}(\mathbf{X},t) = \frac{\partial f_j}{\partial X_j}(\mathbf{X},t),$$

again with the summation convention understood.

2.2.3 The Deformation Gradient

The following derivative is central to all of continuum mechanics.

DEFINITION. *The* **deformation gradient** *is*

$$\mathsf{F}(\mathbf{X},t) = \operatorname{Grad}\chi_\kappa(\mathbf{X},t).$$

Thus the tensor-valued function F is a referential quantity. We sometimes write $\mathsf{F}(\mathbf{X},t) = \operatorname{Grad}\mathbf{x}(\mathbf{X},t)$. At each point $(\mathbf{X},t)$, the deformation gradient has representation

$$\mathsf{F}(\mathbf{X},t) = \frac{\partial x_i}{\partial X_j}(\mathbf{X},t)\,\mathbf{e}_i \otimes \mathbf{e}_j. \tag{2.2.4}$$

EXERCISE 40 *The representation (2.2.4) uses the orthonormal basis* $\{\mathbf{e}_1,\mathbf{e}_2,\mathbf{e}_3\}$ *for* $\mathbb{E}$ *in both the reference configuration* $\boldsymbol{\kappa}(\mathcal{B})$ *and the current configuration* $\chi(\mathcal{B},t)$. *Suppose instead that we use a different orthonormal basis* $\{\hat{\mathbf{e}}_1,\hat{\mathbf{e}}_2,\hat{\mathbf{e}}_3\}$ *for* $\mathbb{E}$ *in the reference configuration. Use the expansions* $\mathbf{x} = x_i\mathbf{e}_i$ *and* $\operatorname{Grad} f = (\partial f/\partial X_j)\hat{\mathbf{e}}_j$, *together with the fact that* $(\operatorname{Grad}\mathbf{f})^\top\mathbf{b} = \operatorname{Grad}(\mathbf{f}\cdot\mathbf{b})$ *for every* $\mathbf{b} = b_i\mathbf{e}_i \in \mathbb{E}$, *to establish the relationship*

$$\mathsf{F}(\mathbf{X},t) = \operatorname{Grad}\mathbf{x} = \frac{\partial x_i}{\partial X_j}\,\mathbf{e}_i \otimes \hat{\mathbf{e}}_j.$$

One way to describe the result of Exercise 40 is to say that F has one "leg" in the current configuration and the other in the reference configuration.

THEOREM 2.2.2 (INVERTIBILITY OF THE DEFORMATION GRADIENT). $\mathsf{F}(\mathbf{X},t)$ *is invertible at each* $(\mathbf{X},t)$, *and*

$$\mathsf{F}^{-1}(\mathbf{X},t) = \operatorname{grad}\chi_\kappa^{-1}(\mathbf{x}(\mathbf{X},t),t).$$

PROOF: Because χ_κ is a diffeomorphism, we need only show that the formula works. With respect to the basis $\{\mathbf{e}_1, \mathbf{e}_2, \mathbf{e}_3\}$,

$$\operatorname{grad} \chi_\kappa^{-1}(\mathbf{x}, t) = \frac{\partial X_k}{\partial x_l}(\mathbf{x}, t)\, \mathbf{e}_k \otimes \mathbf{e}_l.$$

But by the chain rule,

$$\frac{\partial x_j}{\partial X_k}\frac{\partial X_k}{\partial x_l} = \frac{\partial x_j}{\partial x_l} = \delta_{jl},$$

so $\mathsf{F}\mathsf{F}^{-1} = \mathsf{I}$. Similarly, $\mathsf{F}^{-1}\mathsf{F} = \mathsf{I}$. ■

The invertibility of F implies that $\det \mathsf{F}(\mathbf{X}, t) \neq 0$.

EXERCISE 41 *Compute χ_κ^{-1}, F, and F^{-1} for pure rotation and simple shear.*

Two other gradients associated with the motion appear frequently.

DEFINITION. *The* **displacement gradient** *is*

$$\mathsf{H}(\mathbf{X}, t) = \operatorname{Grad} \mathbf{u}(\mathbf{X}, t),$$

a referential quantity. The **velocity gradient** *is*

$$\mathsf{L}(\mathbf{x}, t) = \operatorname{grad} \mathbf{v}(\mathbf{x}, t),$$

a spatial quantity.

The following identity is easy to verify:

$$\mathsf{H} = \mathsf{F} - \mathsf{I}. \qquad (2.2.5)$$

EXERCISE 42 *Assume that all entries of the displacement gradient H are small, in the sense that $|\partial u_i/\partial X_j| < \epsilon$ for some small, positive parameter $\epsilon < 1$. Working in Cartesian coordinates, justify the approximation*

$$\det \mathsf{F} \simeq 1 + \frac{\partial u_1}{\partial X_1} + \frac{\partial u_2}{\partial X_2} + \frac{\partial u_3}{\partial X_3}$$

by ignoring terms proportional to ϵ^2.

THEOREM 2.2.3 (DERIVATIVE OF THE DEFORMATION GRADIENT). *If the deformation is twice continuously differentiable, then the deformation gradient changes according to the equation*

$$\frac{D\mathsf{F}}{Dt} = \mathsf{L}\mathsf{F}. \qquad (2.2.6)$$

PROOF: Since F is a referential quantity,

$$\frac{D\mathsf{F}}{Dt}(\mathbf{X},t) = \frac{\partial \mathsf{F}}{\partial t}(\mathbf{X},t) = \frac{\partial}{\partial t}\operatorname{Grad}\mathbf{x}(\mathbf{X},t) = \operatorname{Grad}\frac{\partial \mathbf{x}}{\partial t}(\mathbf{X},t),$$

the last identity following from the smoothness conditions on $\mathbf{x}(\mathbf{X},t)$. But

$$\frac{\partial \mathbf{x}}{\partial t}(\mathbf{X},t) = \mathbf{v}(\mathbf{x}(\mathbf{X},t),t).$$

Therefore, by the chain rule,

$$\frac{D\mathsf{F}}{Dt}(\mathbf{X},t) = \operatorname{grad}\mathbf{v}(\mathbf{x},t)\operatorname{Grad}\mathbf{x}(\mathbf{X},t) = \mathsf{LF},$$

as claimed. ■

Recall from Exercise 28 that, in any Cartesian coordinate system, the trace of a second-order tensor A is $\operatorname{tr}\mathsf{A} = A_{11} + A_{22} + A_{33}$.

EXERCISE 43 *Show that* $\operatorname{tr}\mathsf{L} = \operatorname{div}\mathbf{v}$.

THEOREM 2.2.4 (DETERMINANT OF THE DEFORMATION GRADIENT). *The determinant of the deformation gradient changes according to the identity*

$$\frac{D}{Dt}\det\mathsf{F} = (\det\mathsf{F})\operatorname{tr}\mathsf{L}. \tag{2.2.7}$$

PROOF: F is a referential quantity, so $D(\det\mathsf{F})/Dt = \partial(\det\mathsf{F})/\partial t$. Applying the identity (1.2.20) yields

$$\frac{D}{Dt}\det\mathsf{F} = (\det\mathsf{F})\operatorname{tr}\left(\frac{D\mathsf{F}}{Dt}\mathsf{F}^{-1}\right).$$

Substituting for $D\mathsf{F}/Dt$ using Equation (2.2.6) yields

$$\frac{D}{Dt}\det\mathsf{F} = (\det\mathsf{F})\operatorname{tr}\left(\mathsf{LFF}^{-1}\right) = (\det\mathsf{F})\operatorname{tr}\mathsf{L},$$

completing the proof. ■

2.2.4 Summary

Differentiating functions associated with motion requires us to distinguish among several descriptions. The two most useful are the referential description, for functions of the referential coordinate $\mathbf{X}$, and the spatial description, for functions of the spatial coordinate $\mathbf{x}$. An important form of time derivative is the material derivative D/Dt, which is the partial derivative with respect to time, holding the particle $\mathbf{X}$ fixed.

One can also differentiate with respect to position variables $\mathbf{X}$ or $\mathbf{x}$, yielding the differential operators Grad and grad, respectively. The identity

$$\frac{D}{Dt} = \frac{\partial}{\partial t} + \mathbf{v} \cdot \operatorname{grad}$$

allows us to apply the concept of material derivative to functions in the spatial description.

Two of the most important derivatives with respect to position are the deformation gradient, $\mathsf{F} = \operatorname{Grad} \boldsymbol{\chi}_\kappa$, and the velocity gradient, $\mathsf{L} = \operatorname{grad} \mathbf{v}$.

2.3 PATHLINES, STREAMLINES, AND STREAKLINES

Three types of curves in $\mathbb{E}$ illustrate applications of differentiation to the analysis of motion. Pathlines, streamlines, and streaklines are all related to the trajectories of particles, but the three concepts are distinct. This section examines these concepts.

2.3.1 Three Types of Arc

DEFINITION. *An* **arc** *in* $\mathbb{E}$ *is a continuously differentiable function of the form* $\gamma\colon [a,b] \subset \mathbb{R} \to \mathbb{E}$.

Thus for any interval $[a,b] \subset \mathbb{R}$ the set

$$\gamma([a,b]) = \left\{ \gamma(s) \mid s \in [a,b] \right\}$$

is a smooth image in $\mathbb{E}$ of a segment of the real line $\mathbb{R}$. The argument s of γ is the **parameter** of the arc. We sometimes take the parameter to be time t, but more generally it is an abstract real variable ranging over the parameter domain $[a,b]$.

DEFINITION. *The* **pathline** *of a particle* $\mathbf{X}$ *is the arc* $\gamma\colon \mathbb{R} \to \mathbb{E}$ *defined by* $\gamma(t) = \mathbf{x}(\mathbf{X}, t)$.

The pathline gives the arc that the particle follows as time progresses. This arc obeys the following differential equation:

$$\gamma'(t) = \mathbf{v}(\gamma(t), t). \qquad \text{Pathline} \qquad (2.3.1)$$

Geometrically, this equation asserts that the velocity $\mathbf{v}(\gamma(t), t)$ at time t is the vector $\gamma'(t)$ tangent to the arc at the point $\mathbf{x} = \gamma(t)$, as Figure 2.9 illustrates. If we select the reference configuration such that $\mathbf{X}$ gives the position of the particle at time t_0, then the pathline of the particle $\mathbf{X}$ solves the initial-value problem consisting of the differential equation (2.3.1) together with the initial condition $\gamma(t_0) = \mathbf{X}$.

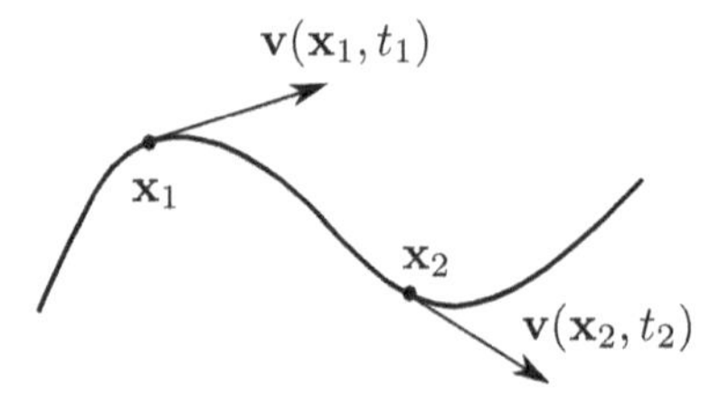

Figure 2.9. The pathline of a particle.

DEFINITION. *A* **streamline** *at time t is an integral curve of the velocity field* $\mathbf{v}(\mathbf{x}, t)$.

In this case, we freeze time at an instant t and consider arcs that are everywhere tangent to the instantaneous velocity field $\mathbf{v}(\mathbf{x}, t)$. Any streamline at time t satisfies a differential equation of the form

$$\gamma'(s) = \mathbf{v}(\gamma(s), t). \qquad \text{Streamline}$$

Here we use the abstract variable s, not time t, as the parameter of the arc, reflecting the fact that particles may follow arcs different from the instantaneous streamlines. One imposes an initial condition by specifying any point $\gamma(s_0)$ in space through which the arc passes at the fixed time t. Figure 2.10 illustrates a streamline through a velocity field at an instant in time.

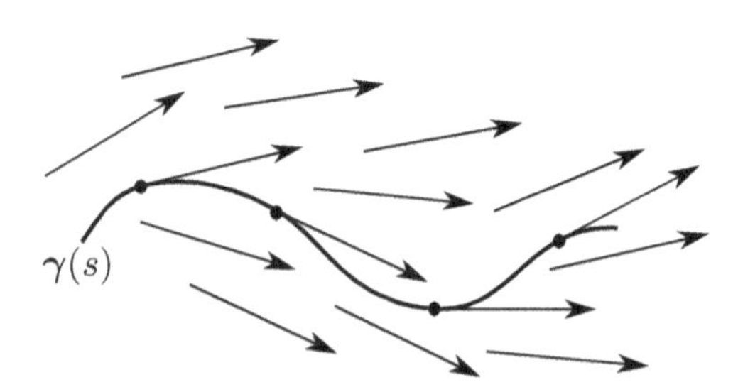

Figure 2.10. A velocity field $\mathbf{v}(\mathbf{x}, t)$ and a streamline at time t.

Streaklines are more complicated.

DEFINITION. *A* **streakline** *at time t through* $\mathbf{y}$ *is the locus at time t of particles that passed through* $\mathbf{y}$ *at earlier times $s \leqslant t$.*

The following equation defines a streakline through $\mathbf{y}$ at time t:

$$\boldsymbol{\gamma}(s) = \mathbf{x}(\mathbf{X}(\mathbf{y}, s), t). \qquad \text{Streakline}$$

Here the parameter s, representing the time at which the particle at $\boldsymbol{\gamma}(s)$ passed through $\mathbf{y}$, parametrizes the arc. The utility of streaklines resides in the fact that they are often easy to observe in fluids. Throughout a certain time interval, one injects a stream or sequence of tagged particles at the fixed point $\mathbf{y}$ and records the arc that the ensemble of particles traces at time t, as illustrated in Figure 2.11. One may consider a plume of aerosol leaving a smokestack to be an approximation of a streakline if its particles do not diffuse too rapidly.

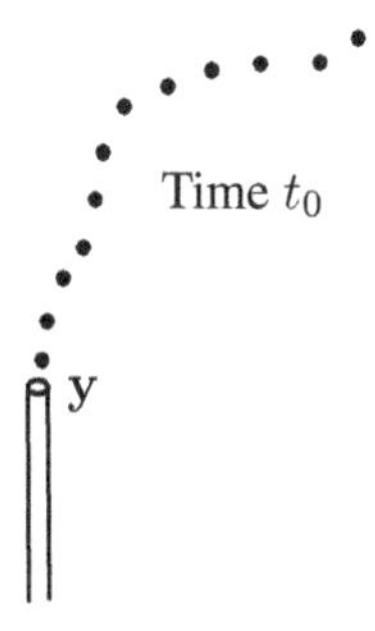

Figure 2.11. A streakline at time t formed by tagged particles injected through a port at $\mathbf{y}$.

2.3.2 An Example

In a given motion, pathlines, streamlines, and streaklines generally differ. In fact, they may be qualitatively different in appearance, as the following example illustrates. Consider a planar motion with spatial velocity field

$$\mathbf{v}(\mathbf{x}, t) = \begin{bmatrix} x_1/(1+t) \\ x_2/(1+t/2) \\ 0 \end{bmatrix} \tag{2.3.2}$$

in Cartesian coordinates. The pathline through a specified point $\mathbf{X} = (X_1, X_2, 0)^\top$ at time $t = 0$ is the solution $\boldsymbol{\gamma}(t) = (\gamma_1(t), \gamma_2(t), \gamma_3(t))^\top$ to the initial-value problem

$$\boldsymbol{\gamma}'(t) = \begin{bmatrix} \gamma_1/(1+t) \\ \gamma_2/(1+t/2) \\ 0 \end{bmatrix}, \qquad \boldsymbol{\gamma}(0) = \begin{bmatrix} X_1 \\ X_2 \\ 0 \end{bmatrix}.$$

One solves this problem by integrating the initial-value problems

$$\frac{\gamma_1'}{\gamma_1} = \frac{1}{1+t}, \qquad \gamma_1(0) = X_1,$$
$$\frac{\gamma_2'}{\gamma_2} = \frac{1}{1+t/2}, \qquad \gamma_2(0) = X_2,$$

getting

$$\gamma(t) = \begin{bmatrix} X_1(1+t) \\ X_2(1+t/2)^2 \\ 0 \end{bmatrix}. \tag{2.3.3}$$

To view this arc in the (x_1, x_2)-plane, solve for $\gamma_2(t)$ in terms of $\gamma_1(t)$: since $t = \gamma_1/X_1 - 1$, $\gamma_2 = \frac{1}{4}X_2(1 + \gamma_1/X_1)^2$. For example, the pathline passing through $(1, 1, 0)^\top$ at $t = 0$ is $\gamma_2 = (1 + \gamma_1)^2/4$. Figure 2.12 shows several pathlines of this form, corresponding to various choices of t_0 and $(X_1, X_2, 0)^\top$.

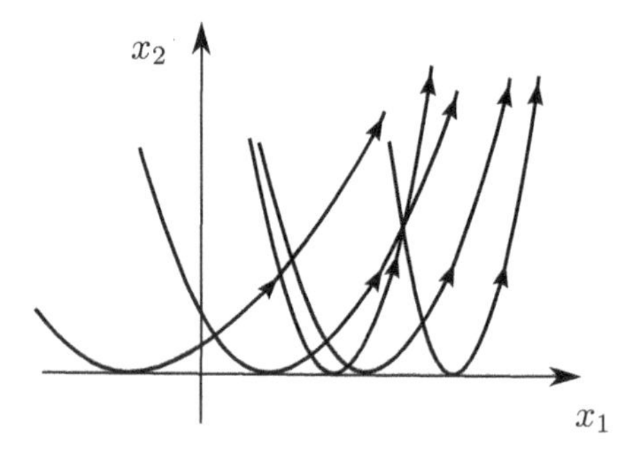

Figure 2.12. Pathlines for the velocity field (2.3.2).

The streamlines at a fixed time t obey initial-value problems of the following form:

$$\gamma'(s) = \begin{bmatrix} \gamma_1/(1+t) \\ \gamma_2/(1+t/2) \\ 0 \end{bmatrix}, \qquad \gamma(0) = \begin{bmatrix} X_1 \\ X_2 \\ 0 \end{bmatrix}.$$

Here the abstract parameter s that measures progress along the arc differs from time t, which remains frozen. The solutions have the form

$$\gamma(s) = \begin{bmatrix} X_1 \exp[s/(1+t)] \\ X_2 \exp[s/(1+t/2)] \\ 0 \end{bmatrix}.$$

Since $s = (1+t)\log(\gamma_1/X_1)$, we can eliminate s to obtain γ_2 in terms of γ_1 as follows:

$$\gamma_2 = X_2 \left(\frac{\gamma_1}{X_1}\right)^{(1+t)/(1+t/2)}.$$

These curves differ from the pathlines. For example, Figure 2.13 shows the streamline passing through $(1,1,0)^\top$ when $t=0$. It is a straight line in the (x_1,x_2)-plane, while the pathline passing through $(1,1,0)^\top$ at $t=0$, also shown in the figure, is concave upward.

EXERCISE 44 *Sketch the streamlines through more general points $(X_1, X_2, 0)$ at a time $t=0$.*

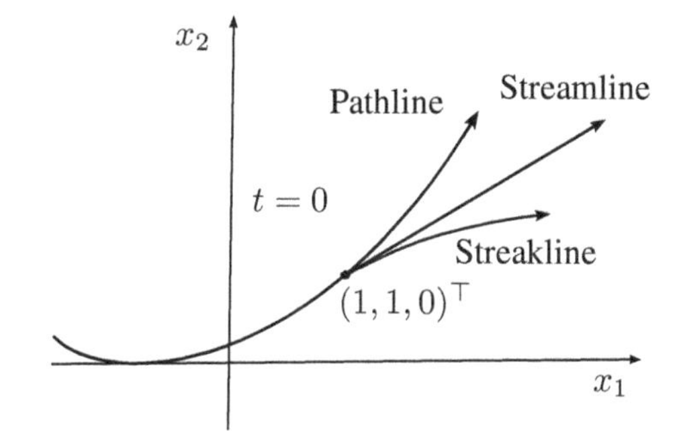

Figure 2.13. Pathline, streamline, and streakline through the point $(1,1,0)^\top$ for the velocity field (2.3.2) at $t=0$.

To find the streakline at time t through the point $\mathbf{y}$, we impose the condition that at some time $s \leqslant t$, the position of a particle now on the streakline was at $\mathbf{y}$. Using the motion given in Equation (2.3.3) and denoting $\mathbf{y} = (y_1, y_2, 0)^\top$, we have the following expression for such a particle's position at time s:

$$\mathbf{x}(\mathbf{y}, s) = \begin{bmatrix} y_1 \\ y_2 \\ 0 \end{bmatrix} = \begin{bmatrix} \xi_1(1+s) \\ \xi_2(1+s/2)^2 \\ 0 \end{bmatrix},$$

where $\boldsymbol{\xi} = (\xi_1, \xi_2, 0)^\top$ stands for the particle's position at time $t=0$. Solving for $\boldsymbol{\xi}$ yields

$$\begin{bmatrix} \xi_1 \\ \xi_2 \\ 0 \end{bmatrix} = \begin{bmatrix} y_1/(1+s) \\ y_2/(1+s/2)^2 \\ 0 \end{bmatrix}.$$

At time t, these particles, each associated with a value of the history parameter s, lie on the arc

$$\gamma(s) = \begin{bmatrix} \xi_1(1+t) \\ \xi_2(1+t/2)^2 \\ 0 \end{bmatrix} = \begin{bmatrix} y_1(1+t)/(1+s) \\ y_2(1+t/2)^2/(1+s/2)^2 \\ 0 \end{bmatrix}.$$

For comparison with the pathlines and streamlines, let us examine this arc at $t = 0$ in the (x_1, x_2)-plane. Since $s = y_1/\gamma_1 - 1$ at $t = 0$, we can solve for γ_2 in terms of γ_1, obtaining

$$\gamma_2 = \frac{4y_2}{(1 + y_1/\gamma_1)^2}.$$

For example, when $(y_1, y_2, 0)^\top = (1, 1, 0)^\top$, $\gamma_2 = 4\gamma_1^2/(\gamma_1 + 1)^2$. As Figure 2.13 illustrates, this arc is concave downward. Figure 2.14 shows this streakline in more detail, indicating how particles following various concave-upward pathlines arrive on the streakline at $t = 0$.

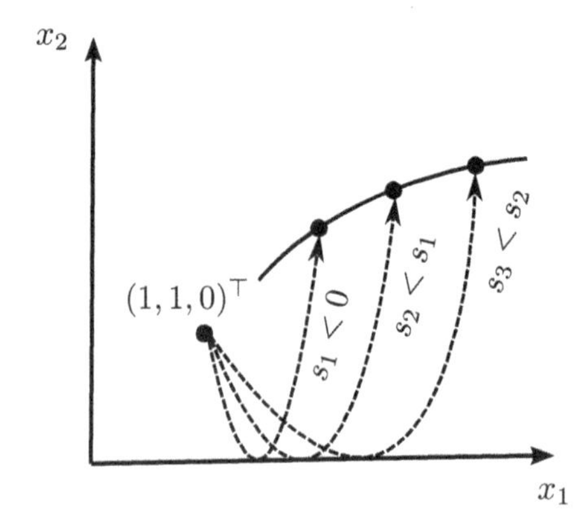

Figure 2.14. A streakline through $(1, 1, 0)^\top$ for the velocity field (2.3.2) at $t = 0$. Shown as dashed curves are pathlines of several particles that lie on the streakline at time $t = 0$, each indexed by the time s at which they passed through $(1, 1, 0)^\top$.

This example shows how qualitatively different pathlines, streamlines, and streaklines can be. For the simple, planar velocity field examined here, pathlines are concave upward; streamlines at $t = 0$ are straight lines, and the streakline is concave downward.

EXERCISE 45 *Consider the planar motion defined by*

$$\mathbf{x}(\mathbf{X}, t) = \begin{bmatrix} X_1(1+t) \\ X_2 + t \\ X_3 \end{bmatrix}.$$

Show that the streakline through $\mathbf{y} = (1, 1, 0)^\top$ *at time* $t_0 = 1$ *is a hyperbola, while the pathlines are straight lines.*

If the motion is steady—that is, $\mathbf{v} = \mathbf{v}(\mathbf{x})$, independent of t—then pathlines and streamlines coincide. In particular, no pathline crosses a streamline in this case. The converse is false. Consider the motion $\mathbf{x}(\mathbf{X}, t) = \mathbf{X} + t^2\mathbf{e}_1$, with velocity $\partial\mathbf{x}/\partial t = 2t\mathbf{e}_1$. Here both the pathlines and streamlines are lines parallel to $\mathbf{e}_1$, but the motion is not steady.

2.3.3 Summary

Pathlines, streamlines, and streaklines define three conceptually different types of arcs that one may observe in materials—especially fluids—under motion. A pathline is simply the arc $\chi_\kappa(\mathbf{X}, t)$ that a *fixed particle* $\mathbf{X}$ traces under the deformation as time progresses. A streamline is an integral curve of the instantaneous tangent field associated with the spatial velocity $\mathbf{v}(\mathbf{x}, t)$ at a *fixed time* t. A streakline is a less elegant entity, mathematically, but it has experimental utility: it is the locus, at time t, of particles that passed through a *fixed position* $\mathbf{y}$ during some earlier time interval.

2.4 INTEGRALS UNDER MOTION

Motion affects line, surface, and volume integrals associated with bodies. To analyze these effects, we investigate how integrals over sets of particles transform under the mapping from the reference configuration $\boldsymbol{\kappa}(\mathcal{B})$ to the deformed configuration $\chi(\mathcal{B}, t)$. We then examine time derivatives of volume integrals over parts of a body — a topic that plays a pivotal role in the development of balance laws in Chapter 4.

2.4.1 Arc, Surface, and Volume Integrals

We turn first to the effects of motion on certain line, surface, and volume integrals. Of interest are integrals taken over sets defined by the fixed collections of particles that they contain.

DEFINITION. *A* **material set** *in* $\mathbb{E}$ *is a set that always contains the same particles from* $\mathcal{B}$.

For example, let $\boldsymbol{\Gamma}\colon [a, b] \to \boldsymbol{\kappa}(\mathcal{B})$ be an arc in the reference configuration of $\mathcal{B}$. At any time t, the deformation χ_κ carries the particles on this arc to a new arc

$\gamma_t \colon [a,b] \to \chi(\mathcal{B},t)$ defined by $\gamma_t(s) = \chi_\kappa(\boldsymbol{\Gamma}(s),t)$, as depicted in Figure 2.15. The image $\gamma_t([a,b])$ is a material set, being the locus at time t of a fixed set of particles—namely, those that reside on $\boldsymbol{\Gamma}([a,b])$ in the reference configuration. We call γ_t a **material arc**.

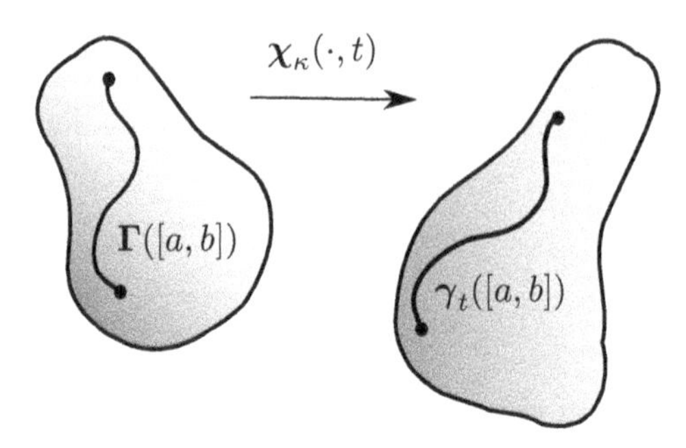

Figure 2.15. A material arc γ_t.

The tangent vector to the material arc γ_t at the point $\gamma_t(s)$ is $\gamma_t'(s)$. Since $\gamma_t(s) = \mathbf{x}(\boldsymbol{\Gamma}(s),t)$, the chain rule gives $\gamma_t'(s) = \operatorname{Grad}\mathbf{x}(\boldsymbol{\Gamma}(s),t)\,\boldsymbol{\Gamma}'(s)$ or, in differential notation,

$$\frac{d\mathbf{x}}{ds} = \mathsf{F}\frac{d\mathbf{X}}{ds}.$$

One often sees this relationship cast in the less formal notation of infinitesimals:

$$d\mathbf{x} = \mathsf{F}\,d\mathbf{X}, \tag{2.4.1}$$

where people refer to $\gamma_t'(s)\,ds = d\mathbf{x}$ and $\boldsymbol{\Gamma}'(s)\,ds = d\mathbf{X}$ as **differential elements of arclength**. Figure 2.16 shows the geometric interpretation.

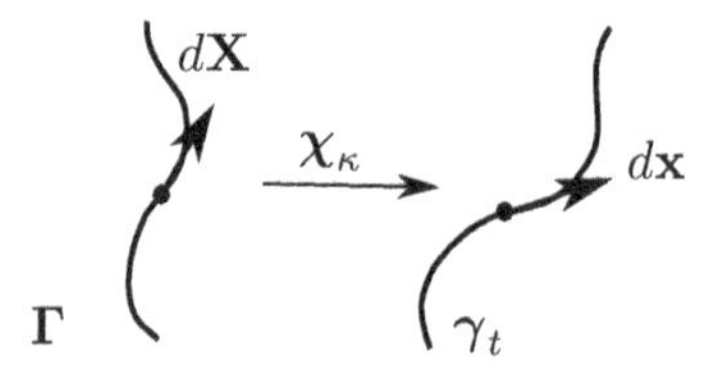

Figure 2.16. Transformation of a material arc element under deformation.

As an application of this relationship, let us convert an integral of a vector field $\boldsymbol{\Psi}(\mathbf{x})$ along a material arc γ_t to an integral along its reference configuration:

$$\begin{aligned}\int_{\gamma_t} \boldsymbol{\Psi}\cdot d\mathbf{x} &= \int_a^b \boldsymbol{\Psi}(\gamma_t(s))\cdot\gamma_t'(s)\,ds \\ &= \int_a^b \boldsymbol{\Psi}(\chi_\kappa(\boldsymbol{\Gamma}(s),t))\cdot\mathsf{F}(\boldsymbol{\Gamma}(s),t)\,\boldsymbol{\Gamma}'(s)\,ds = \int_{\boldsymbol{\Gamma}} \boldsymbol{\Psi}\cdot\mathsf{F}\,d\mathbf{X}.\end{aligned}$$

Next consider material volume elements. Suppose that $\mathcal{P} \subset \mathcal{B}$ occupies a region $\boldsymbol{\kappa}(\mathcal{P}) \subset \mathbb{E}$ in the reference configuration $\boldsymbol{\kappa}\colon \mathcal{B} \to \mathbb{E}$. Then the set $\boldsymbol{\chi}(\mathcal{P}, t)$ is a material region, being the image under $\boldsymbol{\chi}_\kappa(\cdot, t)$ of the set $\boldsymbol{\kappa}(\mathcal{P})$, as shown in Figure 2.17. We assume that $\boldsymbol{\chi}(\mathcal{P}, t)$ and $\boldsymbol{\kappa}(\mathcal{P})$ satisfy the hypotheses needed to apply the change-of-variables theorem (see Appendix B): if Ψ is a bounded, continuous function of position, then

$$\int_{\boldsymbol{\chi}(\mathcal{P},t)} \Psi(\mathbf{x})\, dv = \int_{\boldsymbol{\kappa}(\mathcal{P})} \Psi(\boldsymbol{\chi}_\kappa(\mathbf{X}, t))\, |\det \mathsf{F}(\mathbf{X}, t)|\, dV.$$

Here the elements dv and dV signify volume integration in the spatial and reference coordinates, respectively. Less formally, we write

$$dv = |\det \mathsf{F}|\, dV \tag{2.4.2}$$

to describe how material volume elements transform under motion.

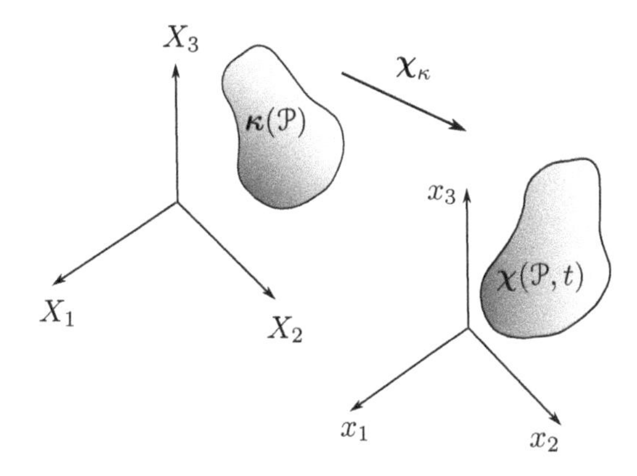

Figure 2.17. Transformation of a material region under deformation.

The scalar triple product furnishes an interesting heuristic for the change-of-variables formula and motivates an additional assumption about the reference configuration. Using the informal language of infinitesimals, consider a volume element in the reference configuration formed by three arc elements $d\mathbf{X}_1, d\mathbf{X}_2, d\mathbf{X}_3$, as sketched in Figure 2.18. By the relationships (1.2.7) and (2.4.1),

$$\begin{aligned} dv &= \left[d\mathbf{x}_1, d\mathbf{x}_2, d\mathbf{x}_3\right] \\ &= \left[\mathsf{F}\, d\mathbf{x}_1, \mathsf{F}\, d\mathbf{x}_2, \mathsf{F}\, d\mathbf{x}_3\right] \\ &= \det \mathsf{F} \left[d\mathbf{X}_1, d\mathbf{X}_2, d\mathbf{X}_3\right] = \det \mathsf{F}\, dV. \end{aligned} \tag{2.4.3}$$

Equation (2.4.3) lends some insight into how the factor $\det \mathsf{F}$ connects volume elements dv in the region under deformation to corresponding volume elements dV in

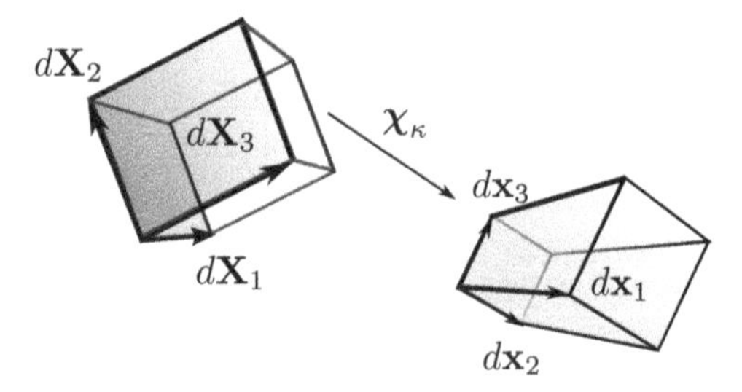

Figure 2.18. Transformation of an infinitesimal material parallelepiped under deformation.

the reference configuration. It also indicates that the signed volume in the reference configuration can differ in sign from the signed volume at time t. This happens when the derivative $\mathsf{F} = \text{Grad}\,\chi_\kappa$ has negative determinant. The change-of-variables theorem ensures nonnegative volumes by using $|\det \mathsf{F}|$ instead of $\det \mathsf{F}$. In this text, we henceforth restrict the choice of reference configuration so that it has the same orientation as the motion, in the sense that every scalar triple product $[\mathbf{X}, \mathbf{Y}, \mathbf{Z}]$ in the reference configuration has the same sign as $[\chi_\kappa(\mathbf{X}, t), \chi_\kappa(\mathbf{Y}, t), \chi_\kappa(\mathbf{Z}, t)]$. (The sign cannot change, since $\det \mathsf{F}$ is continuous in $\mathbf{X}$ and t and never vanishes.) Convenient but not mathematically essential, this restriction implies that $\det \mathsf{F} > 0$, so Equation (2.4.2) becomes

$$dv = \det \mathsf{F}\, dV. \tag{2.4.4}$$

To investigate the transformation of material surfaces, let us begin by reviewing their construction. A **parametrized smooth surface** in the reference configuration $\boldsymbol{\kappa}(\mathcal{B})$ is the image of a connected, two-dimensional region $\Omega \in \mathbb{R}^2$ under a continuously differentiable function $\boldsymbol{\varphi}\colon \Omega \to \boldsymbol{\kappa}(\mathcal{B})$ for which the vector field

$$\frac{\partial \boldsymbol{\varphi}}{\partial \xi_1}(\boldsymbol{\xi}) \times \frac{\partial \boldsymbol{\varphi}}{\partial \xi_2}(\boldsymbol{\xi}) \neq \mathbf{0} \tag{2.4.5}$$

for all $\boldsymbol{\xi} = (\xi_1, \xi_2) \in \Omega$. We call Ω the **parameter domain**. Figure 2.19 shows such a mapping, together with the Cartesian coordinate system $\boldsymbol{\xi} = (\xi_1, \xi_2)$ on the parameter domain. The image set $\boldsymbol{\varphi}(\Omega) \subset \mathbb{E}$ is the surface.

The condition (2.4.5) has geometric significance. The vectors $\boldsymbol{\tau}_1 = \partial\boldsymbol{\varphi}/\partial\xi_1$ and $\boldsymbol{\tau}_2 = \partial\boldsymbol{\varphi}/\partial\xi_2$ are tangent to $\boldsymbol{\varphi}(\Omega)$. The requirement that their cross product be nonzero ensures that the function $\boldsymbol{\varphi}$ maps areas in Ω onto areas in $\boldsymbol{\varphi}(\Omega)$ and that there exists a well-defined unit normal vector field $\mathbf{n}$ on $\boldsymbol{\varphi}(\Omega)$:

$$\mathbf{n}(\boldsymbol{\xi}) = \frac{\dfrac{\partial \boldsymbol{\varphi}}{\partial \xi_1}(\boldsymbol{\xi}) \times \dfrac{\partial \boldsymbol{\varphi}}{\partial \xi_2}(\boldsymbol{\xi})}{\left\| \dfrac{\partial \boldsymbol{\varphi}}{\partial \xi_1}(\boldsymbol{\xi}) \times \dfrac{\partial \boldsymbol{\varphi}}{\partial \xi_2}(\boldsymbol{\xi}) \right\|} \tag{2.4.6}$$

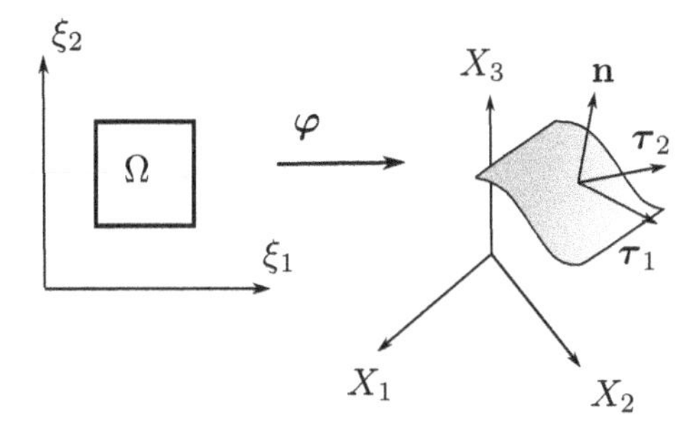

Figure 2.19. Representation of a parametrized smooth surface as an image of a region $\Omega \subset \mathbb{R}^2$.

As Figure 2.20 illustrates, the deformation χ_κ maps the surface $\varphi(\Omega)$ into a set $\chi_\kappa(\varphi(\Omega))$, defining a new parametrized smooth surface $\chi_\kappa \circ \varphi \colon \Omega \to \chi(\mathcal{B}, t)$.

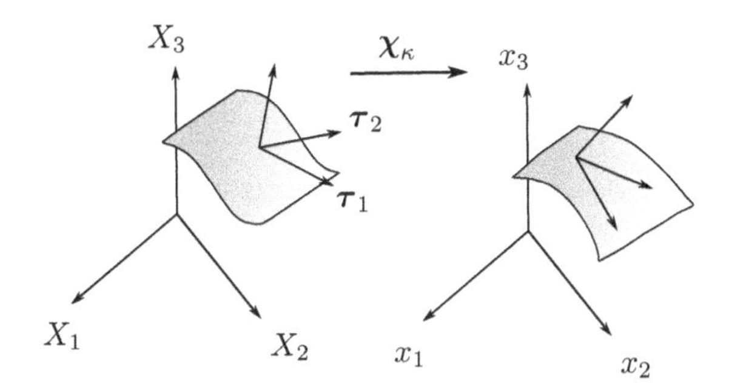

Figure 2.20. Transformation of a material surface under motion.

EXERCISE 46 *Why is*

$$\frac{\partial(\chi_\kappa \circ \varphi)}{\partial \xi_1}(\boldsymbol{\xi}) \times \frac{\partial(\chi_\kappa \circ \varphi)}{\partial \xi_2}(\boldsymbol{\xi}) \neq \mathbf{0}?$$

In the deformed configuration, the integral of a vector field $\boldsymbol{\Psi}$ over the transformed surface $\chi_\kappa \circ \varphi$ is, by definition,

$$\int_{(\chi_\kappa \circ \varphi)(\Omega)} \boldsymbol{\Psi}(\mathbf{x}) \cdot d\boldsymbol{\sigma} = \iint_\Omega \boldsymbol{\Psi}(\chi_\kappa(\varphi(\boldsymbol{\xi}), t)) \cdot d\boldsymbol{\sigma},$$

where

$$d\boldsymbol{\sigma} = \mathbf{n}(\boldsymbol{\xi})\, d\sigma = \left[\frac{\partial(\chi_\kappa \circ \varphi)}{\partial \xi_1}(\boldsymbol{\xi}) \times \frac{\partial(\chi_\kappa \circ \varphi)}{\partial \xi_2}(\boldsymbol{\xi})\right] d\xi_1\, d\xi_2 \tag{2.4.7}$$

serves as shorthand notation for the surface element in the deformed configuration. By the chain rule,

$$\frac{\partial(\boldsymbol{\chi}_\kappa \circ \boldsymbol{\varphi})}{\partial \xi_j}(\boldsymbol{\xi}) = \mathsf{F}(\boldsymbol{\varphi}(\boldsymbol{\xi}), t)\, \frac{\partial \boldsymbol{\varphi}}{\partial \xi_j}(\boldsymbol{\xi}).$$

We use this fact to rewrite the surface integral in the deformed configuration in terms of a surface integral in the reference configuration: by the relationship (1.2.7),

$$\begin{aligned}
\int_{(\boldsymbol{\chi}_\kappa \circ \boldsymbol{\varphi})(\Omega)} \boldsymbol{\Psi} \cdot d\boldsymbol{\sigma} &= \iint_\Omega \boldsymbol{\Psi}^\top \left(\mathsf{F}\frac{\partial \boldsymbol{\varphi}}{\partial \xi_1} \times \mathsf{F}\frac{\partial \boldsymbol{\varphi}}{\partial \xi_2} \right) d\xi_1\, d\xi_2 \\
&= \iint_\Omega (\mathsf{F}\mathsf{F}^{-1}\boldsymbol{\Psi})^\top \left(\mathsf{F}\frac{\partial \boldsymbol{\varphi}}{\partial \xi_1} \times \mathsf{F}\frac{\partial \boldsymbol{\varphi}}{\partial \xi_2} \right) d\xi_1\, d\xi_2 \\
&= \iint_\Omega \left[\mathsf{F}\mathsf{F}^{-1}\boldsymbol{\Psi}, \mathsf{F}\frac{\partial \boldsymbol{\varphi}}{\partial \xi_1}, \mathsf{F}\frac{\partial \boldsymbol{\varphi}}{\partial \xi_2} \right] d\xi_1\, d\xi_2 \\
&= \iint_\Omega \det \mathsf{F}\, (\mathsf{F}^{-1}\boldsymbol{\Psi})^\top \left(\frac{\partial \boldsymbol{\varphi}}{\partial \xi_1} \times \frac{\partial \boldsymbol{\varphi}}{\partial \xi_2} \right) d\xi_1\, d\xi_2.
\end{aligned}$$

The shorthand expression

$$d\boldsymbol{\Sigma} = \left(\frac{\partial \boldsymbol{\varphi}}{\partial \xi_1} \times \frac{\partial \boldsymbol{\varphi}}{\partial \xi_2} \right) d\xi_1\, d\xi_2$$

represents the surface element in the reference configuration. Therefore,

$$\int_{(\boldsymbol{\chi}_\kappa \circ \boldsymbol{\varphi})(\Omega)} \boldsymbol{\Psi} \cdot d\boldsymbol{\sigma} = \int_{\boldsymbol{\varphi}(\Omega)} \boldsymbol{\Psi} \cdot (\det \mathsf{F})\, \mathsf{F}^{-\top})\, d\boldsymbol{\Sigma}.$$

In shorthand, we write

$$d\boldsymbol{\sigma} = (\det \mathsf{F})\mathsf{F}^{-\top}\, d\boldsymbol{\Sigma} \tag{2.4.8}$$

to describe the transformation of material surface elements under motion.

An additional remark about surface integrals will be helpful in future sections. For a scalar field f, we denote by

$$\int_{\boldsymbol{\varphi}(\Omega)} f\, d\sigma$$

an integral of the form

$$\int_\Omega f(\boldsymbol{\varphi}(\xi_1, \xi_2)) \left\| \frac{\partial \boldsymbol{\varphi}}{\partial \xi_1} \times \frac{\partial \boldsymbol{\varphi}}{\partial \xi_2} \right\| d\xi_1\, d\xi_2, \tag{2.4.9}$$

in keeping with the definition (2.4.6) of the unit outward normal vector field $\mathbf{n}$ and the notational convention $d\boldsymbol{\sigma} = \mathbf{n}\, d\sigma$ adopted in Equation (2.4.7). This notation extends to surface integrals of the form

$$\int_{\boldsymbol{\varphi}(\Omega)} \mathbf{f}\, d\sigma,$$

where $\mathbf{f}$ is a vector field.

2.4.2 Reynolds Transport Theorem

A standard calculus problem asks for the derivative a function of t having the form

$$\int_{\alpha(t)}^{\beta(t)} \Psi(x,t)\,dx.$$

Since both the integrand and the one-dimensional region of integration depend on t, one cannot just pass the derivative operator through the integral sign. An analogous problem arises with an integral over a material region. To determine the rate of change of some quantity inside the region, we must differentiate with respect to t. But since the spatial configuration of the region may vary with time, we cannot simply pass the operator d/dt through the integral sign. The following theorem shows what else is needed.

THEOREM 2.4.1 (REYNOLDS TRANSPORT THEOREM). *Suppose that $\mathcal{R}(t)$ is a material region and that $\Psi(\mathbf{x},t)$ has integrable derivatives $\partial\Psi/\partial t$ and* $\operatorname{div}\Psi$. *Then*

$$\frac{d}{dt}\int_{\mathcal{R}(t)} \Psi(\mathbf{x},t)\,dv = \int_{\mathcal{R}(t)} \left\{\frac{\partial\Psi}{\partial t}(\mathbf{x},t) + \operatorname{div}\left[\Psi(\mathbf{x},t)\mathbf{v}(\mathbf{x},t)\right]\right\} dv. \qquad (2.4.10)$$

Applying the divergence theorem (see Appendix B) yields the following more suggestive version.

COROLLARY 2.4.2 (FLUX FORM OF THE REYNOLDS TRANSPORT THEOREM). *If the material region $\mathcal{R}(t)$ is a normal region at each time t, then*

$$\underbrace{\frac{d}{dt}\int_{\mathcal{R}} \Psi\,dv}_{\text{(I)}} = \underbrace{\int_{\mathcal{R}} \frac{\partial\Psi}{\partial t}\,dv}_{\text{(II)}} + \underbrace{\int_{\partial\mathcal{R}} \Psi\mathbf{v}\cdot d\boldsymbol{\sigma}}_{\text{(III)}}\,. \qquad (2.4.11)$$

The terms in Equation (2.4.11) have distinctive physical interpretations:

(I) is the net rate of change of Ψ in $\mathcal{R}$;

(II) is the rate of change of Ψ occurring within $\mathcal{R}$;

(III) is the rate of influx of Ψ across the boundary $\partial\mathcal{R}$.

The term labeled (III) can be confusing unless one keeps in mind that $\mathbf{v}$ denotes the velocity of the particles that define the boundary of the material region $\mathcal{R}(t)$.

Therefore, when $\mathbf{v} \cdot \mathbf{n} > 0$, the region expands outwardly to encompass more of the quantity measured per unit volume by Ψ.

PROOF (OF THEOREM 2.4.1): Begin by changing variables to the reference configuration:

$$\frac{d}{dt} \int_{\mathcal{R}(t)} \Psi(\mathbf{x}, t)\, dv = \frac{d}{dt} \int_{\chi_\kappa^{-1}(\mathcal{R}(t), t)} \Psi(\mathbf{X}, t) \det \mathsf{F}(\mathbf{X}, t)\, dV.$$

The integral on the right is over the region $\mathcal{R}_\kappa = \chi^{-1}(\mathcal{R}(t), t)$ occupied in the reference configuration, which does not vary with t. We can pass the derivative operator d/dt through this integral, getting

$$\frac{d}{dt} \int_{\mathcal{R}(t)} \Psi\, dv = \int_{\mathcal{R}_\kappa} \frac{\partial}{\partial t}(\Psi \det \mathsf{F})\, dV.$$

But the integrand in the reference configuration is a material quantity, for which $\partial/\partial t$ equals D/Dt. By the product rule and Equation (2.2.7),

$$\begin{aligned} \frac{d}{dt} \int_{\mathcal{R}(t)} \Psi\, dv &= \int_{\mathcal{R}_\kappa} \left(\frac{D\Psi}{Dt} \det \mathsf{F} + \Psi \frac{D}{Dt} \det \mathsf{F} \right) dV \\ &= \int_{\mathcal{R}_\kappa} \left(\frac{D\Psi}{Dt} + \Psi \operatorname{tr} \mathsf{L} \right) \det \mathsf{F}\, dV. \end{aligned}$$

Now change variables back to the spatial configuration $\mathcal{R}(t)$ to get

$$\frac{d}{dt} \int_{\mathcal{R}(t)} \Psi\, dv = \int_{\mathcal{R}(t)} \left(\frac{D\Psi}{Dt} + \Psi \operatorname{tr} \mathsf{L} \right) dv.$$

Finally, the identity (2.2.2) and the fact that $\operatorname{tr} \mathsf{L} = \operatorname{div} \mathbf{v}$ yield

$$\frac{d}{dt} \int_{\mathcal{R}(t)} \Psi\, dv = \int_{\mathcal{R}(t)} \left[\frac{\partial \Psi}{\partial t} + \operatorname{div}(\Psi \mathbf{v}) \right] dv,$$

completing the proof. ■

EXERCISE 47 *Show that, for a material region $\mathcal{R}(t)$,*

$$\frac{d}{dt} \int_{\mathcal{R}(t)} dv = \int_{\partial \mathcal{R}(t)} \mathbf{v} \cdot d\boldsymbol{\sigma}. \tag{2.4.12}$$

Interpret the integral on the right.

In Equation (2.4.12) the special case when the time derivative vanishes is noteworthy enough to merit a name.

DEFINITION. *A motion is* **isochoric** *if, for any material region* $\mathcal{R}(t)$, *at any time* t,

$$\int_{\mathcal{R}(t)} dv = \int_{\mathcal{R}_\kappa} dV,$$

where $\mathcal{R}_\kappa$ *denotes the region occupied in the reference configuration.*

Such a motion preserves the volume occupied by any material region, even if the region changes shape.

EXERCISE 48 *Show that the motion is isochoric whenever* $\det \mathsf{F} = 1$ *identically.*

2.4.3 Summary

Using the chain rule and the change-of-variables theorem, one can derive identities relating integrals over material paths, volumes, and surfaces in spatial coordinates to their corresponding quantities in referential coordinates. The change-of-variables theorem also yields the Reynolds transport theorem, which provides a useful identity for computing the time derivative of an integral over a material region.

CHAPTER 3

KINEMATICS II: STRAIN AND ITS RATES

A body undergoes strain when the deformation alters distances between particles. To analyze effects of this type, we use a variety of quantities arising from the deformation gradient $\mathsf{F}(\mathbf{X}, t) = \mathrm{Grad}\, \mathbf{x}(\mathbf{X}, t)$. This chapter examines the properties of this tensor and its time derivatives.

3.1 STRAIN

We begin with the algebraic properties of symmetric tensors, which figure prominently in the algebra of $\mathsf{F}(\mathbf{X}, t)$ and in continuum mechanics more generally. The concepts lead to an important factoring technique for invertible tensors, the polar decomposition. We apply this decomposition to F and then examine several quantities related to F that prove useful in calculations.

Continuum Mechanics: The Birthplace of Mathematical Models, First Edition. M.B. Allen.

3.1.1 Symmetric Tensors

Several important properties of F rest on a central topic in linear algebra:

DEFINITION. *A second-order tensor* $\mathsf{A} \in \mathrm{L}(\mathbb{E})$ *is* **positive definite** *if* $\mathbf{a} \cdot \mathsf{A}\mathbf{a} > 0$ *for every nonzero vector* $\mathbf{a} \in \mathbb{E}$. A *is* **symmetric** *if* $\mathsf{A} = \mathsf{A}^\top$, *that is, if* $\mathbf{a} \cdot \mathsf{A}\mathbf{b} = \mathsf{A}\mathbf{a} \cdot \mathbf{b}$ *for all vectors* $\mathbf{a}, \mathbf{b} \in \mathbb{E}$. *We denote*

$$\begin{aligned} \mathrm{Sym}\,(\mathbb{E}) &= \left\{\mathsf{A} \in \mathrm{L}(\mathbb{E}) \;\middle|\; \mathsf{A}^\top = \mathsf{A}\right\}, \\ \mathrm{SPD}(\mathbb{E}) &= \left\{\mathsf{A} \in \mathrm{Sym}(\mathbb{E}) \;\middle|\; \mathsf{A} \text{ is positive definite}\right\}. \end{aligned}$$

The next two theorems establish important properties of $\mathrm{Sym}(\mathbb{E})$.

THEOREM 3.1.1 (EIGENVALUES OF SYMMETRIC TENSORS). *If* $\mathsf{A} \in \mathrm{Sym}\,(\mathbb{E})$, *then its eigenvalues are real, and there are real eigenvectors associated with each eigenvalue.*

PROOF: *A priori*, we must allow eigenvalues, which are roots of the cubic polynomial $\det(\lambda\mathsf{I} - \mathsf{A})$, to be complex numbers. Similarly, we allow eigenvectors to have real and imaginary vector parts. Let $\lambda = d + iw$ be an eigenvalue of A with associated eigenvector $\mathbf{p} + i\mathbf{q}$, where $\mathbf{p}, \mathbf{q} \in \mathbb{E}$. Then

$$\mathsf{A}(\mathbf{p} + i\mathbf{q}) = (d + iw)(\mathbf{p} + i\mathbf{q}). \tag{3.1.1}$$

Taking complex conjugates of both sides yields

$$\mathsf{A}(\mathbf{p} - i\mathbf{q}) = (d - iw)(\mathbf{p} - i\mathbf{q}). \tag{3.1.2}$$

Now take the inner product of $\mathbf{p} - i\mathbf{q}$ with Equation (3.1.1) and the inner product of $\mathbf{p} + i\mathbf{q}$ with Equation (3.1.2), then subtract and simplify:

$$\begin{aligned} (\mathbf{p} + i\mathbf{q}) \cdot \mathsf{A}(\mathbf{p} - i\mathbf{q}) - (\mathbf{p} - i\mathbf{q}) \cdot \mathsf{A}(\mathbf{p} + i\mathbf{q}) & \\ = (d - iw)(\|\mathbf{p}\|^2 + \|\mathbf{q}\|^2) &- (d + iw)(\|\mathbf{p}\|^2 + \|\mathbf{q}\|^2) \\ = -2iw(\|\mathbf{p}\|^2 + \|\mathbf{q}\|^2).& \end{aligned}$$

The left side of this equation vanishes, because $\mathsf{A} \in \mathrm{Sym}(\mathbb{E})$. On the right side, $\|\mathbf{p}\|^2 + \|\mathbf{q}\|^2 \neq 0$, since $\mathbf{p} + i\mathbf{q} \neq \mathbf{0}$. It follows that $w = 0$ and, hence, that $\lambda \in \mathbb{R}$. To show that there is a real eigenvector associated with λ, observe that if $\mathbf{p} + i\mathbf{q}$ is an eigenvector associated with λ, then so is $\mathbf{p} - i\mathbf{q}$. It follows that

$$\mathsf{A}\mathbf{p} = \mathsf{A}[\tfrac{1}{2}(\mathbf{p} + i\mathbf{q}) + \tfrac{1}{2}(\mathbf{p} - i\mathbf{q})] = \lambda\mathbf{p}.$$

Therefore, $\mathbf{p}$ is a real eigenvector of A associated with the eigenvalue λ. ■

THEOREM 3.1.2 (BASIS ASSOCIATED WITH A SYMMETRIC TENSOR). *If* $\mathsf{A} \in \mathrm{Sym}(\mathbb{E})$, *then* $\mathbb{E}$ *has an orthonormal basis consisting of eigenvectors of* A.

PROOF: Let $\lambda_1, \lambda_2, \lambda_3$ be the eigenvalues of A, listed according to multiplicity. By the previous theorem, each $\lambda_i \in \mathbb{R}$, and there are three cases:

Case 1: $\lambda_1 = \lambda_2 = \lambda_3 = \lambda$. In this case, symmetry implies that $\mathsf{A} = \lambda\mathsf{I}$, and any orthonormal basis for $\mathbb{E}$ consists of eigenvectors of A.

Case 2: $\lambda_1, \lambda_2, \lambda_3$ are distinct. It suffices to show that unit-length eigenvectors $\mathbf{p}_i, \mathbf{p}_j$ associated with distinct eigenvalues λ_i, λ_j are orthogonal, from which it follows that $\{\mathbf{p}_1, \mathbf{p}_2, \mathbf{p}_3\}$ is an orthonormal basis for $\mathbb{E}$. Since A is symmetric,

$$0 = \mathbf{p}_i \cdot \mathsf{A}\mathbf{p}_j - \mathsf{A}\mathbf{p}_i \cdot \mathbf{p}_j = (\lambda_j - \lambda_i)\, \mathbf{p}_i \cdot \mathbf{p}_j.$$

If $\lambda_i \neq \lambda_j$, then this equation shows that $\mathbf{p}_i \cdot \mathbf{p}_j = 0$.

Case 3: $\lambda_1 \neq \lambda_2 = \lambda_3$. Denote the unit-length eigenvectors associated with λ_1, λ_2 by $\mathbf{p}_1, \mathbf{p}_2$, respectively. By the reasoning used in Case 2, $\mathbf{p}_1 \cdot \mathbf{p}_2 = 0$. Pick $\mathbf{p}_3 = \mathbf{p}_1 \times \mathbf{p}_2$. The set $\{\mathbf{p}_1, \mathbf{p}_2, \mathbf{p}_3\}$ constitutes an orthonormal basis for $\mathbb{E}$, so it suffices to show that $\mathbf{p}_3$ is an eigenvector of A. Expanding $\mathsf{A}\mathbf{p}_3$ with respect to the basis (see Exercise 1) and using the symmetry of A gives

$$\begin{aligned}\mathsf{A}\mathbf{p}_3 &= \sum_{i=1}^{3}(\mathsf{A}\mathbf{p}_3 \cdot \mathbf{p}_i)\mathbf{p}_i = \sum_{i=1}^{3}(\mathbf{p}_3 \cdot \mathsf{A}\mathbf{p}_i)\mathbf{p}_i \\ &= (\lambda_1\mathbf{p}_3 \cdot \mathbf{p}_1)\mathbf{p}_1 + (\lambda_2\mathbf{p}_3 \cdot \mathbf{p}_2)\mathbf{p}_2 + (\mathbf{p}_3 \cdot \mathsf{A}\mathbf{p}_3)\mathbf{p}_3.\end{aligned}$$

The first two terms on the right vanish by orthogonality, and $\mathbf{p}_3 \cdot \mathsf{A}\mathbf{p}_3$ is a scalar, so $\mathbf{p}_3$ is an eigenvector of A. ■

EXERCISE 49 *Show that the eigenvalues of any tensor* $\mathsf{A} \in \mathrm{SPD}(\mathbb{E})$ *are positive.*

EXERCISE 50 *Let* $\mathsf{A}, \mathsf{B} \in \mathrm{Sym}(\mathbb{E})$, *and suppose there exists an orthogonal tensor* Q *such that* $\mathsf{A} = \mathsf{Q}\mathsf{B}\mathsf{Q}^\top$. *Show that* A *and* B *have the same eigenvalues.*

EXERCISE 51 *Establish the converse of the result in Exercise 50: assume that* $\mathsf{A}, \mathsf{B} \in \mathrm{Sym}(\mathbb{E})$ *have the same eigenvalues* $\lambda_1, \lambda_2, \lambda_3$, *and find an orthonormal tensor* Q *such that* $\mathsf{A} = \mathsf{Q}\mathsf{B}\mathsf{Q}^\top$. *(Hint: Define* Q *by its action on an orthonormal basis of eigenvectors.)*

EXERCISE 52 *Prove that if* $\{\mathbf{p}_1, \mathbf{p}_2, \mathbf{p}_3\}$ *is an orthonormal basis of eigenvectors of a symmetric tensor, then*

$$\mathsf{I} = \sum_{i=1}^{3} \mathbf{p}_i \otimes \mathbf{p}_i.$$

This identity facilitates the proof of the following theorem:

THEOREM 3.1.3 (SPECTRAL DECOMPOSITION). *Let* $\mathsf{A} \in \mathrm{Sym}(\mathbb{E})$, *and suppose that* $\{\mathbf{p}_1, \mathbf{p}_2, \mathbf{p}_3\}$ *is an orthonormal basis for* $\mathbb{E}$ *consisting of eigenvectors of* A. *Then*

$$\mathsf{A} = \sum_{i=1}^{3} \lambda_i \mathbf{p}_i \otimes \mathbf{p}_i. \tag{3.1.3}$$

PROOF: By Exercise 52,

$$\mathsf{A} = \sum_{i=1}^{3} \mathsf{A}(\mathbf{p}_i \otimes \mathbf{p}_i) = \sum_{i=1}^{3} (\mathsf{A}\mathbf{p}_i) \otimes \mathbf{p}_i = \sum_{i=1}^{3} \lambda_i \mathbf{p}_i \otimes \mathbf{p}_i,$$

finishing the proof. ■

The spectral decomposition (3.1.3) shows that the eigenvalues and eigenvectors determine any tensor $\mathsf{A} \in \mathrm{Sym}(\mathbb{E})$ uniquely: if two tensors have the same spectral decomposition, then they exhibit the same action on vectors in $\mathbb{E}$ and, hence, must be equal. Equation (3.1.3) also reveals that A has matrix representation

$$\mathsf{A} = \begin{bmatrix} \lambda_1 & 0 & 0 \\ 0 & \lambda_2 & 0 \\ 0 & 0 & \lambda_3 \end{bmatrix}$$

with respect to the orthonormal basis $\{\mathbf{p}_1, \mathbf{p}_2, \mathbf{p}_3\}$.

EXERCISE 53 *Expand* $p_A(\lambda)$ *to prove the identities*

$$\begin{aligned} \mathrm{I}_A &= \lambda_1 + \lambda_2 + \lambda_3, \\ \mathrm{II}_A &= \lambda_2\lambda_3 + \lambda_3\lambda_1 + \lambda_1\lambda_2, \\ \mathrm{III}_A &= \lambda_1\lambda_2\lambda_3. \end{aligned} \tag{3.1.4}$$

EXERCISE 54 *Each of the expressions on the right in Equations (3.1.4) is symmetric in the eigenvalues* $\lambda_1, \lambda_2, \lambda_3$ *of the symmetric tensor* A. *Another trio of expressions that are symmetric in the eigenvalues of a symmetric tensor is* $\lambda_1 + \lambda_2 + \lambda_3$, $\lambda_1^2 + \lambda_2^2 + \lambda_3^2$, $\lambda_1^3 + \lambda_2^3 + \lambda_3^3$. *Prove the following representations for this alternative set of invariants for* $\mathsf{A} \in \mathrm{Sym}(\mathbb{E})$*:*

$$\begin{aligned} \lambda_1 + \lambda_2 + \lambda_3 &= \operatorname{tr} \mathsf{A}, \\ \lambda_1^2 + \lambda_2^2 + \lambda_3^2 &= \operatorname{tr} \mathsf{A}^2, \\ \lambda_1^3 + \lambda_2^3 + \lambda_3^3 &= \operatorname{tr} \mathsf{A}^3. \end{aligned} \tag{3.1.5}$$

The following exercise uses the orthonormality of the basis $\{\mathbf{p}_1, \mathbf{p}_2, \mathbf{p}_3\}$ and the concept of spectral decomposition to construct a new tensor associated with $\mathsf{A} \in \mathrm{SPD}(\mathbb{E})$:

EXERCISE 55 *Let* $\mathsf{A} \in \mathrm{SPD}(\mathbb{E})$*, with positive eigenvalues* $\lambda_1, \lambda_2, \lambda_3$ *and real eigenvectors* $\mathbf{p}_1, \mathbf{p}_2, \mathbf{p}_3$ *forming an orthonormal basis for* $\mathbb{E}$*. Show that*

$$\left(\sum_{i=1}^{3} \sqrt{\lambda_i}\mathbf{p}_i \otimes \mathbf{p}_i\right)^2 = \mathsf{A}.$$

Here, $\sqrt{\lambda_i}$ *denotes the positive square root of* λ_i*. Show that the tensor in parentheses belongs to* $\mathrm{SPD}(\mathbb{E})$.

Based on this result, we call the tensor

$$\sqrt{\mathsf{A}} = \sum_{i=1}^{3} \sqrt{\lambda_i}\mathbf{p}_i \otimes \mathbf{p}_i \tag{3.1.6}$$

the **square root** of A. The following theorem guarantees that $\sqrt{\mathsf{A}}$ is the unique tensor in $\mathrm{SPD}(\mathbb{E})$ whose square is A.

THEOREM 3.1.4 (UNIQUENESS OF TENSOR SQUARE ROOTS). *If* $\mathsf{S} \in \mathrm{SPD}(\mathbb{E})$ *and* $\mathsf{S}^2 = \mathsf{A} \in \mathrm{SPD}(\mathbb{E})$*, then* $\mathsf{S} = \sqrt{\mathsf{A}}$.

PROOF: Denote the eigenvalues of A by $\lambda_1, \lambda_2, \lambda_3$, all positive, with associated eigenvectors $\mathbf{p}_1, \mathbf{p}_2, \mathbf{p}_3$ forming an orthonormal basis. Since $\mathsf{A}\mathbf{p}_i = \lambda_i \mathbf{p}_i$ for $i = 1, 2, 3$,

$$(\mathsf{S}^2 - \lambda_i \mathsf{I})\mathbf{p}_i = \left(\mathsf{S} + \sqrt{\lambda_i}\mathsf{I}\right)\left(\mathsf{S} - \sqrt{\lambda_i}\mathsf{I}\right)\mathbf{p}_i = \mathbf{0},$$

that is, $(\mathsf{S} + \sqrt{\lambda_i}\mathsf{I})\mathbf{y} = \mathbf{0}$, where $\mathbf{y} = (\mathsf{S} - \sqrt{\lambda_i}\mathsf{I})\mathbf{p}_i$. It follows that $\mathbf{y} = \mathbf{0}$, since otherwise $-\sqrt{\lambda_i}$ would be a negative eigenvalue of the symmetric, positive definite tensor S. Thus $(\mathsf{S} - \sqrt{\lambda_i}\mathsf{I})\mathbf{p}_i = \mathbf{0}$, that is, $\mathbf{p}_i$ is an eigenvector of S associated with eigenvalue $\sqrt{\lambda_i}$. This argument shows that $\mathsf{S}\mathbf{p}_i = \sqrt{\lambda_i}\mathbf{p}_i = \sqrt{\mathsf{A}}\mathbf{p}_i$ for each basis vector $\mathbf{p}_i$, and hence $\mathsf{S}\mathbf{a} = \sqrt{\mathsf{A}}\mathbf{a}$ for every vector $\mathbf{a} \in \mathbb{E}$. Therefore, $\mathsf{S} = \sqrt{\mathsf{A}}$. ■

EXERCISE 56 *Show that each of the eigenvectors* $\mathbf{p}_j$ *in the orthonormal basis associated with* A *is an eigenvector of* $\sqrt{\mathsf{A}}$ *with corresponding eigenvalue* $\sqrt{\lambda_j}$.

EXERCISE 57 *Let* $\mathsf{A} \in \mathrm{SPD}(\mathbb{E})$ *be as in the previous exercise. Show that*

$$\sum_{i=1}^{3} \lambda_i^{-1}\mathbf{p}_i \otimes \mathbf{p}_i = \mathsf{A}^{-1}.$$

It follows that $\mathrm{SPD}(\mathbb{E}) \subset \mathrm{GL}(\mathbb{E})$. *Is* $\mathrm{SPD}(\mathbb{E})$ *a subgroup of* $\mathrm{GL}(\mathbb{E})$*?*

EXERCISE 58 *Show that, whenever* $\mathsf{A} \in \mathrm{GL}(\mathbb{E})$, $\mathsf{A}^\top\mathsf{A} \in \mathrm{SPD}(\mathbb{E})$.

3.1.2 Polar Decomposition and the Deformation Gradient

An arbitrary tensor $\mathsf{A} \in \mathrm{GL}(\mathbb{E})$ may not be symmetric, but $\mathsf{A}^\top\mathsf{A}$ and $\mathsf{A}\mathsf{A}^\top$ are. In fact, they are symmetric and positive definite. This observation, when coupled with the concept of the square root, yields an important representation for A in terms of symmetric and orthogonal tensors.

THEOREM 3.1.5 (POLAR DECOMPOSITION). *If* $\mathsf{A} \in \mathrm{GL}(\mathbb{E})$, *then there exist linear transformations* $\mathsf{U}, \mathsf{V} \in \mathrm{SPD}(\mathbb{E})$ *and a linear transformation* $\mathsf{R} \in \mathrm{O}(\mathbb{E})$ *such that* $\mathsf{A} = \mathsf{R}\mathsf{U} = \mathsf{V}\mathsf{R}$. *These decompositions are unique.*

These decompositions generalize the polar decomposition $a + ib = r\exp(i\theta)$ of a nonzero complex number.

PROOF: Exercise 58 establishes that the tensor $\mathsf{C} = \mathsf{A}^\top\mathsf{A} \in \mathrm{SPD}(\mathbb{E})$. From this fact it follows that the tensors

$$\mathsf{U} = \sqrt{\mathsf{A}^\top\mathsf{A}}, \quad \mathsf{V} = \sqrt{\mathsf{A}\mathsf{A}^\top} \in \mathrm{SPD}(\mathbb{E}). \tag{3.1.7}$$

Define $\mathsf{R} = \mathsf{A}\mathsf{U}^{-1}$, $\mathsf{Q} = \mathsf{V}^{-1}\mathsf{A}$. Then $\mathsf{Q}, \mathsf{R} \in \mathrm{O}(\mathbb{E})$, since

$$\mathsf{R}^\top\mathsf{R} = \mathsf{U}^{-\top}\mathsf{A}^\top\mathsf{A}\mathsf{U}^{-1} = \mathsf{U}^{-1}\mathsf{A}^\top\mathsf{A}\mathsf{U}^{-1} = \mathsf{U}^{-1}\mathsf{U}^2\mathsf{U}^{-1} = \mathsf{I},$$

and similarly for Q. Therefore, $\mathsf{A} = \mathsf{R}\mathsf{U}$ and $\mathsf{A} = \mathsf{V}\mathsf{Q}$. It remains to show that these decompositions are unique and that $\mathsf{Q} = \mathsf{R}$. To show that the decomposition $\mathsf{A} = \mathsf{R}\mathsf{U}$ is unique, assume that $\mathsf{A} = \mathsf{R}_1\mathsf{U}_1$ is another polar decomposition. Then

$$\mathsf{A}^\top\mathsf{A} = \mathsf{U}_1^\top\mathsf{R}_1^\top\mathsf{R}_1\mathsf{U}_1 = \mathsf{U}_1^2,$$

so $\mathsf{U}_1 = \mathsf{U}$ by the uniqueness of square roots in $\mathrm{SPD}(\mathbb{E})$. Hence, $\mathsf{R}_1 = \mathsf{AU}^{-1} = \mathsf{R}$. Similar reasoning shows that the polar decomposition $\mathsf{A} = \mathsf{VQ}$ is unique. To establish that $\mathsf{Q} = \mathsf{R}$, observe that

$$\mathsf{A} = \mathsf{RU} = \mathsf{VQ} = \mathsf{QQ}^\top \mathsf{VQ},$$

with $\mathsf{Q} \in \mathrm{O}(\mathbb{E})$ and $\mathsf{Q}^\top \mathsf{VQ} \in \mathrm{SPD}(\mathbb{E})$. The uniqueness of polar decompositions now implies that $\mathsf{Q} = \mathsf{R}$ and $\mathsf{Q}^\top \mathsf{VQ} = \mathsf{U}$. ∎

EXERCISE 59 *Show that* U *and* V *defined in Equation (3.1.7) have the same eigenvalues.*

With this background in place, we turn specifically to the deformation gradient $\mathsf{F} = \mathrm{Grad}\, \chi_\kappa$.

COROLLARY 3.1.6 (POLAR DECOMPOSITIONS OF F). *The deformation gradient* F *has polar decompositions*

$$\mathsf{F}(\mathbf{X}, t) = \mathsf{R}(\mathbf{X}, t)\, \mathsf{U}(\mathbf{X}, t) = \mathsf{V}(\mathbf{X}, t)\, \mathsf{R}(\mathbf{X}, t). \tag{3.1.8}$$

PROOF: Since $\det \mathsf{F}(\mathbf{X}, t) \neq 0$, $\mathsf{F}(\mathbf{X}, t) \in \mathrm{GL}(\mathbb{E})$ for each argument $(\mathbf{X}, t)$. ∎

For the remainder of this chapter we use the symbols R, U, and V to denote the tensors in the polar decomposition of F.

EXERCISE 60 *Show that* $\mathsf{R}(\mathbf{X}, t) \in \mathrm{O}^+(\mathbb{E})$ *at each* $(\mathbf{X}, t)$.

To interpret the decompositions (3.1.8) heuristically, consider the effect of the motion on arc elements: from the informal equation (2.4.1),

$$d\mathbf{x} = \mathsf{F}d\mathbf{X} = \begin{cases} \mathsf{RU}\, d\mathbf{X} \\ \mathsf{VR}\, d\mathbf{X}. \end{cases}$$

In the first decomposition, the motion stretches the arc element $d\mathbf{X}$ by a transformation $\mathsf{U} \in \mathrm{SPD}(\mathbb{E})$, having positive real eigenvalues, then rotates it by a transformation $\mathsf{R} \in \mathrm{O}^+(\mathbb{E})$. In the second decomposition, $d\mathbf{X}$ undergoes first rotation by R then stretching by V. We call R the **rotation tensor**, while U and V are the **right** and **left stretch tensors**, respectively. The eigenvalues of U (and, hence, of V) are the **principal stretches**. These numbers quantify the material's extension in directions defined by the orthogonal eigenvectors of U (before rotation) or V (after rotation).

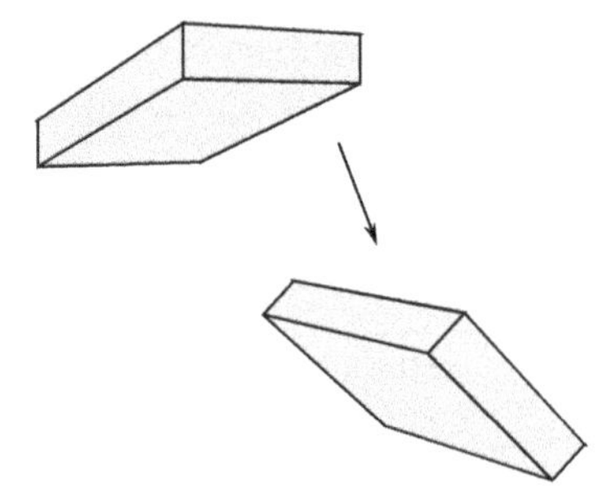

Figure 3.1. Rigid motion.

3.1.3 Examples

Four examples of motion illustrate these concepts. The first is rigid motion, which superposes pure rotation and pure translation:

$$\mathbf{x}(\mathbf{X}, t) = \mathsf{Q}(t)\mathbf{X} + \mathbf{c}(t). \tag{3.1.9}$$

Here $\mathsf{Q}\colon \mathbb{R} \to \mathrm{O}^+(\mathbb{E})$ is a continuous, one-parameter family of proper orthogonal transformations. Figure 3.1 depicts this motion. The deformation gradient for this motion has polar decompositions

$$\mathsf{F} = \operatorname{Grad} \mathbf{x}(\mathbf{X}, t) = \mathsf{Q}(t) = \mathsf{Q}(t)\mathsf{I} = \mathsf{I}\,\mathsf{Q}(t).$$

Since $\mathsf{U} = \mathsf{V} = \mathsf{I}$, the principal stretches are $\lambda_1 = \lambda_2 = \lambda_3 = 1$.

EXERCISE 61 *Show that rigid motion is isochoric.*

The second example is **pure extension**, which has the form

$$\mathbf{x}(\mathbf{X}, t) = \left[\sum_{i=1}^{3} \lambda_i(t)\mathbf{e}_i \otimes \mathbf{e}_i\right]\mathbf{X}.$$

Using matrix representations with respect to $\{\mathbf{e}_1, \mathbf{e}_2, \mathbf{e}_3\}$, we write the polar decomposition of the deformation gradient in this case as

$$\mathsf{F} = \begin{bmatrix} \lambda_1(t) & 0 & 0 \\ 0 & \lambda_2(t) & 0 \\ 0 & 0 & \lambda_3(t) \end{bmatrix} = \mathsf{Z}(t) = \mathsf{I}\mathsf{Z}(t) = \mathsf{Z}(t)\mathsf{I},$$

where both the left and right stretch tensors are $\mathsf{Z}(t)$, and the rotation tensor is I. The body undergoes no rotation with respect to the coordinate axes, and the principal stretches are $\lambda_1(t), \lambda_2(t), \lambda_3(t)$. Figure 3.2 illustrates this motion. When

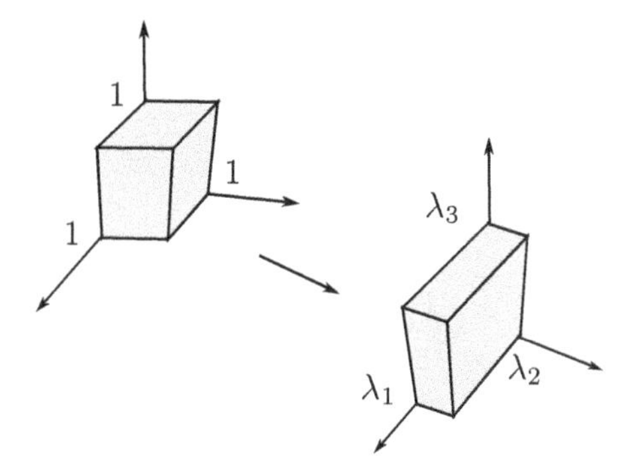

Figure 3.2. Pure extension, with principal stretches $\lambda_1, \lambda_2, \lambda_3$.

$0 < \lambda_i(t) < 1$, the principal stretch in the direction $\mathbf{e}_i$ is a contraction; when $\lambda_i(t) > 1$, it is a dilation.

The third example is simple shear, $\mathbf{x}(\mathbf{X}, t) = [\mathsf{I} + \alpha(t)\mathbf{e}_1 \otimes \mathbf{e}_2]\mathbf{X}$, whose representation in $\mathbb{R}^3$ with respect to $\{\mathbf{e}_1, \mathbf{e}_2, \mathbf{e}_3\}$ is

$$\mathbf{x}(\mathbf{X}, t) = \begin{bmatrix} X_1 + X_2\alpha(t) \\ X_2 \\ X_3 \end{bmatrix}.$$

The polar decomposition for this motion is more complicated than in the previous examples. Again with respect to the orthonormal basis $\{\mathbf{e}_1, \mathbf{e}_2, \mathbf{e}_3\}$,

$$\mathsf{F} = \begin{bmatrix} 1 & \alpha & 0 \\ 0 & 1 & 0 \\ 0 & 0 & 1 \end{bmatrix} = \underbrace{\begin{bmatrix} \cos\beta & \sin\beta & 0 \\ -\sin\beta & \cos\beta & 0 \\ 0 & 0 & 1 \end{bmatrix}}_{\mathsf{R}} \underbrace{\begin{bmatrix} \cos\beta & \sin\beta & 0 \\ \sin\beta & \sec\beta(1+\sin^2\beta) & 0 \\ 0 & 0 & 1 \end{bmatrix}}_{\mathsf{U}}.$$

Here, $\beta = \tan^{-1}(\alpha/2)$. As Figure 3.3 shows, this motion stretches the body and rotates it through an angle β. The principal stretches are

$$\begin{aligned} \lambda_1 &= \sec\beta + \tan\beta, \\ \lambda_2 &= 1, \\ \lambda_3 &= \sec\beta - \tan\beta. \end{aligned}$$

A final example arises from a formal Taylor expansion of the deformation χ_κ about a fixed point $\mathbf{X}_0 \in \boldsymbol{\kappa}(\mathcal{B})$:

$$\chi_\kappa(\mathbf{X}, t) = \chi_\kappa(\mathbf{X}_0, t) + \mathsf{F}(\mathbf{X}_0, t)\,(\mathbf{X} - \mathbf{X}_0) + \cdots.$$

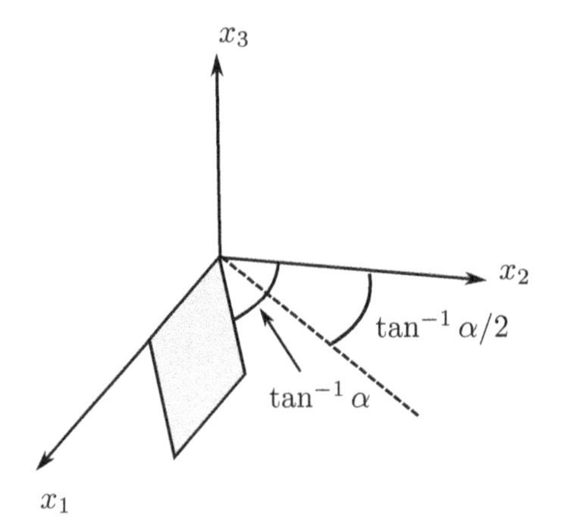

Figure 3.3. Simple shear showing the combined effects of stretch followed by rotation.

Truncating this expansion yields a deformation of the form

$$\chi_\kappa(\mathbf{X}, t) = \mathbf{x}_0(\mathbf{X}_0, t) + \mathsf{F}(\mathbf{X}_0, t)(\mathbf{X} - \mathbf{X}_0). \tag{3.1.10}$$

Here, $\mathbf{x}_0(\mathbf{X}_0, t)$ and $\mathsf{F}(\mathbf{X}_0, t)$ are continuously differentiable functions of t; $\mathsf{F}(\mathbf{X}_0, t)$ has continuously differentiable inverse $\mathsf{F}^{-1}(\mathbf{X}_0, t)$ at each time t; and $\mathbf{X}_0$ stands for a constant vector. This type of motion serves as a linear approximation to more complicated motions appropriate in regions where $\|\mathbf{X} - \mathbf{X}_0\|$ is sufficiently small.

3.1.4 Cauchy–Green and Strain Tensors

The polar decomposition of F into stretch and rotation tensors is important but often computationally inconvenient. Other tensors related to F are typically easier to compute.

DEFINITION. *The* **Cauchy–Green tensors** *are*

$$\begin{aligned} \mathsf{B} &= \mathsf{F}\mathsf{F}^\top && \textbf{left Cauchy–Green tensor}, \\ \mathsf{C} &= \mathsf{F}^\top\mathsf{F} && \textbf{right Cauchy–Green tensor}. \end{aligned} \tag{3.1.11}$$

By the identities (2.2.5) and (3.1.8), these tensors are related to the stretch tensors and to the displacement gradient $\mathsf{H}(\mathbf{X}, t) = \text{Grad}\,\mathbf{u}(\mathbf{X}, t) = \mathsf{F}(\mathbf{X}, t) - \mathsf{I}$ as follows:

$$\begin{aligned} \mathsf{B} &= \mathsf{V}^2 = \mathsf{I} + \mathsf{H} + \mathsf{H}^\top + \mathsf{H}\mathsf{H}^\top; \\ \mathsf{C} &= \mathsf{U}^2 = \mathsf{I} + \mathsf{H} + \mathsf{H}^\top + \mathsf{H}^\top\mathsf{H}. \end{aligned} \tag{3.1.12}$$

It follows that the eigenvalues of B and C are squares of the eigenvalues of V, which has the same eigenvalues as U by Exercise 59.

EXERCISE 62 *Prove the identities (3.1.12).*

EXERCISE 63 *Show that* $\mathsf{B}, \mathsf{C} \in \mathrm{SPD}(\mathbb{E})$.

To see how B and C measure strain, the following exercise examines the effects of the deformation on material spheres. Recall that Equation (3.1.10) serves as an approximation to more general deformations when higher-order terms in the deformation's Taylor expansion are negligible.

EXERCISE 64 *Show that the deformation (3.1.10) maps the sphere*

$$\mathcal{S}_\kappa = \left\{ \mathbf{X} \in \mathbb{E} \,\middle|\, \|\mathbf{X} - \mathbf{X}_0\|^2 = R_\kappa^2 \right\} \subset \boldsymbol{\kappa}(\mathcal{B})$$

into an ellipsoid in $\chi(\mathcal{B}, t)$. *Show that the sphere*

$$\mathcal{S}_t = \left\{ \mathbf{x} \in \mathbb{E} \,\middle|\, \|\mathbf{x} - \mathbf{x}_0\|^2 = R_t^2 \right\} \subset \chi(\mathcal{B}, t)$$

is the image under χ_κ *of an ellipsoid in* $\kappa(\mathcal{B})$. *(The general equation for an ellipsoid in* $\mathbb{E}$ *centered at* $\mathbf{x}_0$ *is* $(\mathbf{x} - \mathbf{x}_0) \cdot \mathsf{A}(\mathbf{x} - \mathbf{x}_0) = R^2$, *where* $\mathsf{A} \in \mathrm{SPD}(\mathbb{E})$. *We call the expression on the left of this equation a* **positive definite quadratic form** *in* $\mathbf{x} - \mathbf{x}_0$.*)*

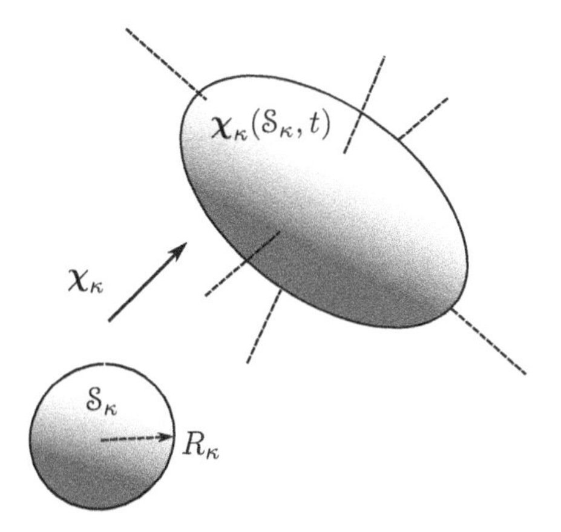

Figure 3.4. Eulerian strain ellipsoid.

The ellipsoids

$$(\mathbf{x} - \mathbf{x}_0) \cdot \mathsf{B}^{-1}(\mathbf{x} - \mathbf{x}_0) = R_\kappa^2, \qquad (\mathbf{X} - \mathbf{X}_0) \cdot \mathsf{C}(\mathbf{X} - \mathbf{X}_0) = R_t^2$$

identified in this exercise are the **Eulerian** and **reciprocal Lagrangian strain ellipsoids**, respectively. Figure 3.4 shows an Eulerian strain ellipsoid. For deformations in which nonzero higher-order terms appear in the Taylor expansion about $\mathbf{X}_0$, the derivations hold in the limit as the radii $R \to 0$; in this sense we regard the strain ellipsoids as infinitesimal.

One way to measure strain is to gauge the extent to which the strain ellipsoids differ from spheres, which are ellipsoids arising from quadratic forms involving I.

DEFINITION. *The* **Eulerian** *and* **Lagrangian strain tensors** *are, respectively,*

$$\mathsf{E}_E = \tfrac{1}{2}(\mathsf{I} - \mathsf{B}^{-1}),$$
$$\mathsf{E}_L = \tfrac{1}{2}(\mathsf{C} - \mathsf{I}).$$

Both tensors are symmetric, and they are nonlinear in the displacement gradient H: from Equations (3.1.12),

$$\mathsf{E}_L = \tfrac{1}{2}(\mathsf{H} + \mathsf{H}^\top + \mathsf{H}^\top \mathsf{H}). \tag{3.1.13}$$

In the next section we investigate a linear approximation to E_L that facilitates analysis of small displacements from the reference configuration.

3.1.5 Strain Invariants

It is possible to distill the information in the Cauchy–Green tensors B and C even further, using information about their invariants. Using concepts reviewed in Section 1.2, we identify the following quantities, which are related to the eigenvalues of B and C but do not require their direct calculation:

DEFINITION. *The* **strain invariants** *are the principal invariants of the Cauchy–Green tensors, denoted as*

$$\mathrm{I}_B = \mathrm{I}_C, \qquad \mathrm{II}_B = \mathrm{II}_C, \qquad \mathrm{III}_B = \mathrm{III}_C.$$

Let us revisit three examples discussed earlier, this time using the Cauchy–Green tensors and strain invariants to characterize the strain. For rigid motion, $\mathbf{x}(\mathbf{X}, t) = \mathsf{Q}(t)\mathbf{X} + \mathbf{c}(t)$. Since $\mathsf{F} = \mathsf{Q}$ is orthogonal in this case, $\mathsf{B} = \mathsf{C} = \mathsf{I}$. Identifying the principal invariants as the coefficients in the characteristic polynomial

$$p_C(\lambda) = \det(\lambda\mathsf{I} - \mathsf{C}) = (\lambda - 1)^3 = \lambda^3 - \mathrm{I}_C\lambda^2 + \mathrm{II}_C\lambda - \mathrm{III}_C$$

gives

$$\mathrm{I}_C = \mathrm{II}_C = 3; \qquad \mathrm{III}_C = 1.$$

For pure extension, $\mathbf{x} = \mathsf{Z}(t)\mathbf{X}$, where $\mathsf{Z}(t)$ has a diagonal matrix representation with respect to $\{\mathbf{e}_1, \mathbf{e}_2, \mathbf{e}_3\}$. The Cauchy–Green tensors are also diagonal in this representation:

$$\mathsf{B} = \mathsf{C} = \mathsf{Z}^2 = \begin{bmatrix} \lambda_1^2 & 0 & 0 \\ 0 & \lambda_2^2 & 0 \\ 0 & 0 & \lambda_3^2 \end{bmatrix},$$

where $\lambda_1(t), \lambda_2(t), \lambda_3(t)$ are the principal stretches. The strain invariants are therefore

$$\begin{aligned} \mathrm{I}_C &= \lambda_1^2 + \lambda_2^2 + \lambda_3^2, \\ \mathrm{II}_C &= \lambda_2^2\lambda_3^2 + \lambda_3^2\lambda_1^2 + \lambda_1^2\lambda_2^2, \\ \mathrm{III}_C &= \lambda_1^2\lambda_2^2\lambda_3^2. \end{aligned}$$

Finally, for simple shear, $\mathbf{x} = [\mathsf{I} + \alpha(t)\mathbf{e}_1 \otimes \mathbf{e}_2]\,\mathbf{X}$, and the Cauchy–Green tensors have the representations

$$\mathsf{B} = \mathsf{F}\mathsf{F}^\top = \begin{bmatrix} 1 & \alpha & 0 \\ 0 & 1 & 0 \\ 0 & 0 & 1 \end{bmatrix} \begin{bmatrix} 1 & 0 & 0 \\ \alpha & 1 & 0 \\ 0 & 0 & 1 \end{bmatrix} = \begin{bmatrix} 1+\alpha^2 & \alpha & 0 \\ \alpha & 1 & 0 \\ 0 & 0 & 1 \end{bmatrix},$$

$$\mathsf{C} = \mathsf{F}^\top\mathsf{F} = \begin{bmatrix} 1 & \alpha & 0 \\ \alpha & 1+\alpha^2 & 0 \\ 0 & 0 & 1 \end{bmatrix}.$$

We compute strain invariants simply by expanding the characteristic polynomial p_C:

$$p_C(\lambda) = \lambda^3 - (3+\alpha^2)\lambda^2 + (3+\alpha^2)\lambda - 1,$$

and hence

$$\mathrm{I}_C = 3 + \alpha^2 = \mathrm{II}_C, \qquad \mathrm{III}_C = 1.$$

3.1.6 Summary

Because the deformation gradient has positive determinant, it admits polar decompositions $\mathsf{F} = \mathsf{R}\mathsf{U} = \mathsf{V}\mathsf{R}$, where R is the orthogonal rotation tensor and U and V are symmetric and positive definite stretch tensors. Related to F are the left and right Cauchy–Green tensors

$$\mathsf{B} = \mathsf{F}\mathsf{F}^\top = \mathsf{V}^2, \qquad \mathsf{C} = \mathsf{F}^\top\mathsf{F} = \mathsf{U}^2,$$

respectively. These tensors have identical principal invariants $\mathrm{I}_B = \mathrm{I}_C$, $\mathrm{II}_B = \mathrm{II}_C$, and $\mathrm{III}_B = \mathrm{III}_C$, which are often easier to calculate than the eigenvalues of F. An examination of the effects of deformation on filaments of material suggests regarding the Lagrangian strain tensor

$$\mathsf{E}_L = \frac{1}{2}(\mathsf{C} - \mathsf{I}) = \frac{1}{2}(\mathsf{H} + \mathsf{H}^\top + \mathsf{H}^\top\mathsf{H})$$

as a measure of strain. The next section examines linear approximations to this tensor.

3.2 INFINITESIMAL STRAIN

In many materials, including certain solids, displacements often amount to small perturbations from rigid motion. Thus, however large the absolute displacements from the body's initial configuration may be, the gradients of these displacements remain small. Examples include the displacements that solid mechanical parts undergo during nondestructive testing, as well as vibrations in segments of the earth's crust subjected to seismic prospecting. Section 5.4 briefly examines the wavelike behavior of such responses. In these cases the analysis of strain uses approximations that linearize the Eulerian and Lagrangian strain tensors introduced in Section 3.1. This section introduces these approximations and their geometric interpretations.

3.2.1 The Infinitesimal Strain Tensor

Recall the displacement gradient, $\mathsf{H}(\mathbf{X},t) = \mathrm{Grad}\,\mathbf{u}(\mathbf{X},t)$, where the vector field $\mathbf{u}(\mathbf{x},t) = \mathbf{x}(\mathbf{X},t) - \mathbf{X}$ is the displacement. From Equations (3.1.13), the Lagrangian strain tensor is

$$\mathsf{E}_L = \tfrac{1}{2}(\mathsf{H} + \mathsf{H}^\top + \mathsf{H}^\top\mathsf{H}).$$

When the components $H_{jk} = \partial u_j/\partial X_k$ are small enough in magnitude, we expect the terms that are linear in H and $\mathsf{H}^\top$ to dominate the quadratic term $\frac{1}{2}\mathsf{H}^\top\mathsf{H}$.

To formalize this reasoning, assume that $|\partial u_j/\partial X_k| < \epsilon$ for some small, positive parameter $\epsilon < 1$, for $j,k = 1,2,3$. We wish to examine the *spatial* gradient of $\mathbf{u} = \mathbf{x} - \mathbf{X}$, the entries of which, in Cartesian coordinates, are

$$\frac{\partial u_j}{\partial x_k} = \frac{\partial x_j}{\partial x_k} - \frac{\partial X_j}{\partial x_k} = \delta_{jk} - F_{jk}^{-1}, \tag{3.2.1}$$

where F_{jk}^{-1} denotes the (j,k)th entry of F^{-1}. Thus the analysis reduces to task of approximating $\mathsf{I} - \mathsf{F}^{-1}$.

EXERCISE 65 *Formally justify the expansion*

$$\begin{aligned}\mathsf{I}-\mathsf{F}^{-1} &= \mathsf{I}-[\mathsf{I}+(\mathsf{F}-\mathsf{I})]^{-1}\\ &= \mathsf{I}-[\mathsf{I}-(\mathsf{F}-\mathsf{I})+(\mathsf{F}-\mathsf{I})^2-+\cdots]\\ &= (\mathsf{F}-\mathsf{I})-(\mathsf{F}-\mathsf{I})^2+(\mathsf{F}-\mathsf{I})^3-+\cdots.\end{aligned}$$

Since $\mathsf{F}-\mathsf{I}$ has entries $\partial u_j/\partial X_k$, the expansion yields, in coordinate form,

$$\frac{\partial u_j}{\partial x_k} = \frac{\partial u_j}{\partial X_k} \underbrace{-\frac{\partial u_j}{\partial X_l}\frac{\partial u_l}{\partial X_k} + \frac{\partial u_j}{\partial X_l}\frac{\partial u_l}{\partial X_m}\frac{\partial u_m}{\partial X_k} - + \cdots}_{\text{Negligible}}.$$

The terms labeled "negligible" are much smaller in magnitude than ϵ, and neglecting them yields the approximation

$$\frac{\partial u_j}{\partial x_k} \simeq \frac{\partial u_j}{\partial X_k}.$$

To this level of approximation,

$$\mathsf{E}_L \simeq \mathsf{E} = \tfrac{1}{2}(\mathsf{H}+\mathsf{H}^\top) \tag{3.2.2}$$

$$= \tfrac{1}{2}(\mathsf{F}+\mathsf{F}^\top) - \mathsf{I}, \tag{3.2.3}$$

where E is the **infinitesimal strain tensor**.

By construction, E is symmetric. Hence, it has real eigenvalues, called **principal strains**, and orthogonal eigenvectors, whose directions define the **principal axes of strain**.

To interpret E geometrically, consider a small material cube of the body, initially having edge length h, with edges parallel to the orthonormal basis vectors $\mathbf{e}_1, \mathbf{e}_2, \mathbf{e}_3$, as drawn in Figure 3.5. Adopt the initial configuration as the reference configuration, labeling a corner of the part as $\mathbf{X}$. Upon deformation, the particles in the cube undergo displacement into a new configuration. At time t, the corner $\mathbf{X}$ occupies the spatial position $\mathbf{x}(\mathbf{X},t) = \mathbf{X}+\mathbf{u}(\mathbf{X},t)$. The adjacent corners $\mathbf{X}+h\mathbf{e}_i$ occupy spatial positions for which we may approximate the displacements using Taylor expansions through first order:

$$\begin{aligned}\mathbf{u}(\mathbf{X}+h\mathbf{e}_i,t) &\simeq \mathbf{u}(\mathbf{X},t) + h\,\mathrm{Grad}\,\mathbf{u}(\mathbf{X},t)\,\mathbf{e}_i\\ &= \mathbf{u}(\mathbf{X},t) + h\frac{\partial u_1}{\partial X_i}\mathbf{e}_1 + h\frac{\partial u_2}{\partial X_i}\mathbf{e}_2 + h\frac{\partial u_3}{\partial X_i}\mathbf{e}_3.\end{aligned}$$

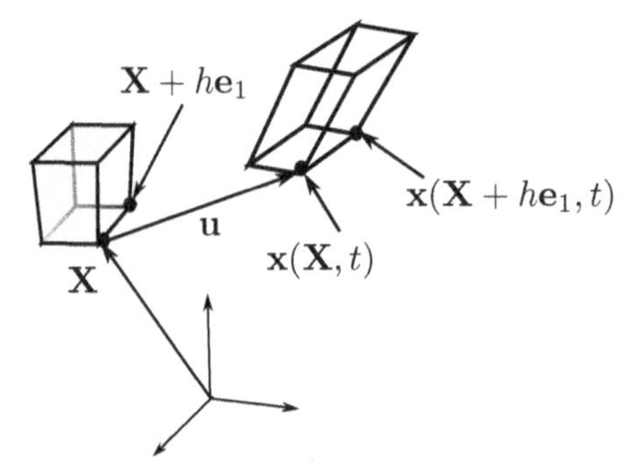

Figure 3.5. Deformation of a cubic part of a body undergoing small displacement.

Now consider the role that the diagonal entries of E,

$$E_{11} = \frac{\partial u_1}{\partial X_1}, \qquad E_{22} = \frac{\partial u_2}{\partial X_2}, \qquad E_{33} = \frac{\partial u_3}{\partial X_3}, \tag{3.2.4}$$

play in Figure 3.5. These entries give the factors by which the deformation extends the material edges $h\mathbf{e}_i$ *in the directions* $\mathbf{e}_i$—that is, longitudinally with respect to the original configuration. Figure 3.6 shows this longitudinal extension for the edge $h\mathbf{e}_1$. We call the diagonal entries of E the **normal strains**.

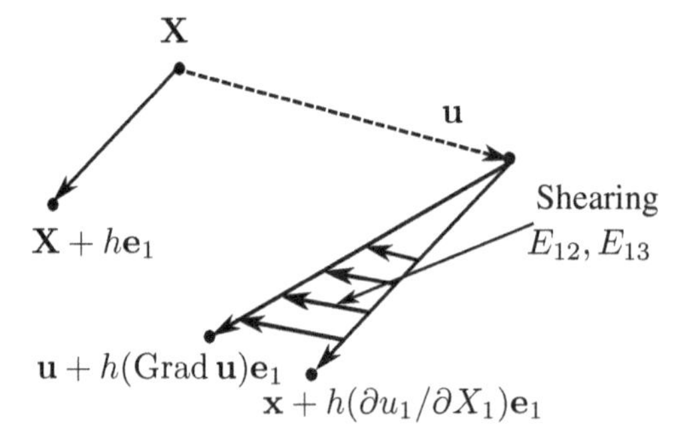

Figure 3.6. Effects of small displacement on a single material edge of a cubic part of the body.

The off-diagonal entries

$$E_{jk} = \frac{1}{2}\left(\frac{\partial u_j}{\partial X_k} + \frac{\partial u_k}{\partial X_j}\right) \tag{3.2.5}$$

of E play a different role. The effects of these entries on material edges involve shearing motion, which changes the angles of the edges with respect to the reference configuration. We call the off-diagonal entries of E the **shear strains**.

3.2.2 Summary

The Lagrangian strain tensor $\mathsf{E}_L = \frac{1}{2}(\mathsf{H} + \mathsf{H}^\top + \mathsf{H}^\top\mathsf{H})$ is nonlinear in the displacement gradient $\mathsf{H} = \text{Grad}\,\mathbf{u}$. When the derivatives $\partial u_i/\partial X_j$ of displacement are small—as may be the case in some motions involving solids—it is useful to neglect the nonlinear term and work with the infinitesimal strain tensor, $\mathsf{E} = \frac{1}{2}(\mathsf{H} + \mathsf{H}^\top)$. This tensor has the computational advantage of being symmetric. In a Cartesian coordinate system its diagonal entries represent longitudinal or normal strain, while the off-diagonal entries capture the effects of shear strain.

3.3 STRAIN RATES

In some materials the time rate of change of strain exerts a greater effect than strain itself. Many fluids behave this way: deformation *per se* from a reference configuration provokes little or no response, but the temporal rate at which the deformation occurs has a significant effect. We explore these responses further in Chapters 5 and 6. For now, we discuss the kinematic aspects of strain rates.

Recall the identity (2.2.6) for the rate of change of the deformation gradient:

$$\frac{D\mathsf{F}}{Dt} = \mathsf{LF},$$

where $\mathsf{L} = \text{grad}\,\mathbf{v}(\mathbf{x}, t)$ is the velocity gradient. Because F is invertible,

$$\mathsf{L} = \frac{D\mathsf{F}}{Dt}\mathsf{F}^{-1}.$$

This observation motivates a focus on L for strain rate analysis.

EXERCISE 66 *Show that, in terms of the stretch and rotation tensors,*

$$\mathsf{L} = \mathsf{R}\frac{D\mathsf{U}}{Dt}\mathsf{U}^{-1}\mathsf{R}^\top + \frac{D\mathsf{R}}{Dt}\mathsf{R}^\top = \frac{D\mathsf{V}}{Dt}\mathsf{V}^{-1} + \mathsf{V}\frac{D\mathsf{R}}{Dt}\mathsf{R}^\top\mathsf{V}^{-1}.$$

The analysis rests on a decomposition of L into tensors that have meaningful geometric interpretations in terms of stretching and rotation rates. As Exercise 66 shows, differentiation of the decompositions $\mathsf{F} = \mathsf{RU} = \mathsf{VR}$ offers little help, because it fails to tease apart the stretching and rotational effects. The following notion furnishes a more productive path:

DEFINITION. *A tensor* $\mathsf{A} \in \mathrm{L}(\mathbb{E})$ *is* **skew** *if* $\mathsf{A}^\top = -\mathsf{A}$.

As we explore in this section, the properties of skew tensors provide insight into the rotational effects associated with time-dependent strain.

EXERCISE 67 *Show that whenever* $\mathsf{Q}\colon \mathbb{R} \to \mathrm{O}(\mathbb{E})$ *is a differentiable function of time, the tensor function* $\mathsf{Q}'\mathsf{Q}^\top$ *is skew.*

For example, the quantity $(D\mathsf{R}/Dt)\mathsf{R}^\top$ appearing in the previous exercise is skew, but the nature of the other terms remains less transparent.

3.3.1 Stretching and Spin Tensors

Decomposing L directly into symmetric and skew parts yields

$$\mathsf{L} = \underbrace{\tfrac{1}{2}(\mathsf{L}+\mathsf{L}^\top)}_{\mathsf{D}} + \underbrace{\tfrac{1}{2}(\mathsf{L}-\mathsf{L}^\top)}_{\mathsf{W}}. \tag{3.3.1}$$

The symmetric part D is the **stretching tensor** or **rate-of-strain tensor**. With respect to the basis $\{\mathbf{e}_1, \mathbf{e}_2, \mathbf{e}_3\}$, it has entries $\frac{1}{2}(\partial v_i/\partial x_j + \partial v_j/\partial x_i)$. The skew part W is the **spin tensor**, having entries $\frac{1}{2}(\partial v_i/\partial x_j - \partial v_j/\partial x_i)$.

EXERCISE 68 *Show that* $\operatorname{tr}(\mathsf{AL}) = \operatorname{tr}(\mathsf{AD})$ *for any symmetric tensor* A.

EXERCISE 69 *Use the identity (2.2.6) to show that*

$$\mathsf{D} = \mathsf{F}^{-\top}\frac{D\mathsf{E}_L}{Dt}\mathsf{F}^{-1}. \tag{3.3.2}$$

The justification for the terminology for D and W may not be immediately apparent. For arbitrary choices of reference configuration, the relationships between D and W and the stretch and rotation tensors introduced in Section 3.1 can seem opaque, as the next exercise suggests.

EXERCISE 70 *Verify the following identities.*

$$\mathsf{D} = \begin{cases} \dfrac{1}{2}\mathsf{R}\left(\dfrac{D\mathsf{U}}{Dt}\mathsf{U}^{-1} + \mathsf{U}^{-1}\dfrac{D\mathsf{U}}{Dt}\right)\mathsf{R}^\top \\[2ex] \dfrac{1}{2}\left(\dfrac{D\mathsf{V}}{Dt}\mathsf{V}^{-1} + \mathsf{V}^{-1}\dfrac{D\mathsf{V}}{Dt} + \mathsf{V}\dfrac{D\mathsf{R}}{Dt}\mathsf{R}^\top\mathsf{V}^{-1} - \mathsf{V}^{-1}\dfrac{D\mathsf{R}}{Dt}\mathsf{R}^\top\mathsf{V}\right), \end{cases}$$

$$\mathsf{W} = \begin{cases} \dfrac{1}{2}\mathsf{R}\left(\dfrac{D\mathsf{U}}{Dt}\mathsf{U}^{-1} - \mathsf{U}^{-1}\dfrac{D\mathsf{U}}{Dt}\right)\mathsf{R}^\top + \dfrac{D\mathsf{R}}{Dt}\mathsf{R}^\top \\[2ex] \dfrac{1}{2}\left(\dfrac{D\mathsf{V}}{Dt}\mathsf{V}^{-1} - \mathsf{V}^{-1}\dfrac{D\mathsf{V}}{Dt} + \mathsf{V}\dfrac{D\mathsf{R}}{Dt}\mathsf{R}^\top\mathsf{V}^{-1} + \mathsf{V}^{-1}\dfrac{D\mathsf{R}}{Dt}\mathsf{R}^\top\mathsf{V}\right). \end{cases}$$

However, with a special choice of reference configuration the terms stretching and spin appear more natural. Fix an instant in time, and choose the reference configuration $\boldsymbol{\kappa}(\mathcal{B})$ to be the configuration of the body at that instant. With this choice, the deformation gradient at the instant in question has its simplest possible polar decomposition: $\mathsf{F} = \mathsf{I} = \mathsf{U} = \mathsf{V} = \mathsf{R}$. Hence, at this instant the relationships just derived also simplify:

$$\mathsf{D} = \left.\frac{D\mathsf{U}}{Dt}\right|_{\mathsf{F}=\mathsf{I}} = \left.\frac{D\mathsf{V}}{Dt}\right|_{\mathsf{F}=\mathsf{I}}$$

$$\mathsf{W} = \left.\frac{D\mathsf{R}}{Dt}\right|_{\mathsf{F}=\mathsf{I}}.$$

Therefore, with respect to this choice of reference configuration, the stretching and spin tensors reduce instantaneously to time derivatives of the stretch tensors U, V and rotation tensor R, respectively

Symmetry of the stretching tensor D implies that it has three real eigenvalues, which we call the **principal stretchings**. Associated with these numbers are three eigenvectors that form an orthonormal basis for $\mathbb{E}$. These mutually orthogonal eigenvectors are the **principal directions of stretching**.

Three examples illustrate these ideas. Consider first pure extension, which has matrix representation

$$\mathbf{x} = \begin{bmatrix} \lambda_1(t) & 0 & 0 \\ 0 & \lambda_2(t) & 0 \\ 0 & 0 & \lambda_3(t) \end{bmatrix} \mathbf{X} = \mathsf{Z}(t)\,\mathbf{X},$$

for differentiable functions $\lambda_1(t), \lambda_2(t), \lambda_3(t)$ that never vanish. The velocity field is

$$\mathbf{v} = \begin{bmatrix} \lambda_1' & 0 & 0 \\ 0 & \lambda_2' & 0 \\ 0 & 0 & \lambda_3' \end{bmatrix} \underbrace{\mathbf{X}}_{\mathsf{Z}^{-1}\mathbf{x}} = \begin{bmatrix} \lambda_1'/\lambda_1 & 0 & 0 \\ 0 & \lambda_2'/\lambda_2 & 0 \\ 0 & 0 & \lambda_3'/\lambda_3 \end{bmatrix} \mathbf{x},$$

so the velocity gradient is

$$\mathsf{L} = \operatorname{grad}\mathbf{v} = \begin{bmatrix} \lambda_1'/\lambda_1 & 0 & 0 \\ 0 & \lambda_2'/\lambda_2 & 0 \\ 0 & 0 & \lambda_3'/\lambda_3 \end{bmatrix}.$$

In this case, $\mathsf{D} = \mathsf{L}$, and $\mathsf{W} = 0$. The principal stretchings are λ_1'/λ_1, λ_2'/λ_2, λ_3'/λ_3, having dimension T^{-1}, and the principal directions of stretching are the standard basis vectors $\mathbf{e}_1, \mathbf{e}_2, \mathbf{e}_3$, as shown in Figure 3.7.

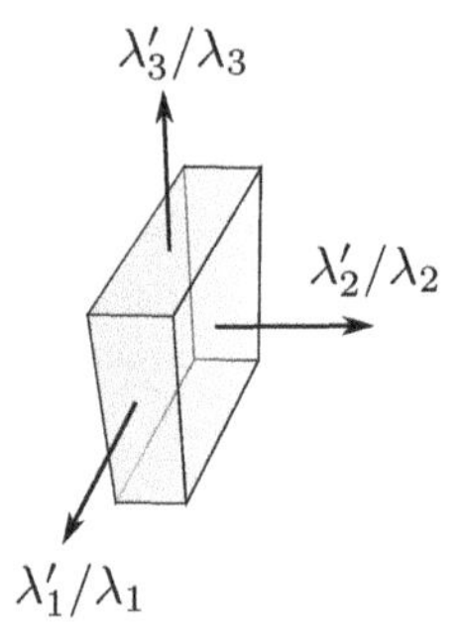

Figure 3.7. Principal stretchings and principal directions of stretching for pure extension.

Next consider two steady, planar vortex flows. For the first, let the velocity have representation

$$\mathbf{v}(\mathbf{x}) = \left(-\frac{x_2}{\|\mathbf{x}\|^2}, \frac{x_1}{\|\mathbf{x}\|^2}, 0\right)^\top,$$

with respect to the basis $\{\mathbf{e}_1, \mathbf{e}_2, \mathbf{e}_3\}$. Figure 3.8 illustrates this motion.

EXERCISE 71 *Show that for this velocity field the velocity gradient has matrix representation*

$$\mathsf{L} = \mathsf{D} = \begin{bmatrix} 2x_1x_2/\|\mathbf{x}\|^4 & (x_2^2 - x_1^2)/\|\mathbf{x}\|^4 & 0 \\ (x_2^2 - x_1^2)/\|\mathbf{x}\|^4 & -2x_1x_2/\|\mathbf{x}\|^4 & 0 \\ 0 & 0 & 0 \end{bmatrix},$$

and $\mathsf{W} = 0$. *Find the principal stretchings and the principal axes of stretching at the point* $\mathbf{x} = (1/2, \sqrt{3}/2, 0)^\top$.

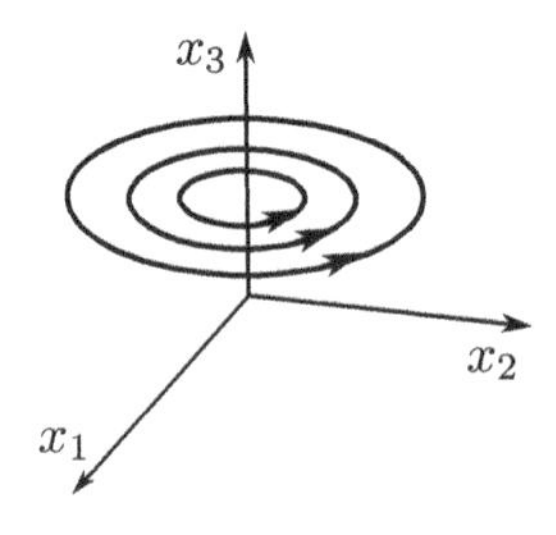

Figure 3.8. Steady, planar vortex flow.

It is striking to see a motion that clearly involves rotation but for which the spin tensor vanishes. To resolve this apparent anomaly, consider a second vortex flow:

EXERCISE 72 *For the velocity field* $\mathbf{v}(\mathbf{x}) = (-x_2, x_1, 0)^\top$, *show that* $\mathsf{D} = 0$ *and* $\mathsf{W} \neq 0$.

The schematic in Figure 3.9, adapted from [50, p. 340–341], suggests the resolution. In the vortex flow illustrated in Figure 3.9(a), filaments of material undergo no rotation as they revolve around the center of the 3.9(a)de vortex. In the vortex flow portrayed in Figure 3.9(b), similar filaments rotate during their revolution, the period of the rotation equalling that of the revolution. The lesson in these two examples is that one cannot analyze the spin of a motion simply by inspecting its pathlines.

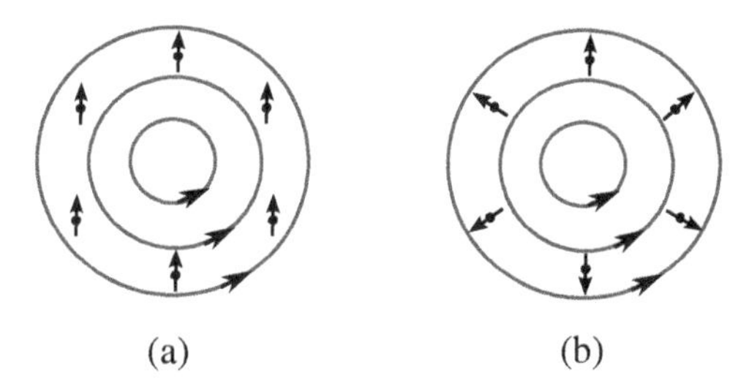

Figure 3.9. Schematic diagrams of material filaments in two different steady vortex flows, showing that the first motion (a) preserves filament orientations, while the second (b) induces spin.

3.3.2 Skew Tensors, Spin, and Vorticity

We now examine properties of skew tensors in general—and the spin tensor W in particular—in more detail. If $\mathsf{S} \in \mathrm{L}(\mathbb{E})$ is skew, then $\mathbf{a} \cdot \mathsf{S}^\top \mathbf{b} = -\mathbf{a} \cdot \mathsf{S}\mathbf{b}$ for all $\mathbf{a}, \mathbf{b} \in \mathbb{E}$. In addition, by definition of the transpose, $\mathbf{a} \cdot \mathsf{S}^\top \mathbf{b} = \mathsf{S}\mathbf{a} \cdot \mathbf{b} = \mathbf{b} \cdot \mathsf{S}\mathbf{a}$. It follows from these two identities that

$$\mathbf{a} \cdot \mathsf{S}\mathbf{b} = -\mathbf{b} \cdot \mathsf{S}\mathbf{a}, \tag{3.3.3}$$

for all $\mathbf{a}, \mathbf{b} \in \mathbb{E}$. Equation (3.3.3) yields two consequences:

1. $\mathbf{a} \cdot \mathsf{S}\mathbf{a} = 0$ for every $\mathbf{a} \in \mathbb{E}$.

2. With respect to any orthonormal basis, S has a matrix representation in which $S_{ij} = -S_{ji}$, and every diagonal entry S_{ii} vanishes.

THEOREM 3.3.1 (EIGENVALUES OF A SKEW TENSOR – PART 1). *If* S *is skew, then it has a zero eigenvalue.*

PROOF: The characteristic polynomial $p_S(\lambda)$ is cubic, so it has at least one real root λ. Let $\mathbf{p} \neq \mathbf{0}$ be an associated eigenvector. Then by Equation (3.3.3), $0 = \mathbf{p} \cdot \mathsf{S}\mathbf{p} = \lambda \|\mathbf{p}\|^2$, which is possible only if $\lambda = 0$. ■

This result leads to an interesting representation for a skew tensor S. Let $\mathbf{p}_1$ be a unit-length eigenvector of S associated with the eigenvalue $\lambda = 0$. Pick vectors $\mathbf{p}_2, \mathbf{p}_3 \in \mathbb{E}$ such that (1) $\{\mathbf{p}_1, \mathbf{p}_2, \mathbf{p}_3\}$ constitutes an orthonormal basis for $\mathbb{E}$ and (2) $[\mathbf{p}_1, \mathbf{p}_2, \mathbf{p}_3] = 1$. Condition (2) guarantees that the basis has positive orientation, that is, that the basis vectors are related by the right-hand rule. With respect to this basis, any tensor has an expansion as a linear combination of dyadic products. In particular, $\mathsf{S} = (\mathbf{p}_i \cdot \mathsf{S}\mathbf{p}_j)\, \mathbf{p}_i \otimes \mathbf{p}_j$.

EXERCISE 73 *Show that, for any skew tensor* S,

$$\mathsf{S} = w\,(\mathbf{p}_3 \otimes \mathbf{p}_2 - \mathbf{p}_2 \otimes \mathbf{p}_3), \tag{3.3.4}$$

where $w = \mathbf{p}_3 \cdot \mathsf{S}\mathbf{p}_2$.

The quantity in parentheses in Equation (3.3.4) is the **exterior product** of $\mathbf{p}_3$ and $\mathbf{p}_2$, often denoted as $\mathbf{p}_3 \wedge \mathbf{p}_2$. It follows from this equation that $\mathbf{p}_3 \cdot \mathsf{S}\mathbf{p}_2 \neq 0$—and similarly that $\mathbf{p}_2 \cdot \mathsf{S}\mathbf{p}_3 \neq 0$—whenever $\mathsf{S} \neq 0$.

The next exercise establishes the promised representation for S.

EXERCISE 74 *With* S *and* w *as above, let* $\mathbf{s} = w\mathbf{p}_1$. *Show the following:*

1. $\mathsf{S}\mathbf{s} = \mathbf{0}$.
2. *For any* $\mathbf{a} \in \mathbb{E}$, $\mathbf{s} \times \mathbf{a} = \mathsf{S}\mathbf{a}$.

Thus we associate with any skew tensor $\mathsf{S} \in \mathrm{L}(\mathbb{E})$ a vector $\mathbf{s} \in \mathbb{E}$ that, acting through the cross product, captures the action of S.

DEFINITION. *The vector* $\mathbf{s}$ *defined in the exercise above is the* **axial vector** *of the skew tensor* S, *and the subspace* $\{c\mathbf{s} \in \mathbb{E} \mid c \in \mathbb{R}\}$ *is the* **axis** *of* S.

EXERCISE 75 *Working with respect to the orthonormal basis* $\{\mathbf{e}_1, \mathbf{e}_2, \mathbf{e}_3\}$, *show that* $\mathbf{a} \times \mathbf{b}$ *is the axial vector of* $\mathbf{b} \otimes \mathbf{a} - \mathbf{a} \otimes \mathbf{b}$.

Tangential to continuum mechanics is a related algebraic question: which Euclidean vector spaces admit a cross product, that is, a binary operation that represents skew tensors as vectors? In an n-dimensional Euclidean vector space, the expansion

$$\mathsf{S} = (\mathbf{e}_i \cdot \mathsf{S}\mathbf{e}_j)\mathbf{e}_i \otimes \mathbf{e}_j$$

of an arbitrary skew tensor S with respect to an orthonormal basis contains up to $n(n-1)/2$ nonzero, independent terms. To represent S as a vector in such a space requires that $n(n-1)/2 = n$. Since this quadratic equation has roots 0 and 3, the representation of skew tensors using vectors acting through a binary operation such as the cross product is possible only when $n = 3$.

One additional algebraic observation characterizes the nonzero eigenvalues of a skew tensor $\mathsf{S} \in \mathrm{L}(\mathbb{E}) \neq \mathsf{0}$.

THEOREM 3.3.2 (EIGENVALUES OF A SKEW TENSOR – PART 2.) *If a nonzero tensor* $\mathsf{S} \in \mathrm{L}(\mathbb{E})$ *is skew, then it has two nonzero imaginary eigenvalues.*

PROOF: Consider the characteristic polynomial $p_S(\lambda) = \lambda^3 - \mathrm{I}_S\lambda^2 + \mathrm{II}_S\lambda - \mathrm{III}_S$ (see Equation (1.2.14)). Using the right-handed orthonormal basis $\{\mathbf{p}_1, \mathbf{p}_2, \mathbf{p}_3\}$ of eigenvectors defined above, the fact that the matrix representation for S with respect to any orthonormal basis has vanishing diagonal entries, and the fact that 0 is an eigenvalue for S, we find that

$$\mathrm{I}_S = \operatorname{tr}\mathsf{S} = 0$$

$$\mathrm{III}_S = \det\mathsf{S} = 0.$$

For the remaining principal invariant, we use an identity from Equations (1.2.17):

$$\begin{aligned}
\mathrm{II}_S &= [\mathsf{S}\mathbf{p}_1, \mathsf{S}\mathbf{p}_2, \mathbf{p}_3] + [\mathsf{S}\mathbf{p}_1, \mathbf{p}_2, \mathsf{S}\mathbf{p}_3] + [\mathbf{p}_1, \mathsf{S}\mathbf{p}_2, \mathsf{S}\mathbf{p}_3] \\
&= [\mathbf{p}_1, w\mathbf{p}_1 \times \mathbf{p}_2, w\mathbf{p}_1 \times \mathbf{p}_3] \\
&= w^2[\mathbf{p}_1, \mathbf{p}_3, -\mathbf{p}_2] = w^2[\mathbf{p}_1, \mathbf{p}_2, \mathbf{p}_3] = w^2.
\end{aligned}$$

Here we have used the representation of S by its axial vector $\mathbf{s} = w\mathbf{p}_1$ and the fact that $\mathsf{S}\mathbf{p}_1 = 0$. Hence, $p_S(\lambda) = \lambda^3 + w^2\lambda$, which has roots $0, \pm iw$. ■

Heuristically, this theorem invites an analogy between the decomposition $\mathsf{L} = \mathsf{D} + \mathsf{W}$ into symmetric and skew parts and the standard Cartesian decomposition $z = d + iw$ of a complex number.

This discussion shows that the spin tensor $\mathsf{W} = \frac{1}{2}(\mathsf{L} - \mathsf{L}^\top)$ has a representation of the following structure with respect to any orthonormal basis:

$$\mathsf{W} = \begin{bmatrix} 0 & -W_{21} & -W_{31} \\ W_{21} & 0 & -W_{32} \\ W_{31} & W_{32} & 0 \end{bmatrix}, \qquad W_{ij} = \frac{1}{2}\left(\frac{\partial v_i}{\partial x_j} - \frac{\partial v_j}{\partial x_i}\right).$$

With respect to such a basis, its axial vector has representation

$$\mathbf{w} = \begin{bmatrix} W_{32} \\ -W_{31} \\ W_{21} \end{bmatrix}.$$

We call this vector the **vorticity**.

The example of simple shear, $\mathbf{x}(\mathbf{X}, t) = \mathsf{I} + \alpha(t)(\mathbf{e}_1 \otimes \mathbf{e}_2)\mathbf{X}$, with α differentiable, illustrates. For this motion, $\mathbf{v}(\mathbf{x}, t) = \alpha'(t)(\mathbf{e}_1 \otimes \mathbf{e}_2)\mathbf{x}$. In matrix form,

$$\mathsf{L} = \operatorname{grad} \mathbf{v} = \alpha'(\mathbf{e}_1 \otimes \mathbf{e}_2) = \begin{bmatrix} 0 & \alpha' & 0 \\ 0 & 0 & 0 \\ 0 & 0 & 0 \end{bmatrix},$$

and

$$\mathsf{D} = \frac{1}{2} \begin{bmatrix} 0 & \alpha' & 0 \\ \alpha' & 0 & 0 \\ 0 & 0 & 0 \end{bmatrix}, \qquad \mathsf{W} = \frac{1}{2} \begin{bmatrix} 0 & \alpha' & 0 \\ -\alpha' & 0 & 0 \\ 0 & 0 & 0 \end{bmatrix}.$$

The eigenvalue–eigenvector pairs of the stretching tensor D are

$$\left(\frac{\alpha'}{2}, \begin{bmatrix} 1 \\ 1 \\ 0 \end{bmatrix} \right), \quad \left(-\frac{\alpha'}{2}, \begin{bmatrix} -1 \\ 1 \\ 0 \end{bmatrix} \right), \quad \left(0, \begin{bmatrix} 0 \\ 0 \\ 1 \end{bmatrix} \right).$$

Since the vorticity is

$$\mathbf{w} = \begin{bmatrix} 0 \\ 0 \\ -\alpha'/2 \end{bmatrix},$$

the axis of spin is the x_3-axis. As Figure 3.10 depicts, over short time intervals one can approximate the action of shear as a superposition of stretching, along the diagonal axes defined by $(1, 1, 0)^\top$ and $(-1, 1, 0)^\top$, and rotation about the axis of spin.

The remaining exercises in this section draw connections between vorticity and classical vector field theory. Recall that, with respect to the basis $\{\mathbf{e}_1, \mathbf{e}_2, \mathbf{e}_3\}$,

$$\operatorname{curl} \mathbf{v} = \left(\frac{\partial v_3}{\partial x_2} - \frac{\partial v_2}{\partial x_3}, \frac{\partial v_1}{\partial x_3} - \frac{\partial v_3}{\partial x_1}, \frac{\partial v_2}{\partial x_1} - \frac{\partial v_1}{\partial x_2} \right)^\top = \varepsilon_{ijk} \frac{\partial v_i}{\partial x_j} \mathbf{e}_k.$$

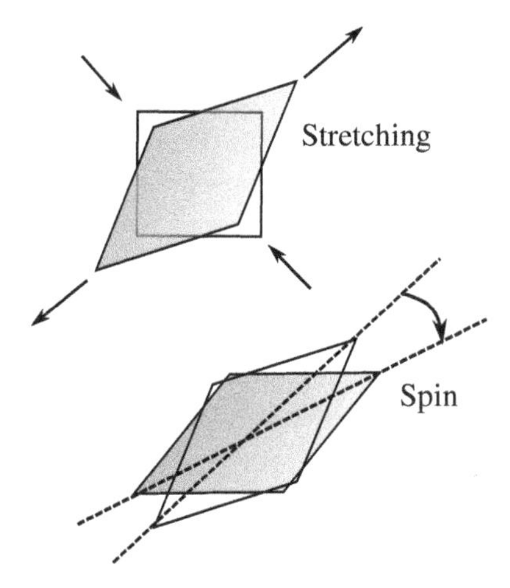

Figure 3.10. Simple shear as a superposition of stretching and spin.

EXERCISE 76 *Show that* $\mathbf{w} = \frac{1}{2}\mathrm{curl}\,\mathbf{v}$. *Thus* $\mathrm{curl}\,\mathbf{v}$ *is the axial vector field of* $\mathrm{grad}\,\mathbf{v} - (\mathrm{grad}\,\mathbf{v})^{\top}$.

EXERCISE 77 *The velocity field* $\mathbf{v}(\mathbf{x}) = (-x_2, x_1, 0)/\|\mathbf{x}\|^{\alpha}$, *where* $\alpha \in \mathbb{R}$ *and* $\mathbf{x} \neq \mathbf{0}$, *generalizes the two vortices examined in Exercises 71 and 72. Compute its vorticity* $\mathbf{w}$. *For what critical value of* α *does* $\mathbf{w} = \mathbf{0}$? *Examine the direction of* $\mathbf{w}$ *for values of* α *greater than and less than this critical value. Sketch diagrams analogous to Figure 3.9 showing the effects of spin in these cases.*

DEFINITION. *The motion is* **irrotational** *if* $\mathbf{w} = \mathbf{0}$.

EXERCISE 78 *A velocity* $\mathbf{v}(\mathbf{x}, t)$ *is a* **gradient field** *if there exists a differentiable scalar potential* $\Phi(\mathbf{x}, t)$ *such that* $\mathbf{v}(\mathbf{x}, t) = \mathrm{grad}\,\Phi(\mathbf{x}, t)$. *Working in Cartesian coordinates, show that any motion for which the velocity is a continuously differentiable gradient field is irrotational.*

It is common to summarize this result as $\mathrm{curl}(\mathrm{grad}\,\Phi) = \mathbf{0}$. A partial converse exists: if $\mathrm{curl}\,\mathbf{v} = \mathbf{0}$ everywhere in $\mathbb{E}$ except possibly at a finite number of exceptional points, then $\mathbf{v}$ is a gradient field, but at the exceptional points the scalar potential is undefined. See [39, Section 8.3].

EXERCISE 79 *Show that, in an isochoric, irrotational motion, the velocity field at time* t *is everywhere orthogonal to surfaces on which the scalar potential* $\Phi(\mathbf{x}, t)$ *is constant. Show that, in this case,* Φ *satisfies the* **Laplace equation**,

$$-\Delta\Phi = -\mathrm{div}(\mathrm{grad}\,\Phi) = 0.$$

3.3.3 Summary

For some materials, such as fluids, the time derivative of strain plays a more important role than strain itself in calculating responses to motion. The identity $D\mathsf{F}/Dt = \mathsf{LF}$ suggests using the velocity gradient tensor $\mathsf{L} = \operatorname{grad}\mathbf{v}$ to quantify strain rates. For this tensor, the most useful decomposition is $\mathsf{L} = \frac{1}{2}(\mathsf{L}+\mathsf{L}^\top) + \frac{1}{2}(\mathsf{L}-\mathsf{L}^\top) = \mathsf{D}+\mathsf{W}$, where the stretching tensor D is symmetric and the spin tensor W is skew.

The real eigenvalues of D are the principal stretchings, and the associated orthonormal basis consisting of eigenvectors of D define the principal directions of stretching.

The skew tensor W has either one or three zero eigenvalues. If there is only one, the remaining eigenvalues are pure imaginary conjugates. In addition, W has an axial vector $\mathbf{w}$, which captures the action of W on any vector $\mathbf{a} \in \mathbb{E}$ as $\mathbf{w} \times \mathbf{a}$. This representation of skew tensors through a binary operation on vectors exists only in three space dimensions. In the case of the spin tensor, we call the axial vector the vorticity, and it obeys the identity $\mathbf{w} = \frac{1}{2}\operatorname{curl}\mathbf{v}$.

3.4 VORTICITY AND CIRCULATION

Vorticity plays a central role in fluid mechanics, having deep connections with mysteries of turbulence. Although exploring these connections would carry us far afield, this section briefly discusses vorticity and its relationship to the geometry of fluid motions. Throughout this section $\gamma\colon [a,b] \to \mathbb{E}$ denotes a continuously differentiable arc whose image $\gamma([a,b]) \subset \chi(\mathcal{B},t)$ and for which $\gamma(a) = \gamma(b)$, that is, γ is **closed**.

3.4.1 Circulation

DEFINITION. *The* **circulation** *around γ is the path integral*

$$C_\gamma(t) = \int_\gamma \mathbf{v}(\mathbf{x},t)\cdot d\mathbf{x} = \int_a^b \mathbf{v}(\gamma(s),t)\cdot\gamma'(s)\,ds.$$

Figure 3.11 shows the geometry.

When γ parametrizes the boundary $\partial\mathcal{S}$ of a parametrized smooth surface $\mathcal{S} \subset \chi(\mathcal{B},t)$, as illustrated in Figure 3.12, the circulation $C_\gamma(t)$ has an interesting inter-

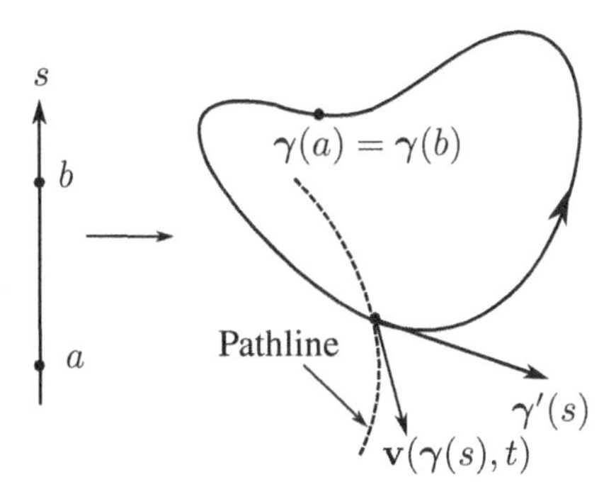

Figure 3.11. Geometry of circulation.

pretation: by Stokes's theorem (see Appendix B),

$$C_\gamma(t) = \int_\gamma \mathbf{v}(\mathbf{x}, t) \cdot d\mathbf{x} = \int_{\mathcal{S}} \operatorname{curl} \mathbf{v} \cdot d\boldsymbol{\sigma}.$$

More explicitly, if $\varphi\colon \Omega \to \mathbb{E}$ is a parametrization of $\mathcal{S}$ with parameter domain $\Omega \subset \mathbb{R}^2$,

$$C_\gamma(t) = \iint_\Omega \operatorname{curl} \mathbf{v}(\mathbf{x}(\xi_1, \xi_2), t) \cdot \left(\frac{\partial \varphi}{\partial \xi_1} \times \frac{\partial \varphi}{\partial \xi_2}\right) d\xi_1\, d\xi_2.$$

From Exercise 76 it follows that

$$C_\gamma(t) = 2 \int_{\mathcal{S}} \mathbf{w} \cdot d\boldsymbol{\sigma} = 2 \times \text{ flux of vorticity across } \mathcal{S}.$$

In particular, irrotational motions, in which $\mathbf{w} = \mathbf{0}$, are circulation–free. We record this result as follows:

THEOREM 3.4.1 (CIRCULATION IN IRROTATIONAL MOTION). *In irrotational motion, the circulation vanishes around any closed path bounding a parametrized smooth surface in the body.*

Of special interest are material arcs, that is, arcs $\gamma(s,t)\colon [a,b] \times \mathbb{R} \to \mathbb{E}$ for which $\gamma(s,t) = \mathbf{x}(\boldsymbol{\Gamma}(s), t)$ for all parameter values s and all times t. Here, $\boldsymbol{\Gamma}\colon [a,b] \to \boldsymbol{\kappa}(\mathcal{B})$ denotes the arc in the reference configuration. Recall from Section 2.4 that the image $\gamma([a,b], t)$ of a material arc is the locus at time t of a fixed set of particles.

DEFINITION. *The motion is* **circulation-preserving** *if $C_\gamma'(t) = 0$ for every time t and for every closed material arc $\gamma(s,t)$ whose image lies in $\chi(\mathcal{B}, t)$.*

To investigate such motions, we establish a relationship between the rate of change of $C_\gamma(t)$ and the velocity field.

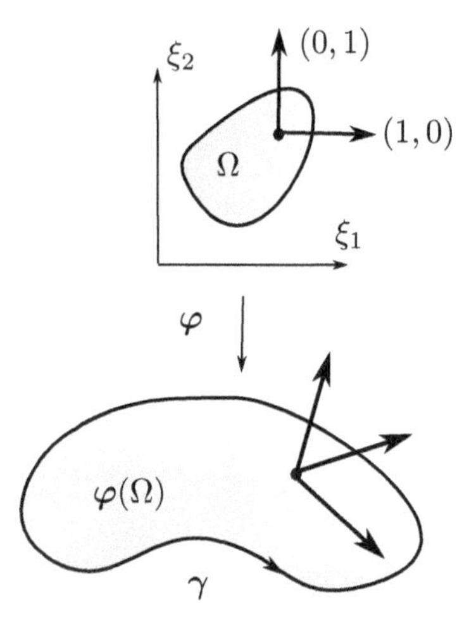

Figure 3.12. The boundary of a parametrized smooth surface $\mathcal{S}$ defined by a mapping $\varphi\colon \Omega \to \mathbb{E}$, where Ω is a parameter domain in $\mathbb{R}^2$.

THEOREM 3.4.2 (RATE OF CHANGE OF CIRCULATION). *If* $\gamma(\cdot, t)\colon [a, b] \to \mathbb{E}$ *is a closed material arc at each time* t *and* $\gamma(s, t)$ *is twice continuously differentiable, then*

$$\frac{dC_\gamma}{dt} = \int_\gamma \frac{D\mathbf{v}}{Dt} \cdot d\mathbf{s}. \tag{3.4.1}$$

Here, by Equation (2.2.3),

$$\frac{D\mathbf{v}}{Dt} = \frac{\partial \mathbf{v}}{\partial t} + (\mathbf{v} \cdot \operatorname{grad})\mathbf{v}$$

is the spatial acceleration.

PROOF: By definition and the product rule,

$$\begin{aligned}
\frac{dC_\gamma}{dt} &= \frac{d}{dt} \int_a^b \mathbf{v}(\gamma(s,t), t) \cdot \frac{\partial \gamma}{\partial s}(s,t)\, ds \\
&= \int_a^b \underbrace{\frac{\partial \mathbf{v}}{\partial t}(\gamma(s,t), t)}_{\text{(I)}} \cdot \frac{\partial \gamma}{\partial s}(s,t)\, ds + \int_a^b \mathbf{v}(\gamma(s,t), t) \cdot \underbrace{\frac{\partial^2 \gamma}{\partial t \partial s}(s,t)}_{\text{(II)}}\, ds.
\end{aligned}$$

Now examine the expressions labeled (I) and (II). Since

$$\frac{\partial \gamma}{\partial t}(s,t) = \frac{\partial \mathbf{x}}{\partial t}(\underbrace{\boldsymbol{\Gamma}(s)}_{\mathbf{X}}, t) = \mathbf{v}(\underbrace{\gamma(s,t)}_{\mathbf{x}(\mathbf{X},t)}, t)$$

and γ is twice continuously differentiable,

$$(\mathrm{II}) = \frac{\partial^2 \gamma}{\partial t \partial s}(s,t) = \frac{\partial^2 \gamma}{\partial s \partial t}(s,t) = \frac{\partial \mathbf{v}}{\partial s}(\gamma(s,t),t).$$

Also, converting to the referential velocity gives

$$(\mathrm{I}) = \frac{\partial}{\partial t}\frac{\partial \mathbf{x}}{\partial t}(\mathbf{\Gamma}(s),t) = \frac{D}{Dt}\frac{\partial \mathbf{x}}{\partial t}(\mathbf{\Gamma}(s),t) = \frac{D\mathbf{v}}{Dt}(\gamma(s,t),t).$$

Substituting these expressions yields

$$\begin{aligned}\frac{dC_\gamma}{dt} &= \int_a^b \frac{D\mathbf{v}}{Dt}(\gamma(s,t),t)\cdot\frac{\partial \gamma}{\partial s}(s,t)\,ds + \int_a^b \mathbf{v}(\gamma(s,t),t)\cdot\frac{\partial \mathbf{v}}{\partial s}(\gamma(s,t),t)\,ds \\ &= \int_\gamma \frac{D\mathbf{v}}{Dt}\cdot d\mathbf{s} + \int_a^b \mathbf{v}(\gamma(s,t),t)\cdot\frac{\partial \mathbf{v}}{\partial s}(\gamma(s,t),t)\,ds.\end{aligned}$$

The second integral on the right simplifies by another application of the product rule:

$$\int_a^b \frac{1}{2}\frac{\partial}{\partial s}\left\|\mathbf{v}(\gamma(s,t),t)\right\|^2 ds = \frac{1}{2}\|\mathbf{v}(\gamma(s,t),t)\|^2\Big|_{s=a}^{s=b} = 0,$$

since γ is closed. Equation (3.4.1) follows. ■

The next result follows immediately.

COROLLARY 3.4.3 (KELVIN'S THEOREM). *If $D\mathbf{v}/Dt$ is the gradient of a continuously differentiable scalar function Φ, then the motion is circulation-preserving.*

PROOF: The path integral of a gradient around a closed arc vanishes:

$$\int_a^b \operatorname{grad}\Phi(\gamma(t))\cdot\gamma'(t)\,dt = \int_a^b \frac{d}{dt}\Phi(\gamma(t))\,dt = \Phi(\gamma(b)) - \Phi(\gamma(a)) = 0,$$

since $\gamma(b) = \gamma(a)$. ■

EXERCISE 80 *Let $\mathbf{v}(\mathbf{x},t)$ be the gradient of a continuously differentiable scalar field $\Phi(\mathbf{x},t)$, so $\mathbf{v} = \operatorname{grad}\Phi$. Show that*

$$\frac{D\mathbf{v}}{Dt} = \operatorname{grad}\left(\frac{\partial \Phi}{\partial t} + \frac{\|\mathbf{v}\|^2}{2}\right),$$

so $D\mathbf{v}/Dt$ is also a gradient field.

DEFINITION. *A* **vortex surface** *is a parametrized smooth surface $\mathcal{S}$ that is everywhere tangent to the vorticity $\mathbf{w}$.*

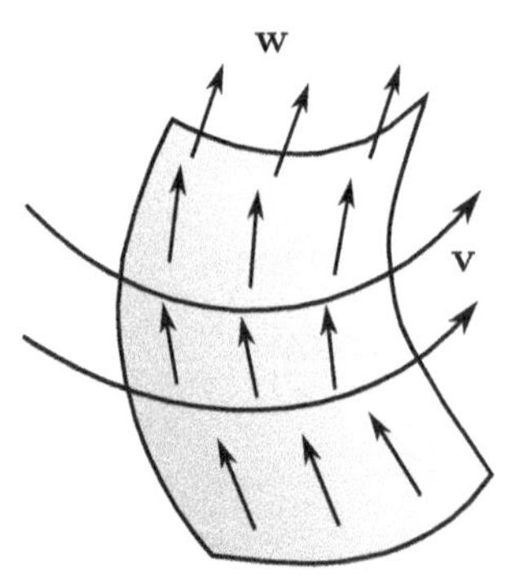

Figure 3.13. A vortex surface.

Figure 3.13 illustrates a vortex surface. Circulation-preserving motions also preserve vortex surfaces, in the sense made precise in the next exercise.

EXERCISE 81 *In a circulation-preserving motion, a material surface that is a vortex surface at some time t_0 is a vortex surface for every time t.*

3.4.2 Summary

The circulation around any closed path γ in a body is

$$C_\gamma(t) = \int_\gamma \mathbf{v}(\mathbf{x}, t) \cdot d\mathbf{x}.$$

When the path parametrizes the boundary of a smooth surface, Stokes's theorem identifies $C_\gamma(t)$ as the flux of vorticity across the surface. Since irrotational motions are vorticity–free, they are circulation–free.

The analysis of circulation around material paths yields an additional result. If $\gamma(s,t)$ is a material arc at each time t, then

$$\frac{dC_\gamma}{dt} = \int_\gamma \frac{D\mathbf{v}}{Dt} \cdot d\mathbf{s}.$$

In particular, if $D\mathbf{v}/Dt$ is a gradient field (which is true when $\mathbf{v}$ itself is a gradient field), then the motion is circulation-preserving.

3.5 OBSERVER TRANSFORMATIONS

Section 1.1 introduces the concept of a frame of reference $(\boldsymbol{f}, t_0)$, which includes two elements: (1) a rule $\boldsymbol{f}\colon \mathbb{X} \to \mathbb{E}$ for associating points in the Euclidean point space $\mathbb{X}$ with position vectors in $\mathbb{E}$ and (2) a choice of instant t_0 to assign to the origin 0 of the time axis $\mathbb{R}$. Different observers may adopt different frames of reference, but their choices do not alter the underlying physics. To respect this principle, mathematical descriptions of the physics must transform in specific ways in response to changes from one frame of reference $(\boldsymbol{f}, t_0)$ to another, $(\hat{\boldsymbol{f}}, \hat{t}_0)$. We close this chapter with a discussion of the kinematic implications of such changes.

3.5.1 Changes in Frame of Reference

Of particular interest are the following types of change in frame of reference:

DEFINITION. *A frame of reference* $(\hat{\boldsymbol{f}}, \hat{t}_0)$ *is related to* $(\boldsymbol{f}, t_0)$ *by an* **observer transformation** *if*

1. *there exists a continuously differentiable, one-parameter family* $\mathsf{Q}\colon \mathbb{R} \to \mathrm{O}(\mathbb{E})$ *of orthogonal tensors and a continuously differentiable, vector-valued function* $\mathbf{c}\colon \mathbb{R} \to \mathbb{E}$ *such that*
$$\hat{\boldsymbol{f}}(P) = \mathsf{Q}(t)\boldsymbol{f}(P) + \mathbf{c}(t)$$
for all points $P \in \mathbb{X}$ *and all times* $t \in \mathbb{R}$*;*

2. *the units of time are the same in the two frames, but the origin of time may be shifted by a constant real number* b*:*
$$\hat{t_0} = t_0 - b.$$

We write the observer transformation as follows:

$$\begin{aligned} \hat{\mathbf{x}} &= \mathsf{Q}(t)\mathbf{x} + \mathbf{c}(t), \\ \hat{t} &= t - b. \end{aligned} \tag{3.5.1}$$

Equations (3.5.1) assert that the difference between the two frames of reference amounts to rigid motion of $(\hat{\boldsymbol{f}}, \hat{t}_0)$ with respect to $(\boldsymbol{f}, t_0)$, as sketched in Figure 3.14. By the chain rule, $D/D\hat{t} = D/Dt$; hence, in differentiating with respect to time, we need not distinguish between the two frames. However, as shown in the following, spatial differentiation is a different matter.

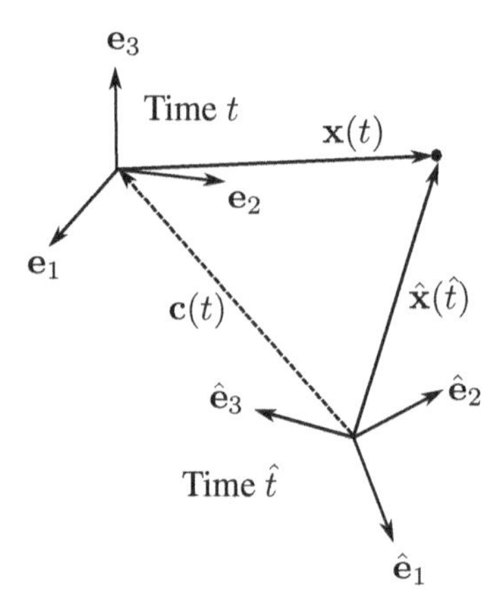

Figure 3.14. Frames of reference in rigid motion with respect to each other. (The coordinate systems used in each frame need not be Cartesian.)

EXERCISE 82 *For this exercise, write* $(\hat{\boldsymbol{f}}, \hat{t}_0) \sim (\boldsymbol{f}, t_0)$ *to indicate that* $(\hat{\boldsymbol{f}}, \hat{t}_0)$ *is related to* $(\boldsymbol{f}, t_0)$ *by an observer transformation. Show that* $\sim$ *is an equivalence relation, that is,*

1. $(\boldsymbol{f}, t_0) \sim (\boldsymbol{f}, t_0)$ *(**reflexivity**).*
2. *If* $(\hat{\boldsymbol{f}}, \hat{t}_0) \sim (\boldsymbol{f}, t_0)$*, then* $(\boldsymbol{f}, t_0) \sim (\hat{\boldsymbol{f}}, \hat{t}_0)$ *(**symmetry**).*
3. *If* $(\hat{\boldsymbol{f}}, \hat{t}_0) \sim (\boldsymbol{f}, t_0)$ *and* $(\check{\boldsymbol{f}}, \check{t}_0) \sim (\hat{\boldsymbol{f}}, \hat{t}_0)$*, then* $(\check{\boldsymbol{f}}, \check{t}_0) \sim (\boldsymbol{f}, t_0)$ *(**transitivity**).*

A special type of observer transformation arises in connection with the laws of motion:

DEFINITION. *The observer transformation (3.5.1) is* **Galilean** *if* $\mathsf{Q}(t)$ *is a constant orthogonal tensor and* $\mathbf{c}(t) = \mathbf{c}_0 + \mathbf{c}_1 t$ *for some constant vectors* $\mathbf{c}_0, \mathbf{c}_1 \in \mathbb{E}$.

Chapter 4 discusses how equations of motion transform under Galilean and non-Galilean observer transformations, and Chapter 6 explores the concept that observer transformations do not affect certain measurable responses of materials to motion. Of interest for the present is how more general quantities transform under Galilean and non-Galilean changes in frame of reference.

Consider a typical scalar-valued function a, a vector-valued function $\mathbf{a}$, and a tensor-valued function A, as observed in the frame of reference $(\boldsymbol{f}, t_0)$. An observer associated with the frame of reference $(\hat{\boldsymbol{f}}, \hat{t}_0)$ sees quantities $\hat{a}$, $\hat{\mathbf{a}}$, and $\hat{\mathsf{A}}$, conceivably having different numerical values. For certain functions a, $\mathbf{a}$, A, one accounts for the effects of an observer transformation simply by accounting for the rotation of one frame of reference with respect to the other.

DEFINITION. *The scalar function* $a(\mathbf{x},t)$, *the vector field* $\mathbf{a}(\mathbf{x},t)$, *and the tensor field* $\mathsf{A}(\mathbf{x},t)$ *are* **objective** *if, for any observer transformation of the form (3.5.1), the following identities hold:*

$$\begin{aligned}\hat{a}(\hat{\mathbf{x}},\hat{t}) &= a(\mathbf{x},t),\\ \hat{\mathbf{a}}(\hat{\mathbf{x}},\hat{t}) &= \mathsf{Q}(t)\mathbf{a}(\mathbf{x},t),\\ \hat{\mathsf{A}}(\hat{\mathbf{x}},\hat{t}) &= \mathsf{Q}(t)\mathsf{A}(\mathbf{x},t)\mathsf{Q}^{\top}(t).\end{aligned}$$

The functions are **Galilean objective** *if these identities hold for every Galilean observer transformation.*

The following theorem motivates this definition.

THEOREM 3.5.1 (PROPERTIES OF OBJECTIVE FUNCTIONS). *If* a, $\mathbf{a}$, *and* A *are objective, then*

1. *The value of* a *remains invariant under observer transformations.*
2. *Geometrically, the vectors* $\mathbf{a}$ *and* $\hat{\mathbf{a}}$ *represent the same translation, in the sense that if* $\mathbf{a} = \mathbf{b}_1 - \mathbf{b}_2$, *then* $\hat{\mathbf{a}} = \hat{\mathbf{b}}_1 - \hat{\mathbf{b}}_2$.
3. *The tensor* A *transforms objective vectors into objective vectors, in the following sense: if* $\hat{\mathbf{a}} = \mathsf{Q}\mathbf{a}$, *then* $\hat{\mathsf{A}}\hat{\mathbf{a}} = \mathsf{Q}(\mathsf{A}\mathbf{a})$.
4. *The scalar function* $\mathbf{a}\cdot\mathsf{A}\mathbf{a}$ *is objective.*

EXERCISE 83 *Prove assertions 2, 3, and 4. (Assertion 1 is part of the definition of objectivity.)*

Equation (3.5.1) shows that spatial position $\mathbf{x}$ is not objective. Neither is the velocity: by the product rule,

$$\begin{aligned}\hat{\mathbf{v}} = \frac{D\hat{\mathbf{x}}}{Dt} &= \mathsf{Q}(t)\frac{D\mathbf{x}}{Dt} + \mathsf{Q}'(t)\mathbf{x} + \mathbf{c}'(t)\\ &= \mathsf{Q}(t)\mathbf{v} + \mathsf{Q}'(t)\mathbf{x} + \mathbf{c}'(t).\end{aligned} \tag{3.5.2}$$

Hence, $\hat{\mathbf{v}} = \mathsf{Q}\mathbf{v}$ only if $\mathsf{Q}' = 0$ and $\mathbf{c}' = \mathbf{0}$, that is, only if the frames of reference are in uniform motion with respect to each other. Thus the velocity is not objective, nor is it Galilean objective.

EXERCISE 84 *Show that*

$$\hat{\mathbf{v}} - \mathsf{Q}(t)\mathbf{v} = \mathsf{Q}'(t)\mathsf{Q}^{\top}(t)[\hat{\mathbf{x}} - \mathbf{c}(t)] + \mathbf{c}'(t). \tag{3.5.3}$$

Recall from Exercise 67 that $\mathsf{Q}'\mathsf{Q}^\top$ is skew, that is, $(\mathsf{Q}'\mathsf{Q}^\top)^\top = -\mathsf{Q}'\mathsf{Q}^\top$. This tensor, having dimension $[\mathrm{T}^{-1}]$, gives the rate and direction of rotation of the hatted frame with respect to the original frame. This fact allows us to rewrite Equation (3.5.3) in a form that will prove useful later. Let $\boldsymbol{\omega}$ be the axial vector for $\mathsf{Q}'\mathsf{Q}^\top$ (see Exercise 74), so that $\mathsf{Q}'\mathsf{Q}^\top\mathbf{a} = \boldsymbol{\omega} \times \mathbf{a}$ for every vector $\mathbf{a} \in \mathbb{E}$. Then

$$\hat{\mathbf{v}} - \mathsf{Q}\mathbf{v} = \boldsymbol{\omega} \times (\hat{\mathbf{x}} - \mathbf{c}) + \mathbf{c}'. \tag{3.5.4}$$

This identity implies that the acceleration is not objective, as the following exercise shows.

EXERCISE 85 *Show that the acceleration transforms as follows under an observer transformation:*

$$\frac{D^2\hat{\mathbf{x}}}{Dt^2} - \mathsf{Q}\frac{D^2\mathbf{x}}{Dt^2} = \mathbf{c}'' + 2\boldsymbol{\omega} \times (\hat{\mathbf{v}} - \mathbf{c}') + \boldsymbol{\omega}' \times (\hat{\mathbf{x}} - \mathbf{c}) - \boldsymbol{\omega} \times [\boldsymbol{\omega} \times (\hat{\mathbf{x}} - \mathbf{c})]. \tag{3.5.5}$$

However, since $\boldsymbol{\omega} = \mathbf{c}'' = \mathbf{0}$ for any Galilean observer transformation, the acceleration is Galilean objective.

Even if we start with an objective vector field $\mathbf{a}(\mathbf{X}, t)$, for which $\hat{\mathbf{a}} = \mathsf{Q}\mathbf{a}$, its material derivative is not generally objective:

$$\frac{D\hat{\mathbf{a}}}{Dt} = \frac{D}{Dt}(\mathsf{Q}\mathbf{a}) = \mathsf{Q}\frac{D\mathbf{a}}{Dt} + \mathsf{Q}'\mathbf{a}.$$

The following theorem shows that yet another important quantity fails to be objective.

THEOREM 3.5.2 (TRANSFORMATION OF THE DEFORMATION GRADIENT). *Under an observer transformation between two frames of reference, using the body's configuration at time t_0 as the reference configuration, the deformation gradient transforms according to the equation*

$$\hat{\mathsf{F}}(\hat{\mathbf{X}}, \hat{t}) = \mathsf{Q}(t)\mathsf{F}(\mathbf{X}, t)\mathsf{Q}^\top(t_0). \tag{3.5.6}$$

PROOF: For any particle X, the spatial positions in the two frames at time t stand in the relationship

$$\hat{\chi}(X, \hat{t}) = \mathsf{Q}(t)\chi(X, t) + \mathbf{c}(t).$$

If the reference configuration $\boldsymbol{\kappa}$ is the configuration occupied by the body at time t_0 in the unhatted frame, then the referential coordinate corresponding to $\mathbf{X} = \boldsymbol{\kappa}(X)$ in the hatted frame is

$$\hat{\mathbf{X}} = \hat{\boldsymbol{\kappa}}(X) = \mathsf{Q}(t_0)\chi(X, t_0) + \mathbf{c}(t_0),$$

or, in terms of the referential coordinate $\mathbf{X}$,

$$\hat{\mathbf{X}} = \mathsf{Q}(t_0)\mathbf{X} + \mathbf{c}(t_0).$$

The deformations in the two frames are then

$$\begin{aligned} \mathbf{x} &= \chi_{\kappa}(\mathbf{X}, t), \\ \hat{\mathbf{x}} &= \chi_{\hat{\kappa}}(\hat{\mathbf{X}}, \hat{t}) = \mathsf{Q}(t)\chi_{\kappa}(\mathbf{X}, t) + \mathbf{c}(t_0). \end{aligned} \tag{3.5.7}$$

Now compute $\hat{\mathsf{F}} = \mathrm{Grad}_{\hat{\mathbf{X}}}\chi_{\hat{\kappa}}$ by applying the chain rule to Equation (3.5.7):

$$\begin{aligned} \mathrm{Grad}_{\hat{\mathbf{X}}}\, \chi_{\hat{\kappa}}(\hat{\mathbf{X}}, \hat{t}) &= \mathrm{Grad}_{\mathbf{X}}\, [\mathsf{Q}(t)\chi_{\kappa}(\mathbf{X}, t) + \mathbf{c}(t)]\ \mathrm{Grad}_{\hat{\mathbf{X}}}\mathbf{X} \\ &= \mathsf{Q}(t)\, \mathrm{Grad}_{\mathbf{X}}\chi_{\kappa}\mathrm{Grad}_{\hat{\mathbf{X}}}\mathbf{X}. \end{aligned}$$

Upon recognizing that $\mathrm{Grad}_{\mathbf{X}}\, \chi_{\kappa} = \mathsf{F}$ and $\mathrm{Grad}_{\hat{\mathbf{X}}}\mathbf{X} = \mathsf{Q}^{\top}(t_0)$ we obtain Equation (3.5.6). ■

EXERCISE 86 *Verify that* $\mathrm{Grad}_{\hat{\mathbf{X}}}\mathbf{X} = \mathsf{Q}^{\top}(t_0)$ *in the previous proof.*

By now one may wonder if anything is objective. The next few observations may be reassuring.

EXERCISE 87 *Using the hypotheses of the previous theorem, show that* $\det \hat{\mathsf{F}} = \det \mathsf{F}$. *Therefore, the scalar field* $\det \mathsf{F}$ *is objective.*

THEOREM 3.5.3 (OBJECTIVITY OF GRADIENTS). *If a differentiable scalar field* $a(\mathbf{x}, t)$ *is objective, so is its spatial gradient.*

PROOF: We work in Cartesian coordinates with the summation convention. By the chain rule and the fact that $\hat{a}(\hat{\mathbf{x}}, \hat{t}) = a(\mathbf{x}, t)$,

$$(\mathrm{grad}_{\hat{\mathbf{x}}}\hat{a})_i = \frac{\partial \hat{a}}{\partial \hat{x}_i} = \frac{\partial x_j}{\partial \hat{x}_i}\frac{\partial \hat{a}}{\partial x_j} = \frac{\partial x_j}{\partial \hat{x}_i}\frac{\partial a}{\partial x_j}.$$

But by Equations (3.5.1), $x_j = Q^{\top}_{ji}[\hat{x}_i - c_i(t)]$, so $\partial x_j / \partial \hat{x}_i = Q^{\top}_{ji} = Q_{ij}$. It follows that

$$(\mathrm{grad}_{\hat{\mathbf{x}}}\hat{a})_i = Q_{ij}\frac{\partial a}{\partial x_j} = (\mathsf{Q}\, \mathrm{grad}_{\mathbf{x}} a)_i,$$

that is,

$$\mathrm{grad}_{\hat{\mathbf{x}}}\hat{a}(\hat{\mathbf{x}}, \hat{t}) = \mathsf{Q}(t)\, \mathrm{grad}_{\mathbf{x}} a(\mathbf{x}, t),$$

which completes the proof. ■

EXERCISE 88 *Working in Cartesian coordinates, show that if* $\mathsf{A}(\mathbf{x},t)$ *is a differentiable, objective tensor field, then* $\operatorname{div}\mathsf{A}(\mathbf{x},t)$ *is an objective vector field.*

EXERCISE 89 *One way to represent a surface in* $\mathbb{E}$ *is as a level set*

$$\left\{\mathbf{x}\in\mathbb{E} \mid f(\mathbf{x},t)=0\right\},$$

where $f\colon \mathbb{E}\times\mathbb{R}\to\mathbb{R}$ *is a continuously differentiable, objective scalar field such that* $\operatorname{grad} f$ *never vanishes. Show that the corresponding unit-length vector field*

$$\mathbf{n}(\mathbf{x},t)=\frac{\operatorname{grad} f(\mathbf{x},t)}{\|\operatorname{grad} f(\mathbf{x},t)\|}$$

normal to the surface at each point $\mathbf{x}$ *and instant* t *is objective.*

The following fact figures prominently in fluid mechanics.

THEOREM 3.5.4 (OBJECTIVITY OF THE STRETCHING TENSOR). *The stretching tensor* D *is objective.*

PROOF: First, calculate the velocity gradient $\hat{\mathsf{L}}=\operatorname{grad}\hat{\mathbf{v}}$ in the hatted frame. Cartesian coordinates and index notation make the task simpler: by the chain rule applied to Equation (3.5.2),

$$\begin{aligned}\hat{L}_{kl}=\frac{\partial\hat{v}_k}{\partial\hat{x}_l}&=\frac{\partial}{\partial\hat{x}_l}(Q_{km}v_m+Q'_{km}x_m+c'_k)\\&=Q_{km}\frac{\partial v_m}{\partial x_n}\frac{\partial x_n}{\partial\hat{x}_l}+Q'_{km}\frac{\partial x_m}{\partial\hat{x}_l}.\end{aligned}$$

But the relationship $\hat{x}_l=Q_{ln}x_n+c_l$ implies that $x_n=Q_{ln}^{\top}(\hat{x}_l-c_l)$, so $\partial x_n/\partial\hat{x}_l=Q_{ln}^{\top}$. Hence,

$$\hat{L}_{kl}=Q_{km}\frac{\partial v_m}{\partial x_n}Q_{ln}^{\top}+Q'_{km}Q_{lm}^{\top},$$

that is, $\hat{\mathsf{L}}=\mathsf{QLQ}^{\top}+\mathsf{Q}'\mathsf{Q}^{\top}$. Taking the symmetric part yields

$$\hat{\mathsf{D}}=\tfrac{1}{2}(\hat{\mathsf{L}}+\hat{\mathsf{L}}^{\top})=\tfrac{1}{2}\mathsf{Q}(\mathsf{L}+\mathsf{L}^{\top})\mathsf{Q}^{\top}+\tfrac{1}{2}(\mathsf{Q}'\mathsf{Q}^{\top}+\mathsf{Q}\mathsf{Q}'^{\top}).$$

The last term on the right vanishes:

$$\tfrac{1}{2}(\mathsf{Q}'\mathsf{Q}^{\top}+\mathsf{Q}\mathsf{Q}'^{\top})=\tfrac{1}{2}(\mathsf{Q}\mathsf{Q}^{\top})'=\mathsf{I}'=0.$$

It follows that $\hat{\mathsf{D}}=\mathsf{QDQ}^{\top}$. ∎

EXERCISE 90 *Show that the spin tensor* W *is Galilean objective but not objective:* $\hat{\mathsf{W}} = \mathsf{QWQ}^\top + \mathsf{Q}'\mathsf{Q}^\top$.

EXERCISE 91 *Show that the left Cauchy–Green tensor* B *is objective.*

THEOREM 3.5.5 (OBJECTIVITY IN POLAR DECOMPOSITIONS). *In the polar decompositions* $\mathsf{F} = \mathsf{RU} = \mathsf{VR}$ *and* $\hat{\mathsf{F}} = \hat{\mathsf{R}}\hat{\mathsf{U}} = \hat{\mathsf{V}}\hat{\mathsf{R}}$ *of the deformation gradient,*

1. $\hat{\mathsf{R}} = \mathsf{QR}$, *so* R *is not objective;*
2. $\hat{\mathsf{U}} = \mathsf{U}$, *so* U *is not objective;*
3. $\hat{\mathsf{V}} = \mathsf{QVQ}^\top$, *so* V *is objective.*

EXERCISE 92 *Use Equation (3.5.6), together with the uniqueness of polar decompositions, to prove Theorem 3.5.5.*

3.5.2 Summary

An observer transformation is a change in frame of reference that allows the observers to be in rigid motion with respect to each other and to assign different origins to the time axis:

$$\hat{\mathbf{x}} = \mathsf{Q}(t)\mathbf{x} + \mathbf{c}(t),$$
$$\hat{t} = t - b.$$

Scalar, vector, and tensor fields are objective if they transform according to specific rules under observer transformations: scalars remain unchanged in value; the changes in vectors are attributable to the relative rotation of the two frames; and the changes in tensors—regarded as linear transformations—are those required to transform objective vector fields in to objective vector fields. In symbols,

$$\hat{a}(\hat{\mathbf{x}}, \hat{t}) = a(\mathbf{x}, t),$$
$$\hat{\mathbf{a}}(\hat{\mathbf{x}}, \hat{t}) = \mathsf{Q}(t)\mathbf{a}(\mathbf{x}, t),$$
$$\hat{\mathsf{A}}(\hat{\mathbf{x}}, \hat{t}) = \mathsf{Q}(t)\mathsf{A}(\mathbf{x}, t)\mathsf{Q}^\top(t).$$

Not all quantities of interest in continuum mechanics are objective; for example, the deformation gradient F is not objective. However, the gradient of any objective scalar field is objective, as are the stretching tensor D, the left Cauchy–Green tensor B, and the left stretch tensor V.

CHAPTER 4

BALANCE LAWS

While kinematics provides the essential descriptive language of continuum mechanics, the equations governing the motions of bodies arise from five fundamental laws:

- Mass balance
- Momentum balance
- Angular momentum balance
- Energy balance
- Entropy inequality.

The first four furnish most of the partial differential equations used to model fluid- and solid-mechanical systems. As Chapter 6 explores, the last law imposes thermodynamic constraints on the equations used to distinguish different types of materials.

Continuum Mechanics: The Birthplace of Mathematical Models, First Edition. M.B. Allen.

In their primitive forms, the balance laws take the forms of axioms involving integrals over configurations. As shown in Section 2.4, such integrals over material sets are amenable to special analysis. If $\boldsymbol{\chi}\colon \mathcal{B} \times \mathbb{R} \to \mathbb{E}$ is a motion on a body $\mathcal{B}$ and $\mathcal{P}$ is a part of $\mathcal{B}$, then the material set $\boldsymbol{\chi}(\mathcal{P}, t) \subset \mathbb{E}$ is the region occupied by $\mathcal{P}$ at time t. It has volume

$$\text{volume}\,(\boldsymbol{\chi}(\mathcal{P},t)) = \int_{\boldsymbol{\chi}(\mathcal{P},t)} dv = \int_{\boldsymbol{\kappa}(\mathcal{P})} \det \mathsf{F}\, dV.$$

Here $\boldsymbol{\kappa}\colon \mathcal{B} \to \mathbb{E}$ denotes the reference configuration, and $\mathsf{F}(\mathbf{X}, t)$ is the deformation gradient introduced in Section 2.2. The last identity follows from the change-of-variables theorem. We assume throughout the exposition that $\boldsymbol{\kappa}(\mathcal{B})$ is bounded and that $\boldsymbol{\chi}(\mathcal{B}, t)$ remains bounded for all t.

Notations such as

$$\int_{\boldsymbol{\chi}(\mathcal{P},t)} \cdots\, dv, \quad \int_{\boldsymbol{\kappa}(\mathcal{P})} \cdots\, dV, \quad \int_{\partial\boldsymbol{\chi}(\mathcal{P},t)} \cdots\, d\sigma, \quad \int_{\partial\boldsymbol{\kappa}(\mathcal{P})} \cdots\, d\Sigma$$

appear throughout this chapter. Despite their ornateness, these notations have the virtue of explicitly indicating both the volumes or surfaces in $\mathbb{E}$ over which the integration is to be performed and whether the volumes or surfaces pertain to current or reference configurations. We assume that all sets over which we integrate are sufficiently tame to justify the use of the integral theorems reviewed in Appendix B.

4.1 MASS BALANCE

Introduction of mass into mechanics requires an axiom.

AXIOM (MASS BALANCE). *Associated with any body $\mathcal{B}$ is a set function called* **mass**, *defined on the parts of $\mathcal{B}$. This function has the following properties.*

1. *The quantity* mass $(\mathcal{P})$ *has a fixed, nonnegative value for any part $\mathcal{P} \subset \mathcal{B}$.*

2. *For every configuration $\boldsymbol{\chi}\colon \mathcal{B} \to \mathbb{E}$, there exists an objective scalar function $\rho_{\boldsymbol{\chi}}\colon \boldsymbol{\chi}(\mathcal{B}) \to \mathbb{R}$, called the* **mass density**, *such that*

$$\text{mass}\,(\mathcal{P}) = \int_{\boldsymbol{\chi}(\mathcal{P})} \rho_{\boldsymbol{\chi}}(\mathbf{x})\, dv,$$

   for every part $\mathcal{P} \subset \mathcal{B}$.

3. *As an extension of the second observation, associated with the motion $\boldsymbol{\chi}(X, t)$ is an objective, one-parameter family $\rho(\cdot, t)\colon \boldsymbol{\chi}(\mathcal{B}, t) \to \mathbb{R}$ of integrable functions*

of $\mathbf{x}$ *such that*

$$\text{mass}\,(\mathcal{P}) = \int_{\boldsymbol{\chi}(\mathcal{P},t)} \rho(\mathbf{x},t)\,dv,$$

for all parts $\mathcal{P} \subset \mathcal{B}$. *We refer to* $\rho(\mathbf{x},t)$ *as the* **mass density** *at* $(\mathbf{x},t)$.

Combining the first and third properties yields the following integral law.

THEOREM 4.1.1 (GLOBAL MASS BALANCE). *For any part* $\mathcal{P} \subset \mathcal{B}$,

$$\frac{d}{dt}\,\text{mass}(\mathcal{P}) = \frac{d}{dt}\int_{\boldsymbol{\chi}(\mathcal{P},t)} \rho(\mathbf{x},t)\,dv = 0. \tag{4.1.1}$$

4.1.1 Local Forms of Mass Balance

In the sections that follow, we deduce from integral forms of the balance laws the local or differential forms common in applications. In doing so, we use a standard argument.

LEMMA 4.1.2 (DUBOIS-REYMOND LOCALIZATION PRINCIPLE). *If the function* $f\colon \boldsymbol{\chi}(\mathcal{B},t) \to \mathbb{R}$ *is continuous and*

$$\int_{\boldsymbol{\chi}(\mathcal{P},t)} f\,dv = 0 \tag{4.1.2}$$

for all $\mathcal{P} \subset \mathcal{B}$, *then* $f = 0$ *identically.*

EXERCISE 93 *Prove this lemma. What can we conclude when we replace* $=$ *by* $\geqslant$ *in Equation (4.1.2)?*

Not all of the integrands that we encounter are continuous in every application of interest. Fluid flows with shocks are among the most well-known examples of motions involving discontinuities. Section 4.6 examines an approach to localization that remains valid under relaxed continuity assumptions.

The remainder of this section explores two local versions of the mass balance. The first takes the referential point of view.

THEOREM 4.1.3 (LOCAL MASS BALANCE, REFERENTIAL VERSION). *If the densities $\rho_\kappa(\mathbf{X})$, associated with the reference configuration, and $\rho(\mathbf{x}, t)$, associated with the motion, are both continuous, then*

$$\rho \det \mathsf{F} = \rho_\kappa. \tag{4.1.3}$$

PROOF: By the global mass balance (4.1.1),

$$\int_{\boldsymbol{\kappa}(\mathcal{P})} \rho_\kappa(\mathbf{X})\, dV = \int_{\boldsymbol{\chi}(\mathcal{P}, t)} \rho(\mathbf{x}, t)\, dv.$$

But by the change-of-variables theorem B.3.1,

$$\int_{\boldsymbol{\chi}(\mathcal{P}, t)} \rho(\mathbf{x}, t)\, dv = \int_{\boldsymbol{\kappa}(\mathcal{P})} \rho(\mathbf{x}(\mathbf{X}, t), t) \det \mathsf{F}(\mathbf{X}, t)\, dV.$$

Combining these two identities yields

$$\int_{\boldsymbol{\kappa}(\mathcal{P})} (\rho_\kappa - \rho \det \mathsf{F})\, dV = 0.$$

Since this equation holds for an arbitrary part $\mathcal{P} \subset \mathcal{B}$, the localization principle 4.1.2 implies that $\rho_\kappa - \rho \det \mathsf{F} = 0$. ■

The second, more familiar version of the mass balance, sometimes misleadingly called the equation of continuity, refers to the spatial view.

THEOREM 4.1.4 (LOCAL MASS BALANCE, SPATIAL VERSION). *If $\rho(\mathbf{x}, t)$ and $\mathbf{v}(\mathbf{x}, t)$ are continuously differentiable, then*

$$\frac{\partial \rho}{\partial t} + \operatorname{div}(\rho \mathbf{v}) = 0. \tag{4.1.4}$$

PROOF: The argument follows a pattern that appears again in the proofs of other local balance laws. By the Reynolds transport theorem 2.4.1,

$$0 = \frac{d}{dt} \int_{\boldsymbol{\chi}(\mathcal{P}, t)} \rho\, dv = \int_{\boldsymbol{\chi}(\mathcal{P}, t)} \left[\frac{\partial \rho}{\partial t} + \operatorname{div}(\rho \mathbf{v}) \right] dv.$$

This equation holds for all parts $\mathcal{P} \subset \mathcal{B}$, so Equation (4.1.4) follows by the localization principle. ■

One often sees this form of the local mass balance written in terms of the material derivative:

$$\frac{D\rho}{Dt} + \rho \operatorname{div} \mathbf{v} = 0.$$

This more suggestive form, which follows from the product rule for $\operatorname{grad}(\rho\mathbf{v})$, indicates that the density following particle trajectories changes in response to the divergence of the velocity field.

DEFINITION. *A body is* **incompressible** *if*

$$\frac{D\rho}{Dt}(\mathbf{x}(\mathbf{X}, t), t) = 0,$$

for all particles $\mathbf{X}$ *and all times* t.

EXERCISE 94 *Show that, for an incompressible body,*

(a) *The velocity field* $\mathbf{v}$ *is* **solenoidal**, *that is,* $\operatorname{div} \mathbf{v} = 0$.

(b) *The motion is isochoric.*

We use the term incompressible to describe a property of the material. The term isochoric refers to a property of the motion. A body can undergo isochoric motion without being incompressible.

EXERCISE 95 *If* $\mathbf{v}(\mathbf{x}) = \operatorname{curl} \mathbf{\Psi}(\mathbf{x})$ *for a twice continuously differentiable function* $\mathbf{\Psi}$*, then* $\operatorname{div} \mathbf{v}(\mathbf{x}) = 0$. *(See Equation (1.2.25). The shorthand for this assertion is a classic vector-field identity:* $\operatorname{div}(\operatorname{curl} \mathbf{\Psi}) = 0$.*)*

Paralleling the discussion of the vector-field identity $\operatorname{curl}(\operatorname{grad} \Phi) = \mathbf{0}$ in Section 3.3, a partial converse holds. If $\mathbf{v}$ is continuously differentiable and solenoidal everywhere in $\mathbb{E}$, then there exists a continuously differentiable vector field $\mathbf{\Psi}$ such that $\mathbf{v} = \operatorname{curl} \mathbf{\Psi}$. For more discussion, see [39, Section 8.3].

In certain applications it is useful conceptually to replace the distributed mass of a part $\mathcal{P}$ of a body by a surrogate point mass. The following definition identifies the correct position of the point.

DEFINITION. *The* **center of mass** *of* $\mathcal{P} \subset \mathcal{B}$ *is*

$$\overline{\mathbf{x}}(\mathcal{P}, t) = \frac{1}{\operatorname{mass}(\mathcal{P})} \int_{\chi(\mathcal{P}, t)} \rho \mathbf{x} \, dv.$$

In other words, $\overline{\mathbf{x}}$ is the spatial moment of the mass density. We use this concept in the next section.

4.1.2 Summary

The global mass balance asserts that

$$\text{mass}\,(\mathcal{P}) = \int_{\chi(\mathcal{P},t)} \rho(\mathbf{x},t)\,dv$$

remains constant for all parts $\mathcal{P} \subset \mathcal{B}$. Here, ρ stands for the mass density. The localization principle yields the following local forms of this law:

$$\begin{aligned} \rho \det \mathsf{F} &= \rho_\kappa \quad \text{(referential form)} \\ \frac{\partial \rho}{\partial t} + \operatorname{div}(\rho\,\mathbf{v}) &= 0 \quad \text{(spatial form).} \end{aligned}$$

Rewriting the spatial form of the mass balance as

$$\frac{D\rho}{Dt} + \rho \operatorname{div} \mathbf{v} = 0$$

shows that for an incompressible body—one for which $D\rho/Dt = 0$—the velocity field satisfies the condition $\operatorname{div} \mathbf{v} = 0$.

4.2 MOMENTUM BALANCE

The second fundamental balance law of continuum mechanics governs a quantity that is familiar from classical physics.

DEFINITION. *For any body $\mathcal{B}$, the* **momentum** *of a part $\mathcal{P} \subset \mathcal{B}$ at time t is*

$$\mathbf{M}(\mathcal{P},t) = \int_{\chi(\mathcal{P},t)} \rho \mathbf{v}\,dv.$$

It is possible to express this quantity in a form that resembles the momentum of a discrete particle.

EXERCISE 96 *Using the center of mass defined in Section 4.1, show that, for any part $\mathcal{P} \subset \mathcal{B}$,*

$$\mathbf{M}(\mathcal{P},t) = \text{mass}\,(\mathcal{P}) \frac{d\overline{\mathbf{x}}}{dt}(\mathcal{P},t).$$

The momentum balance amounts to a version of Newton's second law: the time rate of change of momentum equals the net force. In the context of continua, a **system of forces** is a pair

$$(\mathbf{b}(\mathbf{x},t), \mathbf{t}(\mathbf{x},t,\mathbf{n}))$$

of vector fields having the following interpretations.

- The vector field $\mathbf{b}\colon \chi(\mathcal{B}, t) \times \mathbb{R} \to \mathbb{E}$ is the **specific body force**, giving the total **body force** acting on any part $\mathcal{P} \subset \mathcal{B}$ as follows:

$$\mathbf{F}_b(\mathcal{P}, t) = \int_{\chi(\mathcal{P}, t)} \rho \, \mathbf{b} \, dv.$$

  This expression accounts for the forces that act on particles in the interior of the region $\chi(\mathcal{P}, t)$. The most familiar body force is gravity.

- The vector field $\mathbf{t}\colon \chi(\mathcal{B}, t) \times \mathbb{R} \times \mathbb{E} \to \mathbb{E}$ is the **stress vector**, giving the total **contact force** (or **traction**) acting on the surface of any part $\mathcal{P} \subset \mathcal{B}$ as follows:

$$\mathbf{F}_c(\mathcal{P}, t) = \int_{\partial\chi(\mathcal{P}, t)} \mathbf{t}(\mathbf{x}, t, \mathbf{n}(\mathbf{x}, t)) \, d\sigma.$$

  Here $d\sigma$ denotes the element of surface area; see Equation (2.4.9). Also, $\mathbf{n}(\mathbf{x}, t)$ denotes the unit outward normal vector field for the surface $\partial\chi(\mathcal{P}, t)$, as shown in Figure 4.1. The expression $\mathbf{F}_c$ accounts for forces exerted by matter outside $\mathcal{P}$ via contact with particles on the bounding surface $\partial\chi(\mathcal{P}, t)$.

More general systems of forces are possible; see [53, Section I.5].

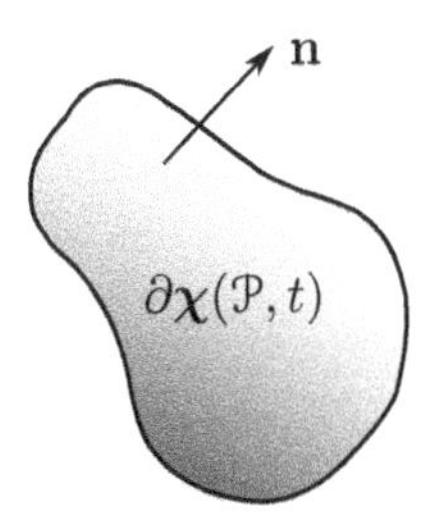

Figure 4.1. The bounding surface of a part $\mathcal{P}$ of a body, showing the unit outward normal vector $\mathbf{n}$.

The following axiom makes explicit the connection with Newton's second law:

AXIOM (MOMENTUM BALANCE). *The specific body force* $\mathbf{b}(\mathbf{x}, t)$ *and contact force* $\mathbf{t}(\mathbf{x}, t, \mathbf{n})$ *are objective vector fields, and there exists a frame of reference in which for any part* $\mathcal{P} \subset \mathcal{B}$,

$$\frac{d}{dt}\mathbf{M}(\mathcal{P}, t) = \mathbf{F}_c(\mathcal{P}, t) + \mathbf{F}_b(\mathcal{P}, t).$$

This axiom yields the **global momentum balance,**

$$\frac{d}{dt}\int_{\chi(\mathcal{P},t)} \rho\mathbf{v}\,dv = \int_{\partial\chi(\mathcal{P},t)} \mathbf{t}\,d\sigma + \int_{\chi(\mathcal{P},t)} \rho\mathbf{b}\,dv. \tag{4.2.1}$$

EXERCISE 97 *Show that*

$$\frac{d}{dt}\int_{\chi(\mathcal{P},t)} \rho\mathbf{v}\,dv = \int_{\chi(\mathcal{P},t)} \rho\frac{D\mathbf{v}}{Dt}\,dv,$$

where

$$\frac{D\mathbf{v}}{Dt} = \frac{\partial\mathbf{v}}{\partial t} + (\mathbf{v}\cdot\mathrm{grad})\mathbf{v} = \frac{\partial\mathbf{v}}{\partial t} + \mathsf{L}\mathbf{v}.$$

(Recall that $(\mathbf{v}\cdot\mathrm{grad})\mathbf{v} = v_j v_{i,j}\mathbf{e}_i$ *in Cartesian coordinates.)*

4.2.1 Analysis of Stress

At this point the vector field $\mathbf{t}$ remains a bit mysterious. For a glimpse of its action consider the part $\mathcal{P}$ drawn in Figure 4.2. The figure shows a plane tangent to the surface $\partial\chi(\mathcal{P},t)$ at a point $\mathbf{x}$ and the unit vector $\mathbf{n}$ normal to the tangent plane at $\mathbf{x}$. The stress vector $\mathbf{t}(\mathbf{x},t,\mathbf{n})$ acting on $\mathcal{P}$ at $\mathbf{x}$ is the sum of a normal component $\mathbf{t}_\perp$ and a tangential component $\mathbf{t}_\parallel$. The **normal stress** $\mathbf{t}_\perp = (\mathbf{t}\cdot\mathbf{n})\mathbf{n}$ is the force per unit area acting in a direction perpendicular to the tangent plane. If $\mathbf{t}\cdot\mathbf{n} > 0$, then the direction of the force is outward, and the stress on $\mathcal{P}$ at $\mathbf{x}$ is **tensile.** If $\mathbf{t}\cdot\mathbf{n} < 0$, then the force is directed inward, and the stress on $\mathcal{P}$ at $\mathbf{x}$ is **compressive**. The **shear stress** $\mathbf{t}_\parallel = \mathbf{t} - \mathbf{t}_\perp$ is the tangential force per unit area on $\mathcal{P}$ at $\mathbf{x}$.

To be more precise about the nature of $\mathbf{t}$ requires more subtle analysis. A fundamental theorem in continuum mechanics asserts that $\mathbf{t}(\mathbf{x},t,\mathbf{n}) = \mathsf{T}(\mathbf{x},t)\mathbf{n}(\mathbf{x},t)$, where the tensor field $\mathsf{T}(\mathbf{x},t)$ is independent of the unit normal vector $\mathbf{n}(\mathbf{x},t)$. Figure 4.3 illustrates unit normal vectors associated with two different parts of $\mathcal{B}$, the boundaries of whose configurations intersect at time t. The advantage of referring to the tensor T instead of the vector $\mathbf{t}$ is that T depends only on the position $\mathbf{x}$ and time t and not on the orientations of planes tangent to conceivable material regions. We call T the **stress tensor.**

The theorem requires two intermediate lemmas. In these lemmas, let δ be an upper bound on the set of distances

$$\left\{\|\mathbf{x}-\mathbf{y}\| \mid \mathbf{x},\mathbf{y}\in\mathcal{S}\right\}.$$

Such a bound exists because $\chi(\mathcal{P},t) \subset \chi(\mathcal{B},t)$, which we have assumed to be bounded. The arguments involve estimates of integrals and limits as $\delta \to 0$ for all t.

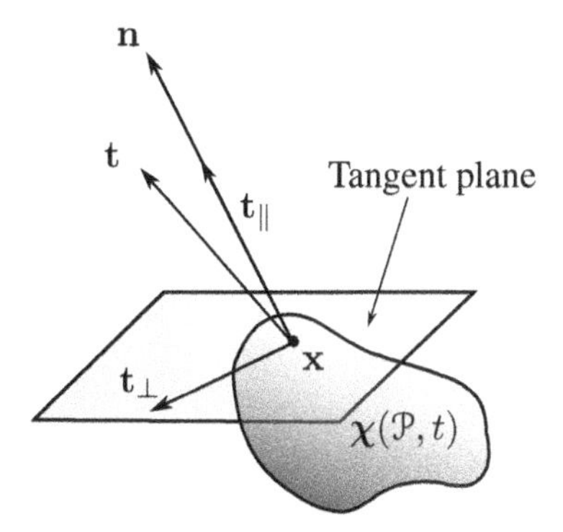

Figure 4.2. A configuration of a part $\mathcal{P}$ of a body showing the plane tangent to the surface $\partial\chi(\mathcal{P}, t)$ and the tangential and normal components of the stress vector acting via contact with the surface.

For the remainder of this subsection we use the temporary shorthand C to denote a generic positive constant, possibly depending on t but not on δ. The value of C may change from one expression to another.

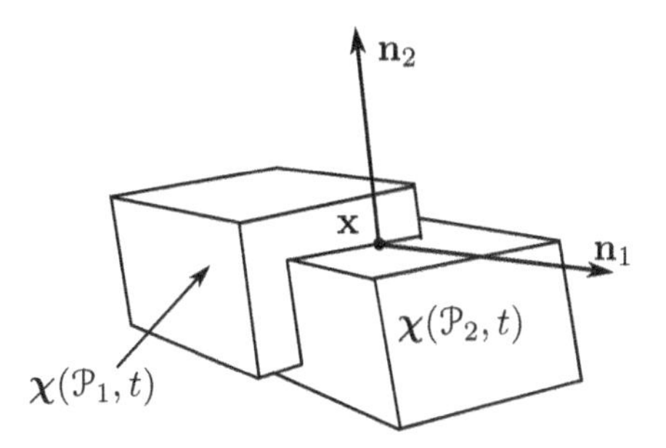

Figure 4.3. Parts of a body whose configurations at time t have bounding surfaces intersecting at a common point $\mathbf{x}$, where their outward unit normal vectors differ in direction.

LEMMA 4.2.1 (LIMIT OF THE STRESS INTEGRAL). *Suppose that* $\mathbf{t}$ *is continuous as a function of* $\mathbf{x}$ *and that* $\rho D\mathbf{v}/Dt$ *and* $\rho\mathbf{b}$ *are continuous and bounded in* $\mathbf{x}$. *At every time* t,

$$\lim_{\delta \to 0} \frac{1}{\delta^2} \int_{\partial\chi(\mathcal{P}, t)} \mathbf{t}\, d\sigma = 0.$$

PROOF: By the global momentum balance (4.2.1),

$$\left\|\int_{\partial\chi(\mathcal{P},t)} \mathbf{t}\,d\sigma\right\| = \left\|\int_{\chi(\mathcal{P},t)} \rho\left(\frac{D\mathbf{v}}{Dt} - \mathbf{b}\right)\,dv\right\|$$

$$\leqslant \int_{\chi(\mathcal{P},t)} \left\|\rho\frac{D\mathbf{v}}{Dt} - \rho\mathbf{b}\right\|\,dv \leqslant C\int_{\chi(\mathcal{P},t)} dv,$$

the last inequality following from the boundedness of $\rho D\mathbf{v}/Dt$ and $\rho\mathbf{b}$. Therefore, by the boundedness of $\chi(\mathcal{P},t)$,

$$\left\|\int_{\partial\chi(\mathcal{P},t)} \mathbf{t}\,d\sigma\right\| \leqslant C\,\delta^3.$$

Multiplying by δ^{-2} and letting $\delta \to 0$ finishes the proof. ∎

LEMMA 4.2.2 (CAUCHY'S LEMMA). *Under the hypotheses of Lemma 4.2.1,*

$$\mathbf{t}(\mathbf{x},t,\mathbf{n}) = -\mathbf{t}(\mathbf{x},t,-\mathbf{n}),$$

for any unit vector $\mathbf{n}$.

EXERCISE 98 *Prove Lemma 4.2.2, arguing as in Lemma 4.2.1 for the wafer-shaped material region drawn in Figure 4.4.*

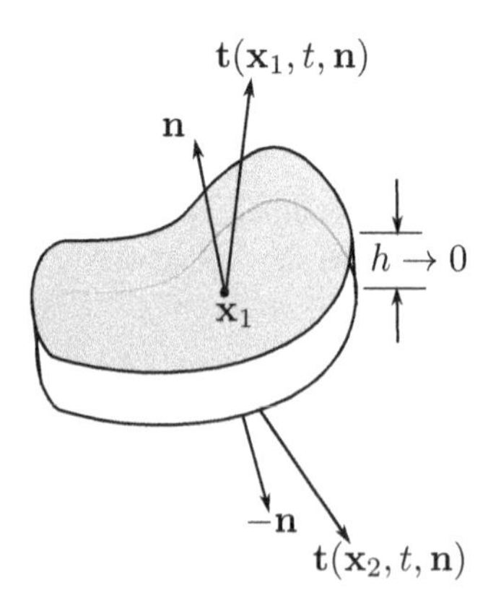

Figure 4.4. Wafer-shaped material region used in the proof of Lemma 4.2.2.

We now address the main theorem.

THEOREM 4.2.3 (CAUCHY'S THEOREM). *Under the hypotheses of Lemmas 4.2.1 and 4.2.2, there exists a time-dependent tensor field* $\mathsf{T}\colon \chi(\mathcal{P},t) \times \mathbb{R} \to \mathrm{L}(\mathbb{E})$ *such that*

$$\mathbf{t}(\mathbf{x},t,\mathbf{n}) = \mathsf{T}(\mathbf{x},t)\,\mathbf{n}, \tag{4.2.2}$$

for all unit vectors $\mathbf{n} \in \mathbb{E}$.

PROOF: It suffices to establish an expression giving $\mathsf{T}(\mathbf{x},t)$ as a linear combination of dyadic products. The argument unfolds by cases. For the first case, assume that $\mathbf{n}$ has nonnegative coordinates with respect to the standard orthonormal basis $\{\mathbf{e}_1, \mathbf{e}_2, \mathbf{e}_2\}$. Construct a tetrahedron in the positive octant, with three faces parallel to the coordinate planes and a face orthogonal to $\mathbf{n}$, containing the point $\mathbf{x}$, as shown in Figure 4.5. The face orthogonal to $\mathbf{n}$ has sides of length $\alpha\delta, \beta\delta, \gamma\delta$, where $\alpha, \beta, \gamma \geqslant 0$ and δ is a positive scale parameter that we eventually allow to tend to zero. By the global momentum balance (4.2.1),

$$\begin{aligned}\int_{\partial\chi(\mathcal{P},t)} \mathbf{t}\,d\sigma &= \sum_{j=1}^{3} \int_{\text{face } j} \mathbf{t}(\boldsymbol{\xi},t,-\mathbf{e}_j)\,d\sigma + \int_{\text{face } \mathbf{n}} \mathbf{t}(\boldsymbol{\xi},t,\mathbf{n})\,d\sigma \\ &= \int_{\chi(\mathcal{P},t)} \rho\left(\frac{D\mathbf{v}}{Dt} - \mathbf{b}\right) dv.\end{aligned}$$

For shorthand, denote the volume integral on the right as $I(\mathcal{P},t)$. Borrowing notation and reasoning from Lemma 4.2.1, we find that $\|I(\mathcal{P},t)\| \leqslant C\delta^3$. Furthermore, by the mean value theorem for integrals, there exist points $\boldsymbol{\xi}_1, \boldsymbol{\xi}_2, \boldsymbol{\xi}_3, \boldsymbol{\xi}_n$ on the respective faces of the tetrahedron such that

$$\int_{\partial\chi(\mathcal{P},t)} \mathbf{t}\,d\sigma = \sum_{j=1}^{3} \mathbf{t}(\boldsymbol{\xi}_j,t,-\mathbf{e}_j) \int_{\text{face } j} d\sigma \;+\; \mathbf{t}(\boldsymbol{\xi}_n,t,\mathbf{n}) \int_{\text{face } \mathbf{n}} d\sigma.$$

But the area of face $\mathbf{n}$ is proportional to δ^2, so

$$\int_{\text{face } \mathbf{n}} d\sigma = C\delta^2, \qquad \int_{\text{face } j} d\sigma = (\mathbf{e}_j \cdot \mathbf{n})C\delta^2.$$

(See Exercise 99.) It follows that

$$\mathbf{t}(\boldsymbol{\xi}_n,t,\mathbf{n}) + \sum_{j=1}^{3} \mathbf{t}(\boldsymbol{\xi}_j,t,-\mathbf{e}_j)(\mathbf{e}_j \cdot \mathbf{n}) = \frac{I(\mathcal{P},t)}{C\delta^2} \leqslant C\delta.$$

Letting $\delta \to 0$ forces $\boldsymbol{\xi}_j \to \mathbf{x}$ for $j = 1, 2, 3$ and $\boldsymbol{\xi}_n \to \mathbf{x}$, yielding

$$\mathbf{t}(\mathbf{x},t,\mathbf{n}) = -\left[\sum_{j=1}^{3} \mathbf{t}(\mathbf{x},t,-\mathbf{e}_j) \otimes \mathbf{e}_j\right] \mathbf{n}.$$

The stress vectors appearing in the sum on the right have expansions in terms of the basis $\{\mathbf{e}_1, \mathbf{e}_2, \mathbf{e}_3\}$:

$$\mathbf{t}(\mathbf{x}, t, -\mathbf{e}_j) = \sum_{j=1}^{3} t_i(\mathbf{x}, t, -\mathbf{e}_j)\mathbf{e}_i = -\sum_{i=1}^{3} t_i(\mathbf{x}, t, \mathbf{e}_j)\,\mathbf{e}_i, \tag{4.2.3}$$

where the negative sign in front of the last sum comes from an application of Lemma 4.2.2. Denoting $t_i(\mathbf{x}, t, \mathbf{e}_j) = T_{ij}(\mathbf{x}, t)$, we obtain

$$\mathbf{t}(\mathbf{x}, t, \mathbf{n}) = \left[\sum_{i=1}^{3}\sum_{j=1}^{3} T_{ij}(\mathbf{x}, t)(\mathbf{e}_i \otimes \mathbf{e}_j)\right] \mathbf{n}, \tag{4.2.4}$$

establishing the theorem for this case. T is the expression in square brackets.

The remaining cases involve orientations of $\mathbf{n}$ pointing into other octants and tetrahedrons in which the remaining faces have normal vectors $\pm\mathbf{e}_j$. Lemma 4.2.2 shows how the corresponding stress vectors $\mathbf{t}(\mathbf{x}, t, \pm\mathbf{e}_j)$ are related to the expression in Equation (4.2.3). The upshot is that the representation (4.2.4) holds for all orientations of $\mathbf{n}$.

The argument so far formally defines the stress tensor T on unit vectors. A standard construction extends the domain of T, by linearity, to all of $\mathbb{E}$. For arbitrary $\mathbf{a} \in \mathbb{E}$, defining

$$\mathsf{T}\mathbf{a} = \begin{cases} \|\mathbf{a}\|\mathsf{T}(\mathbf{a}/\|\mathbf{a}\|), & \text{if} \quad \mathbf{a} \neq \mathbf{0}, \\ \mathbf{0}, & \text{if} \quad \mathbf{a} = \mathbf{0}, \end{cases}$$

completes the proof. ∎

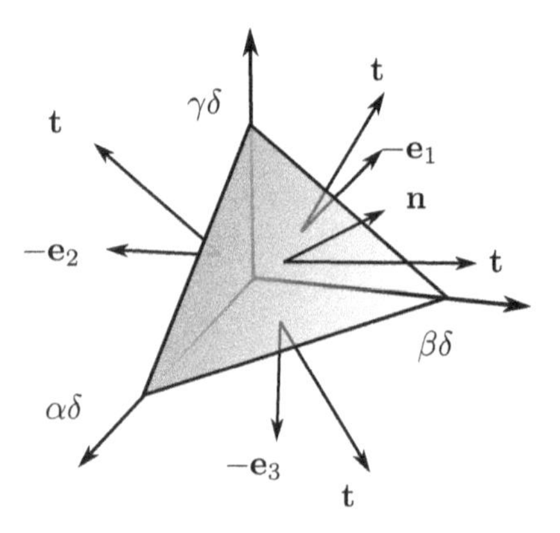

Figure 4.5. Tetrahedron used in the proof of Cauchy's theorem on stress.

EXERCISE 99 *In Figure 4.5, verify that the area of face j and the area of face* $\mathbf{n}$ *stand in the ratio* $\mathbf{e}_j \cdot \mathbf{n}$.

EXERCISE 100 *Show that the tensor field* T *is objective. (Recall from Exercise 89 that the surface normal vector field* $\mathbf{n}$ *is objective.)*

To gain further insight into the action of T, consider the cubic material region sketched in Figure 4.6. If the faces lie parallel to the coordinate planes, then all normal and tangential components of $\mathbf{t}$ lie parallel to the orthonormal basis vectors $\mathbf{e}_1, \mathbf{e}_2, \mathbf{e}_3$. With respect to this basis, the entry T_{ij} of the stress tensor T gives the force per unit area, acting on the face perpendicular to $\mathbf{e}_j$, acting in the direction of $\mathbf{e}_i$.

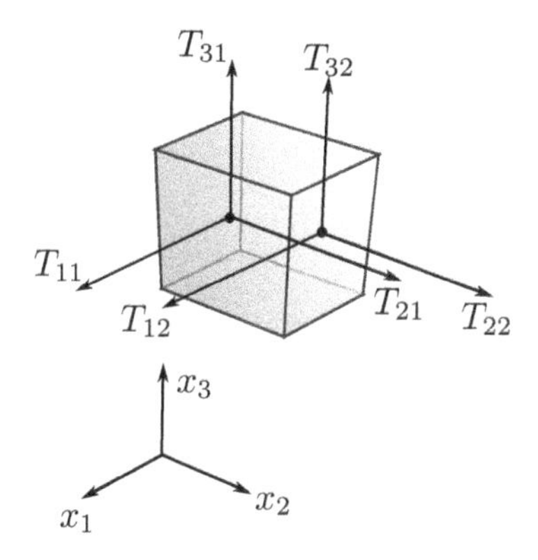

Figure 4.6. Cubic material volume showing the actions of the stress components T_{ij} with respect to the preferred basis for $\mathbb{E}$.

Theorem 4.2.3 facilitates reduction of the global momentum balance (4.2.1) to local form. By substituting the stress tensor into the integral representing the contact forces, we obtain

$$\int_{\chi(\mathcal{P},t)} \rho \frac{D\mathbf{v}}{Dt}\, dv = \int_{\partial\chi(\mathcal{P},t)} \mathsf{T}\mathbf{n}\, d\sigma + \int_{\chi(\mathcal{P},t)} \rho\mathbf{b}\, dv.$$

Further reduction requires use of the divergence theorem in the following form, reviewed in Appendix B:

$$\int_{\partial\chi(\mathcal{P},t)} \mathsf{T}\mathbf{n}\, d\sigma = \int_{\chi(\mathcal{P},t)} \operatorname{div}\mathsf{T}\, dv. \tag{4.2.5}$$

Here, $\operatorname{div}\mathsf{T}$ denotes the divergence of the tensor field T, defined in Equation (1.2.24). Making the substitution (4.2.5), we obtain a volume-integral formulation of the global momentum balance:

$$\int_{\chi(\mathcal{P},t)} \rho \frac{D\mathbf{v}}{Dt}\, dv = \int_{\chi(\mathcal{P},t)} (\operatorname{div}\mathsf{T} + \rho\mathbf{b})\, dv. \tag{4.2.6}$$

Finally, applying the localization principle yields the **local momentum balance** in the following form.

COROLLARY 4.2.4 (CAUCHY'S FIRST LAW OF MOTION). *If the integrands in Equation (4.2.6) are continuous, then*

$$\rho \frac{D\mathbf{v}}{Dt} = \operatorname{div}\mathsf{T} + \rho\mathbf{b}. \tag{4.2.7}$$

EXERCISE 101 *Briefly discuss the similarity between Equation (4.2.7) and Newton's second law of motion, force = mass × acceleration.*

4.2.2 Inertial Frames of Reference

The axiom leading to the momentum balance equation

$$\rho \frac{D^2\mathbf{x}}{Dt^2} = \operatorname{div}\mathsf{T} + \rho\mathbf{b} \tag{4.2.8}$$

guarantees that there exists a frame of reference in which it holds. We have no reason to expect that the equation holds in a different frame of reference obtained through an arbitrary observer transformation: although all expressions in the right side of the equation are objective, Exercise 85 shows that $D^2\mathbf{x}/Dt^2$ is not. Shortly we examine how Equation (4.2.8) transforms under general observer transformations.

Still, there is a special class of observer transformations that preserve the form of Equation (4.2.8). To identify this class, recall from Exercise 85 that $D^2\mathbf{x}/Dt^2$ is Galilean objective. That is, it transforms as an objective vector field under the special case of observer transformations having the form

$$\hat{\mathbf{x}} = \mathsf{Q}\mathbf{x} + \mathbf{c}(t), \tag{4.2.9}$$

where $\mathsf{Q} \in \mathsf{O}(\mathbb{E})$ is constant and $\mathbf{c}'' = \mathbf{0}$.

EXERCISE 102 *Show that, under an observer transformation of the form (4.2.9),*

$$\hat{\rho} \frac{D^2\hat{\mathbf{x}}}{Dt^2} = \operatorname{div}_{\hat{\mathbf{x}}} \hat{\mathsf{T}} + \hat{\rho}\hat{\mathbf{b}}. \tag{4.2.10}$$

Thus the form of Cauchy's first law remains unchanged under Galilean transformations.

DEFINITION. *A frame of reference is* **inertial** *if it can be obtained, by a Galilean transformation, from a frame of reference in which Equation (4.2.1) holds.*

In a noninertial frame, Cauchy's first law takes a different form. To see how the momentum balance equation transforms under arbitrary observer transformations, recall Equation (3.5.5), which shows how the acceleration transforms under such transformations. Using this equation together with the identity

$$\rho\,\mathsf{Q}\frac{D^2\mathbf{x}}{Dt^2} = \mathsf{Q}\,\mathrm{div}\,\mathsf{T} + \rho\,\mathsf{Q}\,\mathbf{b}$$

yields

$$\begin{aligned}\hat{\rho}\frac{D^2\hat{\mathbf{x}}}{Dt^2} &= \mathrm{div}_{\hat{\mathbf{x}}}\,\hat{\mathsf{T}} + \hat{\rho}\hat{\mathbf{b}} \\ &\quad + \underbrace{\hat{\rho}\mathbf{c}''}_{\text{(I)}} + \underbrace{2\hat{\rho}\boldsymbol{\omega}\times(\hat{\mathbf{v}}-\mathbf{c}')}_{\text{(II)}} + \underbrace{\hat{\rho}\boldsymbol{\omega}'\times(\hat{\mathbf{x}}-\mathbf{c})}_{\text{(III)}} - \underbrace{\hat{\rho}\boldsymbol{\omega}\times[\boldsymbol{\omega}\times(\hat{\mathbf{x}}-\mathbf{c})]}_{\text{(IV)}}.\end{aligned} \tag{4.2.11}$$

The labeled terms on the right have the following interpretations:

(I) is the **translational inertia**, the apparent force that compensates for acceleration of the noninertial frame.

(II) represents the **Coriolis effect**, the apparent force attributable to rotation about an axis not collinear with $\hat{\mathbf{v}}$.

(III) is sometimes called the **Euler effect**, an apparent force attributable to variations in the noninertial frame's rotation rate.

(IV) is the **centrifugal effect**, an apparent force compensating for the force required to keep particles rotating with respect to the inertial frame.

Among the most commonly used noninertial frames of reference are those fixed to the Earth. For such applications it is usually possible to choose the origin to coincide with that of an inertial frame. In these cases the translational inertia vanishes. Also, since the Earth's rotation rate is nearly constant at $\|\boldsymbol{\omega}\| = 1\ \mathrm{day}^{-1} \simeq 1.157 \times 10^{-5}\ \mathrm{s}^{-1}$, it is common to neglect the Euler force in these settings. Although the centrifugal effect tends to be quite small in most applications of this type, the Coriolis effect deserves a comment.

For illustration, consider two simple thought experiments. Both involve a rigid disk rotating with constant angular velocity $\boldsymbol{\omega}$ about an axis perpendicular to the disk and passing through its center. In the first, shown in Figure 4.7, a particle moves radially outward from the center O of the plate, always aiming toward a point P fixed to the moving edge of the disk. As viewed in an inertial frame of reference shown in Figure 4.7(a), the particle follows the curved path γ_1 required to stay aimed at its moving target P. In a noninertial frame of reference attached to the rotating disk, shown in Figure 4.7(b), the particle appears to follow a radial line $\hat{\gamma}_1$ between O and P. A force acting in the direction $\boldsymbol{\omega} \times \hat{\gamma}_1$ is required to keep the particle accelerating toward its target instead of drifting to the right of P.

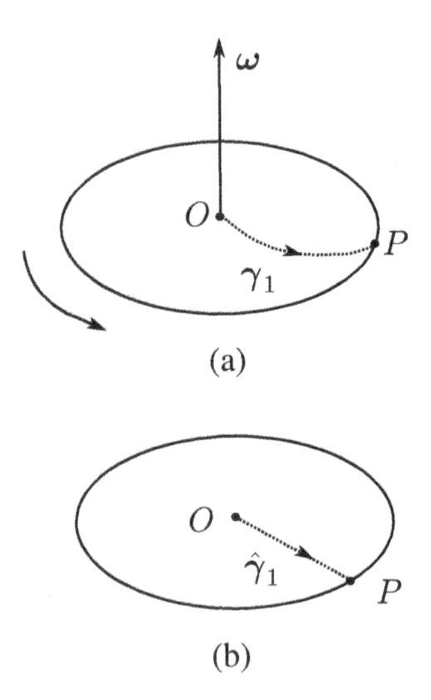

Figure 4.7. Rotating disk with a particle moving from the center O toward a point P attached to the edge. The path γ_1 in (a) shows how the particle's trajectory appears in an inertial frame of reference; $\hat{\gamma}_1$ in (b) shows how it appears in a noninertial frame of reference rotating with the disk.

In the second thought experiment, the particle undergoes uniform motion starting at the center O. In the inertial frame of reference, shown in Figure 4.8(a), the particle follows a straight line γ_2 passing through O. In the noninertial frame of reference rotating with the disk, shown in Figure 4.8(b), the particle appears to undergo deflection to the right, corresponding to an apparent force in the direction $-\boldsymbol{\omega} \times \hat{\gamma}_2'$. Thus, to observers in the rotating frame of reference, the Coriolis effect appears to be a force that, unless counteracted, deflects particle trajectories to the right when the rotation rate $\boldsymbol{\omega}$ points upward.

We encounter the Coriolis effect and the centrifugal force again in Section 5.3.

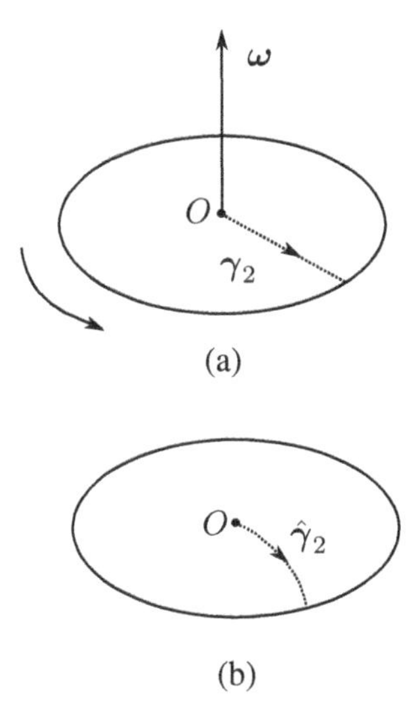

Figure 4.8. Rotating disk with a particle undergoing uniform motion from the center O. The path γ_2 in (a) shows how the particle's trajectory appears in an inertial frame of reference; $\hat{\gamma}_2$ in (b) shows how it appears in a noninertial frame of reference rotating with the disk.

4.2.3 Momentum Balance in Referential Coordinates

The local form (4.2.7) of Cauchy's first law applies to spatial coordinates. For certain applications, especially in solid mechanics, it is useful to have an analogous local momentum balance, written in referential coordinates and governing the reference density $\rho_\kappa(\mathbf{X})$ and the displacement $\mathbf{u}(\mathbf{X}, t)$. To derive this form of Cauchy's first law, start with the global form:

$$\underbrace{\int_{\chi(\mathcal{P},t)} \rho \frac{D\mathbf{v}}{Dt}\, dv}_{\text{(I)}} = \underbrace{\int_{\partial\chi(\mathcal{P},t)} \mathsf{T}\mathbf{n}\, d\sigma}_{\text{(II)}} + \underbrace{\int_{\chi(\mathcal{P},t)} \rho\mathbf{b}\, dv}_{\text{(III)}}.$$

Let us convert the terms (I), (II), and (III), one by one, to referential coordinates. For the first term, the identities $\rho_\kappa = \rho \det \mathsf{F}$ and $\partial\mathbf{u}/\partial t = \partial\mathbf{x}/\partial t$ yield

$$\begin{aligned}
\text{(I)} &= \int_{\boldsymbol{\kappa}(\mathcal{P})} \rho(\mathbf{x}(\mathbf{X},t),t) \frac{D}{Dt}\frac{\partial \mathbf{x}}{\partial t}(\mathbf{X},t)\, \det \mathsf{F}(\mathbf{X},t)\, dV \\
&= \int_{\boldsymbol{\kappa}(\mathcal{P})} \rho_\kappa(\mathbf{X}) \frac{\partial^2 \mathbf{u}}{\partial t^2}(\mathbf{X},t)\, dV.
\end{aligned}$$

Similarly, for the third term we get

$$\text{(III)} = \int_{\boldsymbol{\kappa}(\mathcal{P})} \rho_\kappa(\mathbf{X})\, \mathbf{b}(\mathbf{x}(\mathbf{X},t),t)\, dV.$$

For the second term the change of variables proceeds as follows, using the result (2.4.8) for surface integrals:

$$\begin{aligned}(\mathrm{II}) &= \int_{\partial\boldsymbol{\kappa}(\mathcal{P})} \mathsf{T}(\mathbf{x}(\mathbf{X},t),t)\ \det\mathsf{F}(\mathbf{X},t)\,\mathsf{F}^{-\top}(\mathbf{X},t)\,\mathbf{n}(\mathbf{X})\,d\Sigma \\ &= \int_{\partial\boldsymbol{\kappa}(\mathcal{P})} \mathsf{T}_{\kappa}(\mathbf{X},t)\,\mathbf{n}(\mathbf{X})\,d\Sigma,\end{aligned}$$

where

$$\mathsf{T}_{\kappa} = (\det\mathsf{F})\mathsf{T}\mathsf{F}^{-\top}. \tag{4.2.12}$$

This tensor, known as the **first Piola–Kirchhoff stress**, gives the stress $\mathsf{T}_{\kappa}(\mathbf{X},t)$ in the reference configuration that corresponds to the stress $\mathsf{T}(\mathbf{x},t)$ in the current spatial configuration. (Section 4.3 introduces a second Piola–Kirchhoff stress.) Applying the divergence theorem in the form B.2.2 to the last integral over $\partial\boldsymbol{\kappa}(\mathcal{P})$ yields

$$(\mathrm{II}) = \int_{\boldsymbol{\kappa}(\mathcal{P})} \mathrm{Div}\,\mathsf{T}_{\kappa}\,dV.$$

All of the terms in the global momentum balance now involve volume integrals over the reference configuration $\boldsymbol{\kappa}(\mathcal{P})$. Applying the localization principle yields the referential form of the local momentum balance,

$$\rho_{\kappa}\frac{\partial^2\mathbf{u}}{\partial t^2} = \mathrm{Div}\,\mathsf{T}_{\kappa} + \rho_{\kappa}\mathbf{b}. \tag{4.2.13}$$

We encounter this equation again in Chapter 5.

4.2.4 Summary

The global momentum balance generalizes Newton's second law:

$$\frac{d}{dt}\int_{\boldsymbol{\chi}(\mathcal{P},t)} \rho\mathbf{v}\,dv = \mathbf{F}_b + \mathbf{F}_c,$$

where $\mathbf{F}_b$ stands for the body force and $\mathbf{F}_c$ represents the contact force acting on surfaces. Analyzing the contact force

$$\mathbf{F}_c = \int_{\partial\boldsymbol{\chi}(\mathcal{P},t)} \mathbf{t}(\mathbf{x},t,\mathbf{n})\,d\sigma$$

yields a representation of the stress vector $\mathbf{t}$ as

$$\mathbf{t}(\mathbf{x},t,\mathbf{n}) = \mathsf{T}(\mathbf{x},t)\mathbf{n},$$

where $\mathsf{T}(\mathbf{x}, t)$ is the stress tensor. Localizing the global momentum balance using this representation yields Cauchy's first law,

$$\rho \frac{D\mathbf{v}}{Dt} = \operatorname{div} \mathsf{T} + \rho \mathbf{b}, \qquad \text{(spatial form)}$$

where $\mathbf{b}$ is the body force per unit mass. The corresponding referential equation is

$$\rho_\kappa \frac{\partial^2 \mathbf{u}}{\partial t^2} = \operatorname{Div} \mathsf{T}_\kappa + \rho_\kappa \mathbf{b}, \qquad \text{(referential form)}$$

where $\mathsf{T}_\kappa = (\det \mathsf{F})\mathsf{T}\mathsf{F}^{-\mathsf{T}}$ is the first Piola–Kirchhoff stress.

The momentum balance axiom guarantees only that there exists a frame of reference in which Cauchy's first law holds. Because the acceleration is not objective, Cauchy's first law holds only in a restricted class of frames of reference, namely, inertial frames, obtainable from the first through Galilean transformations. In more general frames of reference it is necessary to use a form of Cauchy's first law that incorporates apparent forces attributable to the frames' rotation and acceleration.

4.3 ANGULAR MOMENTUM BALANCE

In classical discrete mechanics, the angular momentum of a particle having mass m, located at position $\mathbf{x}$ with velocity $\mathbf{v}$ as drawn in Figure 4.9, is $m\mathbf{x} \times \mathbf{v}$. The quantity $\mathbf{x} \times \mathbf{v}$ is the **moment** of the velocity with respect to the origin O. An analogous definition applies to continua.

DEFINITION. *The* **angular momentum** *(or* **moment of momentum***) of a part $\mathcal{P} \subset \mathcal{B}$ about the origin O is*

$$\mathbf{L}(\mathcal{P}, t) = \int_{\chi(\mathcal{P}, t)} \rho \mathbf{x} \times \mathbf{v} \, dv.$$

A fundamental law of classical mechanics asserts that this quantity is conserved, in the sense that it changes only in response to applied moments of force, or **torques**. We develop the version that holds for continua.

The development parallels that given for the momentum balance in Section 4.2. We define a **system of torques** as a pair

$$\begin{aligned} &\mathbf{x} \times \mathsf{T} \qquad \textbf{moment of stress}, \\ &\mathbf{x} \times \mathbf{b} \qquad \textbf{moment of body force}. \end{aligned} \tag{4.3.1}$$

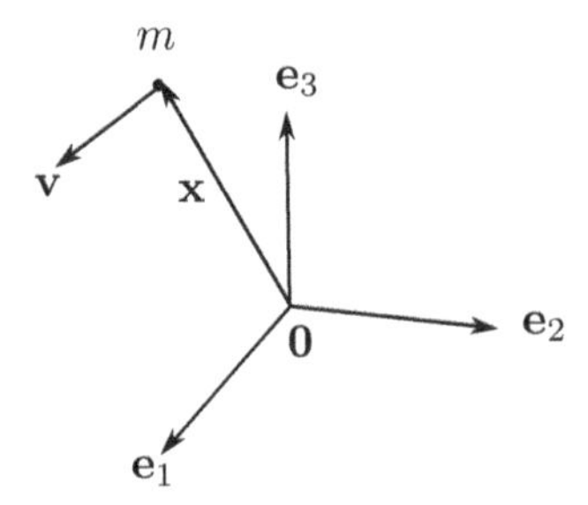

Figure 4.9. A discrete particle having angular momentum $\mathbf{x} \times m\mathbf{v}$.

Here, the product $\mathbf{x} \times \mathsf{T}$ denotes a linear transformation defined by the action $(\mathbf{x} \times \mathsf{T})\mathbf{a} = \mathbf{x} \times (\mathsf{T}\mathbf{a})$ for any $\mathbf{a} \in \mathbb{E}$. With respect to the orthonormal basis $\{\mathbf{e}_1, \mathbf{e}_2, \mathbf{e}_3\}$,

$$\mathbf{x} \times \mathsf{T} = \varepsilon_{ijk} x_j T_{kl} (\mathbf{e}_i \otimes \mathbf{e}_l)$$

$$= \begin{bmatrix} x_2T_{31} - x_3T_{21} & x_2T_{32} - x_3T_{22} & x_2T_{33} - x_3T_{23} \\ x_3T_{11} - x_1T_{31} & x_3T_{12} - x_1T_{32} & x_3T_{13} - x_1T_{33} \\ x_1T_{21} - x_2T_{11} & x_1T_{22} - x_2T_{12} & x_1T_{23} - x_2T_{13} \end{bmatrix}.$$

In terms of the system of torques, the **total torque** on $\mathcal{P}$ is

$$\mathbf{G}(\mathcal{P}, t) = \underbrace{\int_{\partial\boldsymbol{\chi}(\mathcal{P},t)} \mathbf{x} \times \mathsf{T}\, d\sigma}_{\text{Surface torque}} + \underbrace{\int_{\boldsymbol{\chi}(\mathcal{P},t)} \rho\mathbf{x} \times \mathbf{b}\, dv}_{\text{Body torque}}.$$

The system of torques (4.3.1) furnishes a complete description only if all torques are **simple**, meaning that they are moments of forces. It is possible to postulate other types of torques, called **couples**, and to allow bodies to have intrinsic spin. Materials subject to these additional types of torque are called **polar materials**. Eringen [18, Chapter 3] describes how to incorporate couples into the angular momentum balance. We consider only nonpolar materials.

The following axiom governs angular momentum:

Axiom (Angular Momentum Balance). *In any inertial frame, the rate of change of angular momentum balances the torques:*

$$\frac{d\mathbf{L}}{dt}(\mathcal{P}, t) = \mathbf{G}(\mathcal{P}, t).$$

More explicitly,

$$\frac{d}{dt}\int_{\chi(\mathcal{P},t)} \rho\mathbf{x}\times\mathbf{v}\,dv = \int_{\partial\chi(\mathcal{P},t)} \mathbf{x}\times\mathsf{T}\,d\boldsymbol{\sigma} + \int_{\chi(\mathcal{P},t)} \rho\mathbf{x}\times\mathbf{b}\,dv. \tag{4.3.2}$$

EXERCISE 103 *Derive the local version,*

$$\rho\frac{D}{Dt}(\mathbf{x}\times\mathbf{v}) = \operatorname{div}(\mathbf{x}\times\mathsf{T}) + \rho\mathbf{x}\times\mathbf{b}.$$

EXERCISE 104 *Prove the identity*

$$\frac{D}{Dt}(\mathbf{x}\times\mathbf{v}) = \mathbf{x}\times\frac{D\mathbf{v}}{Dt},$$

which reduces the local angular momentum balance to

$$\rho\mathbf{x}\times\frac{D\mathbf{v}}{Dt} = \operatorname{div}(\mathbf{x}\times\mathsf{T}) + \rho\mathbf{x}\times\mathbf{b}. \tag{4.3.3}$$

4.3.1 Symmetry of the Stress Tensor

As the next theorem shows, the local version of the angular momentum balance further reduces to a simple fact about the stress tensor. The theorem employs a version of the product rule:

$$\frac{\partial}{\partial x_l}(\varepsilon_{ijk}a_j A_{kl}) = \varepsilon_{ijk}\frac{\partial a_j}{\partial x_l}A_{kl} + \varepsilon_{ijk}a_j\frac{\partial A_{kl}}{\partial x_l},$$

which we write as $\operatorname{div}(\mathbf{a}\times\mathsf{A}) = (\operatorname{grad}\mathbf{a})\times\mathsf{A} + \mathbf{a}\times\operatorname{div}\mathsf{A}$.

THEOREM 4.3.1 (CAUCHY'S SECOND LAW OF MOTION). *If Cauchy's first law of motion (4.2.7) holds, then the angular momentum balance (4.3.3) is equivalent to the symmetry of the stress tensor,*

$$\mathsf{T} = \mathsf{T}^\top. \tag{4.3.4}$$

PROOF: It suffices to show that, with respect to the basis $\{\mathbf{e}_1, \mathbf{e}_2, \mathbf{e}_3\}$, the angular momentum balance reduces to $T_{ij} = T_{ji}$ for $i, j = 1, 2, 3$. Using the identity $\operatorname{div}(\mathbf{x}\times\mathsf{T}) = (\operatorname{grad}\mathbf{x})\times\mathsf{T} + \mathbf{x}\times\operatorname{div}\mathsf{T}$, we cast Equation 4.3.3 as

$$\underbrace{\mathbf{x}\times\left(\rho\frac{D\mathbf{v}}{Dt} - \operatorname{div}\mathsf{T} - \rho\mathbf{b}\right)}_{\text{(I)}} - \underbrace{(\operatorname{grad}\mathbf{x})\times\mathsf{T}}_{\text{(II)}} = \mathbf{0}.$$

The quantity labeled (I) vanishes by the Cauchy's first law of motion, and in Cartesian coordinates the quantity labeled (II) is $\partial x_j/\partial x_l = \delta_{jl}$. Hence, the local angular momentum balance reduces to

$$\varepsilon_{ijk}\delta_{jl}T_{lk} = \varepsilon_{ijk}T_{jk} = 0.$$

In array form, this identity states that

$$\begin{bmatrix} T_{23} - T_{32} \\ T_{31} - T_{13} \\ T_{12} - T_{21} \end{bmatrix} = \mathbf{0}.$$

Therefore, the angular momentum balance (4.3.3) holds if and only if $T_{ij} = T_{ji}$. ■

While symmetry of the stress tensor is a fundamental result, one should bear in mind two caveats. First, for polar materials, the local angular momentum balance takes a form different from Equation (4.3.3), and the argument just given no longer holds. In such materials, the stress tensor is not generally symmetric. Second, even for nonpolar materials, the first Piola–Kirchhoff stress tensor, which gives the stress corresponding to T in referential coordinates, fails to be symmetric in general.

EXERCISE 105 *Show that the first Piola–Kirchhoff stress tensor obeys the modified symmetry relationship* $(\mathsf{F}^{-1}\mathsf{T}_\kappa)^\top = \mathsf{F}^{-1}\mathsf{T}_\kappa$.

This exercise shows that the tensor

$$\mathsf{S}_\kappa = \mathsf{F}^{-1}\mathsf{T}_\kappa = (\det \mathsf{F})\mathsf{F}^{-1}\mathsf{T}\mathsf{F}^{-\top}, \tag{4.3.5}$$

called the **second Piola–Kirchhoff stress**, is symmetric.

EXERCISE 106 *Referring to the result in Equation (3.5.6), show that, under an observer transformation, the second Piola–Kirchhoff stress transforms according to the relationship* $\hat{\mathsf{S}}_\kappa = \mathsf{Q}_0\mathsf{S}_\kappa\mathsf{Q}_0^\top$, *where* Q_0 *denotes the rotation tensor of the observer transformation at the instant in time associated with the reference configuration. Interpret this result in terms of objectivity.*

4.3.2 Summary

The global angular momentum balance resembles the global momentum balance, except that we use a system of torques

$$\begin{aligned} &\mathbf{x} \times \mathsf{T} && \text{moment of stress,} \\ &\mathbf{x} \times \mathbf{b} && \text{moment of body force} \end{aligned}$$

that represent spatial moments of the forces appearing in the momentum balance. Through use of the localization principle, elimination of terms that are redundant with the momentum balance, and further simplification, the global form of the angular momentum balance

$$\frac{d}{dt}\int_{\boldsymbol{\chi}(\mathcal{P},t)} \rho\mathbf{x}\times\mathbf{v}\,dv = \int_{\partial\boldsymbol{\chi}(\mathcal{P},t)} \mathbf{x}\times\mathsf{T}\,d\boldsymbol{\sigma} + \int_{\boldsymbol{\chi}(\mathcal{P},t)} \rho\mathbf{x}\times\mathbf{b}\,dv$$

reduces to the following simple statement about the stress tensor for nonpolar materials:

$$\mathsf{T} = \mathsf{T}^\top.$$

4.4 ENERGY BALANCE

The fourth fundamental balance law, the energy balance, expresses a continuum analog of the first law of thermodynamics: energy is conserved. The complete statement of this principle involves four new quantities associated with a body $\mathcal{B}$. We introduce these quantities one by one.

Internal Energy. There exists a nonnegative set function $E(\cdot,t)$, called **internal energy**, and an objective scalar function $\varepsilon(\mathbf{x},t)$, called the **specific internal energy**, such that

$$E(\mathcal{P},t) = \int_{\boldsymbol{\chi}(\mathcal{P},t)} \rho(\mathbf{x},t)\varepsilon(\mathbf{x},t)\,dv,$$

for every $\mathcal{P}\subset\mathcal{B}$. Physically, internal energy corresponds to the energy stored in degrees of freedom observable at the molecular scale, such as the energy associated with random molecular motion or stored as a result of applying strain.

Kinetic Energy. The **total kinetic energy** of a part $\mathcal{P}\subset\mathcal{B}$ is

$$K(\mathcal{P},t) = \int_{\boldsymbol{\chi}(\mathcal{P},t)} \tfrac{1}{2}\rho\mathbf{v}\cdot\mathbf{v}\,dv.$$

This definition generalizes the kinetic energy $\frac{1}{2}m\mathbf{v}\cdot\mathbf{v}$ of a discrete particle.

The **total energy** of $\mathcal{P}$ is $E(\mathcal{P},t)+K(\mathcal{P},t)$.

Rate of Working. Stress and body forces can do work on a body, thereby adding to its energy. In many applications we regard work as an "orderly" form of energy, affecting the visible motion of the body. The **rate of working**, or **mechanical power**, associated with the system of forces acting on $\mathcal{P}$ is

$$W(\mathcal{P},t) = \underbrace{\int_{\partial\boldsymbol{\chi}(\mathcal{P},t)} \mathsf{T}\mathbf{v}\cdot\mathbf{n}\,d\sigma}_{\text{(I)}} + \underbrace{\int_{\boldsymbol{\chi}(\mathcal{P},t)} \rho\mathbf{v}\cdot\mathbf{b}\,dv}_{\text{(II)}}.$$

The term labeled (I) represents the rate of working of contact forces or tractions on $\mathcal{P}$, while the term labeled (II) represents the rate of working of body forces on $\mathcal{P}$.

Rate of Heating. We also allow additions to the energy of a body via transfer of heat. Specifically, we associate with the body a time-dependent, objective vector field $\mathbf{q}$ called the **heat flux**. This field gives the net rate of heating of $\mathcal{P} \subset \mathcal{B}$ via heat transfer across the surface $\partial\chi(\mathcal{P}, t)$ as

$$-\int_{\partial\chi(\mathcal{P},t)} \mathbf{q}(\mathbf{x}, t) \cdot \mathbf{n}(\mathbf{x}, t)\, d\sigma. \tag{4.4.1}$$

This concept bears three remarks. First, heat serves as a continuum model of the random motion of molecules, which we regard as a "disorderly" form of energy, in contrast with work. Second, at points on $\partial\chi(\mathcal{P}, t)$ where $\mathbf{q} \cdot \mathbf{n} > 0$, as shown in Figure 4.10, heat energy flows outward from $\mathcal{P}$. Therefore, the integral (4.4.1) represents the net rate of heat flow into $\mathcal{P}$.

Third, by analogy with the analysis of stress, one can introduce the net rate of heating by postulating that it has the form

$$-\int_{\partial\chi(\mathcal{P},t)} q(\mathbf{x}, t, \mathbf{n})\, d\sigma,$$

where $q(\mathbf{x}, t, \mathbf{n})$ is a scalar heat flux density. From this arguably more fundamental starting point one then deduces the existence of the heat flux vector $\mathbf{q}(\mathbf{x}, t)$ by an argument paralleling the tetrahedron-based proof used for the stress tensor T in Section 4.2. We bypass this step.

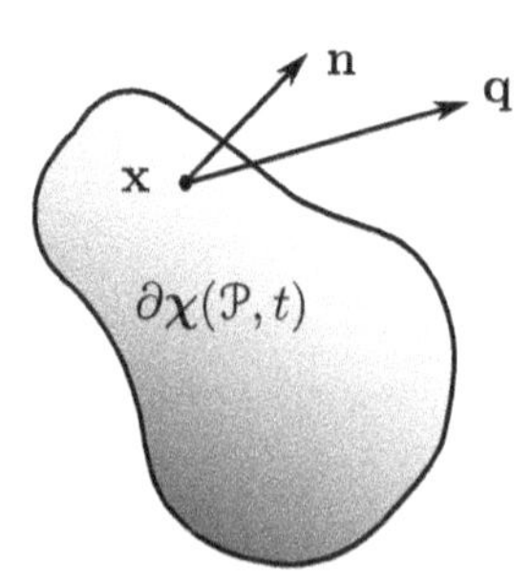

Figure 4.10. A heat flux vector $\mathbf{q}$ directed outward from $\mathcal{P}$.

Heat Supply. Finally, we introduce an objective scalar function $r(\mathbf{x}, t)$ called the **heat supply**. This function gives the rate of heating interior to $\chi(\mathcal{P}, t)$ as a result of external influences as follows:

$$\int_{\chi(\mathcal{P},t)} \rho(\mathbf{x}, t)\, r(\mathbf{x}, t)\, dv. \tag{4.4.2}$$

The heat supply accounts for the addition of energy via such phenomena as the interaction of electromagnetic waves with the material (radiative heat transfer) and Joule heating associated with electric currents.

The quantities $\mathbf{q}(\mathbf{x},t)$ and $r(\mathbf{x},t)$ form a **system of heatings**, giving the **total rate of heating** as

$$Q(\mathbf{x},t) = -\int_{\partial\boldsymbol{\chi}(\mathcal{P},t)} \mathbf{q}\cdot\mathbf{n}\,d\sigma + \int_{\boldsymbol{\chi}(\mathcal{P},t)} \rho\, r\, dv.$$

The following axiom prescribes a relationship among the total kinetic energy, the total internal energy, the rate of working, and the total rate of heating:

AXIOM (FIRST LAW OF THERMODYNAMICS). *In any inertial frame,*

$$\frac{d}{dt}[E(\mathcal{P},t)+K(\mathcal{P},t)] = W(\mathcal{P},t)+Q(\mathcal{P},t). \tag{4.4.3}$$

EXERCISE 107 *Identify the physical dimensions of each of the quantities appearing in $E(\mathcal{P},t)$, $K(\mathcal{P},t)$, $W(\mathcal{P},t)$, and $Q(\mathcal{P},t)$.*

In words, the rate of change of the total (internal and kinetic) energy equals the rate of addition of work and heat. Substituting for each of the terms in this equation gives the **global energy balance**,

$$\begin{aligned}\frac{d}{dt}\int_{\boldsymbol{\chi}(\mathcal{P},t)} \rho\left(\varepsilon+\tfrac{1}{2}\mathbf{v}\cdot\mathbf{v}\right)dv &= \int_{\partial\boldsymbol{\chi}(\mathcal{P},t)} (\mathsf{T}\mathbf{v}-\mathbf{q})\cdot\mathbf{n}\,d\sigma \\ &\quad+\int_{\boldsymbol{\chi}(\mathcal{P},t)} \rho(\mathbf{b}\cdot\mathbf{v}+r)\,dv. \end{aligned}\tag{4.4.4}$$

The divergence theorem and the localization principle yield the local energy balance,

$$\rho\frac{D}{Dt}(\varepsilon+\underbrace{\tfrac{1}{2}\mathbf{v}\cdot\mathbf{v}}_{\text{(M)}}) = \underbrace{\operatorname{div}(\mathsf{T}\mathbf{v})}_{\text{(M)}} - \operatorname{div}\mathbf{q} + \underbrace{\rho\,\mathbf{b}\cdot\mathbf{v}}_{\text{(M)}} + \rho r. \tag{4.4.5}$$

We discuss the terms labeled (M) shortly.

In elementary formulations of the conservation of energy one often finds reference to a quantity called potential energy. No such term appears explicitly in Equation (4.4.5). For the concept of potential energy to make sense we must adopt an additional assumption about the nature of the specific body force $\mathbf{b}$. Suppose that this force is **conservative**, that is, there exists a scalar potential $\Phi(\mathbf{x}, t)$ such that $\mathbf{b} = -\operatorname{grad} \Phi$. In terms of the material derivative,

$$-\mathbf{v} \cdot \operatorname{grad} \Phi = -\rho \frac{D\Phi}{Dt} + \rho \frac{\partial \Phi}{\partial t},$$

and hence the local energy balance (4.4.5) takes the form

$$\rho \frac{D}{Dt}\left(\varepsilon + \Phi + \tfrac{1}{2}\mathbf{v} \cdot \mathbf{v}\right) = \operatorname{div}(\mathsf{T}\mathbf{v}) - \operatorname{div} \mathbf{q} + \rho r + \rho \frac{\partial \Phi}{\partial t}. \tag{4.4.6}$$

We call Φ the **potential energy**. For some body forces, it is reasonable to assume the existence of a time-independent potential Φ. For example, it is common to approximate the effect of gravity near Earth's surface using the model $\mathbf{b}(\mathbf{x}) = -g \operatorname{grad}(\mathbf{e}_3 \cdot \mathbf{x})$, where g is a constant. In this case, Equation (4.4.6) balances the rate of addition of work and heat against the rate of change of internal, kinetic, and potential energy.

4.4.1 Thermal Energy Balance

We now simplify Equation (4.4.5) by culling information already contained in the momentum balance equation. Taking the inner product of the velocity with each of the terms in the local momentum balance (4.2.7) yields the scalar equation

$$\rho \mathbf{v} \cdot \frac{D\mathbf{v}}{Dt} = \mathbf{v} \cdot \operatorname{div} \mathsf{T} + \rho \mathbf{v} \cdot \mathbf{b}. \tag{4.4.7}$$

By manipulating terms in this equation, it is possible to match them with terms labeled (M) in Equation (4.4.5).

Start with the term involving $D\mathbf{v}/Dt$. By the product rule,

$$\rho \mathbf{v} \cdot \frac{D\mathbf{v}}{Dt} = \rho \frac{D}{Dt}\left(\tfrac{1}{2}\mathbf{v} \cdot \mathbf{v}\right),$$

which is precisely the term corresponding to the rate of change of kinetic energy.

Next, consider the term involving the stress tensor. The product rule yields

$$\mathbf{v} \cdot \operatorname{div} \mathsf{T} = \operatorname{div}(\mathsf{T}\mathbf{v}) - \mathsf{T} : \operatorname{grad} \mathbf{v},$$

where the operation : is the double-dot product defined in Equation (1.2.12):

$$\mathsf{T} : \operatorname{grad} \mathbf{v} = \mathsf{T} : \mathsf{L} = \operatorname{tr}(\mathsf{T}^{\top}\mathsf{L}).$$

EXERCISE 108 *Show that*

$$\mathsf{T} : \mathsf{L} = \operatorname{tr}(\mathsf{TL}) = \operatorname{tr}(\mathsf{TD}).$$

Substituting these expressions into Equation (4.4.7) gives

$$\underbrace{\frac{1}{2}\rho\frac{D}{Dt}(\mathbf{v}\cdot\mathbf{v})}_{\text{(I)}} = \underbrace{\operatorname{div}(\mathsf{T}\mathbf{v})}_{\text{(II)}} - \underbrace{\operatorname{tr}(\mathsf{TD})}_{\text{(III)}} + \underbrace{\rho\mathbf{b}\cdot\mathbf{v}}_{\text{(IV)}}. \tag{4.4.8}$$

This is the **mechanical energy balance**. The terms, all of which arise from terms in the momentum balance, have the following physical interpretations:

(I) is the rate of change of kinetic energy;

(II) is the rate of work done by stress;

(III) is the rate of internal energy increase by compression and shearing;

(IV) is the rate of work done by body forces.

Subtracting Equation (4.4.8) from Equation (4.4.5) clears the terms containing redundant information from the momentum balance, leaving the **thermal energy balance**:

$$\underbrace{\rho\frac{D\varepsilon}{Dt}}_{\text{(V)}} = \underbrace{\operatorname{tr}(\mathsf{TD})}_{\text{(VI)}} - \underbrace{\operatorname{div}\mathbf{q}}_{\text{(VII)}} + \underbrace{\rho r}_{\text{(VIII)}}. \tag{4.4.9}$$

In this equation,

(V) is the rate of change of internal energy;

(VI) is the rate of internal energy increase by compression and shearing;

(VII) is the rate of heat flow;

(VIII) is the rate of external heat supply.

EXERCISE 109 *The* **stress power** *associated with a part* $\mathcal{P}$ *of a body is the integral*

$$\int_{\boldsymbol{\chi}(\mathcal{P},t)} \mathsf{T} : \mathsf{L}\, dv = \int_{\boldsymbol{\chi}(\mathcal{P},t)} \mathsf{T} : \mathsf{D}\, dv = \int_{\boldsymbol{\chi}(\mathcal{P},t)} \operatorname{tr}(\mathsf{TD})\, dv.$$

Using the definition (4.2.12) of the first Piola–Kirchhoff stress T_κ *and the identity (2.2.6), prove that*

$$\int_{\boldsymbol{\chi}(\mathcal{P},t)} \mathsf{T} : \mathsf{D}\, dv = \int_{\boldsymbol{\kappa}(\mathcal{P})} \mathsf{T}_\kappa : \frac{D\mathsf{F}}{Dt}\, dV.$$

EXERCISE 110 *Using the definition (4.3.5) of the second Piola–Kirchhoff stress* S_k *and the identity (3.3.2), prove that*

$$\int_{\chi(\mathcal{P},t)} \mathsf{T} : \mathsf{D}\, dv = \int_{\boldsymbol{\kappa}(\mathcal{P})} \mathsf{S}_\kappa : \frac{D\mathsf{E}_L}{Dt}\, dV,$$

where E_L *is the Lagrangian stress tensor defined in Section 3.1.*

4.4.2 Summary

The global energy balance asserts that energy is conserved in the following sense:

$$\frac{d}{dt}[K(\mathcal{P},t) + E(\mathcal{P},t)] = W(\mathcal{P},t) + Q(\mathcal{P},t).$$

Here E stands for the total internal energy of the part $\mathcal{P}$, K denotes the total kinetic energy, W represents the total rate of working done by body forces and contact forces, and Q is the total rating heating, attributable to heat flux across the surface of $\mathcal{P}$ and heat supply. Applying the divergence theorem and the localization principle to this global law yields a local form,

$$\rho\frac{D}{Dt}\left(\varepsilon + \tfrac{1}{2}\mathbf{v}\cdot\mathbf{v}\right) = \operatorname{div}(\mathsf{T}\mathbf{v}) - \operatorname{div}\mathbf{q} + \rho\,\mathbf{b}\cdot\mathbf{v} + \rho\, r.$$

Taking the inner product of $\mathbf{v}$ with terms in the momentum balance yields another equation whose terms also have the dimension of power:

$$\tfrac{1}{2}\rho\frac{D}{Dt}(\mathbf{v}\cdot\mathbf{v}) = \operatorname{div}(\mathsf{T}\mathbf{v}) - \operatorname{tr}(\mathsf{TD}) + \rho\,\mathbf{b}\cdot\mathbf{v}.$$

Subtracting this mechanical energy balance from the local form of the total energy balance yields the thermal energy balance,

$$\rho\frac{D\varepsilon}{Dt} = \operatorname{tr}(\mathsf{TD}) - \operatorname{div}\mathbf{q} + \rho\, r.$$

4.5 ENTROPY INEQUALITY

The entropy inequality is a formal statement of the second law of thermodynamics. Although many treatments simply posit it as an axiom, we begin with an attempt at motivation. It is a risky venture. There are at least as many points of view in print on this subject as there are authors writing about it. Truesdell [51, Lecture 1] offers an unusually lively discussion. Notwithstanding differences of opinion about the philosophic origins of the entropy inequality, it plays a central role in the analysis of constitutive relations, as illustrated in Chapters 6 and 7.

4.5.1 Motivation

Begin by rewriting the energy balance (4.4.3) for a part $\mathcal{P}$ of a body as follows:

$$Q(\mathcal{P}, t) = \frac{d}{dt}\left[E(\mathcal{P}, t) + K(\mathcal{P}, t)\right] - W(\mathcal{P}, t). \tag{4.5.1}$$

As in Section 4.4, Q represents the total rate of heating in $\mathcal{P}$, E represents the total internal energy of $\mathcal{P}$, K stands for the total kinetic energy, and W is the mechanical power. The case when both sides of Equation (4.5.1) are positive—that is, when $Q > 0$—corresponds heuristically to a positive rate of increase in "orderly" forms of energy, represented by the terms on the right side of Equation (4.5.1). The case $Q(\mathcal{P}, t) < 0$ corresponds to the loss of "orderly" forms of energy to heat flux out of $\mathcal{P}$.

Experience suggests that there is an upper bound on the rate at which a process can extract "orderly" energy from heat. We write this statement provisionally as follows:

$$Q(\mathcal{P}, t) = \int_{\boldsymbol{\chi}(\mathcal{P}, t)} (-\operatorname{div}\mathbf{q} + \rho r)\, dv \leqslant B, \tag{4.5.2}$$

where B stands for the assumed bound. In contrast, we do not postulate a lower bound on Q. Other than the energy balance itself, no *a priori* bound constrains the rate at which "orderly" energy can be dissipated into heat.

To be more specific about B, we associate with $\mathcal{P}$ a quantifiable capacity to transfer heat to other parts of the body. As a gauge of this capacity, we assign to the body an objective scalar function $\theta(\mathbf{x}, t)$, called the **temperature**, subject to the constraint that $\theta > 0$. For a given part $\mathcal{P}$ the upper bound B increases with this heat transfer capacity. That is, we expect B to be an increasing function of θ.

The reasoning takes a simple form in the case when θ is spatially uniform: we write the upper bound on $Q(\mathcal{P}, t)$ as

$$B = \theta(t)\frac{dH}{dt}(\mathcal{P}, t),$$

for some function $H(\mathcal{P}, t)$. Thus the inequality (4.5.2) becomes

$$\frac{dH}{dt}(\mathcal{P}, t) \geqslant \frac{Q(\mathcal{P}, t)}{\theta(t)}. \tag{4.5.3}$$

To accommodate spatial variations in temperature $\theta(\mathbf{x}, t)$, we extend the inequality (4.5.3) by adopting the following axiom:

AXIOM (SECOND LAW OF THERMODYNAMICS). *For any body there exists a function $H(\mathcal{P}, t)$, called the* **entropy**, *having the following properties.*

1. *There exists an objective scalar function $\eta(\mathbf{x}, t)$, called the* **specific entropy**, *giving*

$$H(\mathcal{P}, t) = \int_{\chi(\mathcal{P}, t)} \rho(\mathbf{x}, t)\, \eta(\mathbf{x}, t)\, dv,$$

for every $\mathcal{P} \subset \mathcal{B}$.

2. *dH/dt bounds the rate of conversion of heat into other forms of energy in a manner that depends upon the temperature $\theta(\mathbf{x}, t)$:*

$$\frac{dH}{dt}(\mathcal{P}, t) = \frac{d}{dt} \int_{\chi(\mathcal{P}, t)} \rho\, \eta\, dv \geqslant \int_{\chi(\mathcal{P}, t)} \left[-\mathrm{div} \left(\frac{\mathbf{q}}{\theta} \right) + \frac{\rho\, r}{\theta} \right] dv.$$

4.5.2 Clausius–Duhem Inequality

Localizing the second law of thermodynamics yields the **Clausius–Duhem inequality**,

$$\rho \frac{D\eta}{Dt} \geqslant -\mathrm{div} \left(\frac{\mathbf{q}}{\theta} \right) + \frac{\rho\, r}{\theta}. \tag{4.5.4}$$

EXERCISE 111 *Rewrite this inequality as*

$$\rho \frac{D\eta}{Dt} + \frac{1}{\theta} (\mathrm{div}\, \mathbf{q} - \rho\, r) - \frac{1}{\theta^2} \mathbf{q} \cdot \mathrm{grad}\, \theta \geqslant 0.$$

Then use the local energy balance (4.4.9) to obtain the inequality

$$\rho \left(\theta \frac{D\eta}{Dt} - \frac{D\varepsilon}{Dt} \right) + \mathrm{tr}\, (\mathsf{TD}) - \left(\frac{\mathbf{q}}{\theta} \right) \cdot \mathrm{grad}\, \theta \geqslant 0.$$

The grouping of material derivatives in the result of Exercise 111 motivates the following definition.

DEFINITION. *The* **specific Helmholtz free energy** *is*

$$\psi(\mathbf{x}, t) = \varepsilon(\mathbf{x}, t) - \eta(\mathbf{x}, t)\, \theta(\mathbf{x}, t).$$

EXERCISE 112 *Using the Clausius–Duhem inequality and the definition of $\psi(\mathbf{x}, t)$, derive the equivalent inequality*

$$-\rho \left(\frac{D\psi}{Dt} + \eta \frac{D\theta}{Dt} \right) + \mathrm{tr}\, (\mathsf{TD}) - \left(\frac{\mathbf{q}}{\theta} \right) \cdot \mathrm{grad}\, \theta \geqslant 0. \tag{4.5.5}$$

This form of the entropy inequality is especially useful in **isothermal** processes, for which $D\theta/Dt = 0$ and $\text{grad}\,\theta = \mathbf{0}$. For such processes, the Clausius–Duhem inequality reduces to the following simple form:

$$-\rho\frac{D\psi}{Dt} + \text{tr}\,(\mathsf{TD}) \geqslant 0. \tag{4.5.6}$$

EXERCISE 113 *From the inequality (4.5.5), derive the following form of the entropy inequality for referential coordinates:*

$$-\rho_\kappa\left(\frac{D\psi}{Dt} + \eta\frac{D\theta}{Dt}\right) + \mathsf{S}_\kappa : \frac{D\mathsf{E}_L}{Dt} - \frac{\rho_\kappa}{\rho\theta}\mathbf{q}\cdot\text{grad}\,\theta \geqslant 0,$$

where S_κ denotes the second Piola–Kirchhoff stress. See Exercise 110.

4.5.3 Summary

The entropy inequality arises from an upper bound on the rate of conversion of heat into more orderly forms of energy. In its local form, this version of the second law of thermodynamics is the Clausius–Duhem inequality,

$$\rho\frac{D\eta}{Dt} \geqslant -\text{div}\left(\frac{\mathbf{q}}{\theta}\right) + \frac{\rho\,r}{\theta},$$

where θ denotes the temperature. The inequality often appears in a slightly different form: after applying the thermal energy balance and defining the specific Helmholtz free energy

$$\psi(\mathbf{x},t) = \varepsilon(\mathbf{x},t) - \eta(\mathbf{x},t)\,\theta(\mathbf{x},t),$$

we have

$$-\rho\left(\frac{D\psi}{Dt} + \eta\frac{D\theta}{Dt}\right) + \text{tr}\,(\mathsf{TD}) - \left(\frac{\mathbf{q}}{\theta}\right)\cdot\text{grad}\,\theta \geqslant 0.$$

This form of the entropy inequality simplifies the analysis of isothermal processes, in which $D\theta/Dt = 0$ and $\text{grad}\,\theta = \mathbf{0}$.

4.6 JUMP CONDITIONS

Although the localization principle requires integrands to be continuous, in some cases local forms of the balance laws hold under relaxed hypotheses. In this section we derive local equations that hold in cases where quantities appearing in the global balance laws suffer jump discontinuities on reasonably well behaved sets. The reasoning yields **jump conditions** in lieu of balance equations. Subsequent sections do not depend on the material presented in this section.

For efficiency, we first cast the global balance laws into a generic form, then derive the jump condition for that form. The global mass, momentum, angular momentum, and energy balances all amount to special cases of the following template for a part $\mathcal{P}$ of a body:

$$\frac{d}{dt}\int_{\chi(\mathcal{P},t)} \rho\,\Psi\,dv = \int_{\partial\chi(\mathcal{P},t)} \mathbf{\Pi}\cdot\mathbf{n}\,d\sigma + \int_{\chi(\mathcal{P},t)} \rho\,\varpi\,dv. \tag{4.6.1}$$

Here, Ψ, $\mathbf{\Pi}$, and ϖ signify generic quantities having the following interpretations:

Ψ represents the amount per unit mass of the quantity being balanced;

$\mathbf{\Pi}$ represents influx of the balanced quantity through the boundary $\partial\chi(\mathcal{P},t)$;

ϖ represents the external supply of the balanced quantity.

The tensor order of $\mathbf{\Pi}$ is always one higher than that of Ψ and ϖ.

EXERCISE 114 *Assuming these generic quantities are sufficiently smooth, apply the Reynolds transport theorem, the divergence theorem, and the localization principle to Equation (4.6.1) to obtain the local differential balance law,*

$$\rho\frac{D\Psi}{Dt} + \Psi\left(\frac{D\rho}{Dt} + \rho\,\mathrm{div}\,\mathbf{v}\right) = \mathrm{div}\,\mathbf{\Pi} + \rho\,\varpi. \tag{4.6.2}$$

The local balance laws derived in Sections 4.1 through 4.4 are instances Equation (4.6.2). For example, to obtain the local mass balance we set $\Psi = 1$, $\mathbf{\Pi} = \mathbf{0}$, and $\varpi = 0$, getting

$$\frac{D\rho}{Dt} + \rho\,\mathrm{div}\,\mathbf{v} = 0.$$

For the local momentum balance, $\Psi = \mathbf{v}$, $\mathbf{\Pi} = \mathsf{T}$, and $\varpi = \mathbf{b}$, and hence

$$\rho\frac{D\mathbf{v}}{Dt} + \mathbf{v}\left(\frac{D\rho}{Dt} + \rho\,\mathrm{div}\,\mathbf{v}\right) = \mathrm{div}\,\mathsf{T} + \rho\,\mathbf{b}.$$

Since the expression in parentheses vanishes by the local mass balance, this equation reduces to Cauchy's first law.

EXERCISE 115 *Assign values to Ψ, $\mathbf{\Pi}$, and ϖ to obtain the local angular momentum balance.*

4.6.1 Singular Surfaces

Now consider Equation (4.6.1) in cases when a surface $\mathcal{S}(t)$ of discontinuity is present. For the analysis to work, it is necessary to restrict the nature of the surface, as Figure 4.11 illustrates. We assume that $\mathcal{S}(t)$, not necessarily a material set, obeys the following hypotheses.

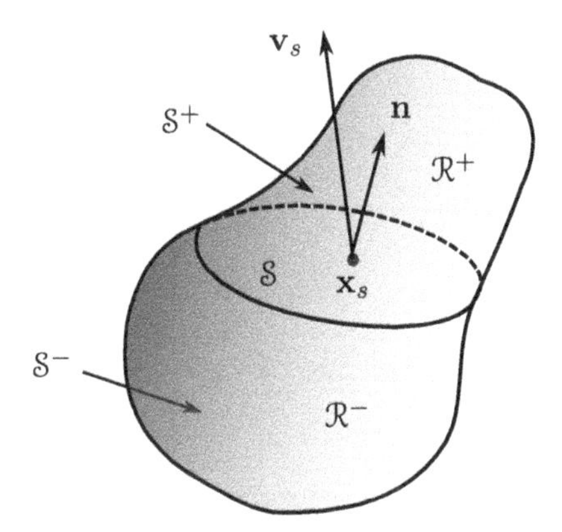

Figure 4.11. A singular surface $\mathcal{S}$ partitioning a material region $\mathcal{R}(t)$ into two subregions.

DEFINITION. *A* **singular surface** *is a smooth, orientable surface $\mathcal{S}$, not necessarily a material set, that satisfies the following five conditions:*

1. *$\mathcal{S}(t)$ has a continuous unit-length normal vector field $\mathbf{n}(\mathbf{x}_S, t)$ at every point $\mathbf{x}_S \in \mathcal{S}$.*

2. *The velocity $\mathbf{v}_S(\mathbf{x}_S, t)$ of the surface is continuously differentiable at every point $\mathbf{x}_S \in \mathcal{S}(t)$.*

3. *$\mathcal{S}(t)$ partitions a material region $\mathcal{R}(t)$ into two disjoint regions $\mathcal{R}^+(t)$ and $\mathcal{R}^-(t)$, so that*

$$\mathcal{R} = \mathcal{R}^+ \cup \mathcal{S} \cup \mathcal{R}^-.$$

We decompose the boundaries of $\mathcal{R}^+(t)$ and $\mathcal{R}^-(t)$ as

$$\partial\mathcal{R}^+ = \mathcal{S}^+ \cup \mathcal{S}, \qquad \partial\mathcal{R}^- = \mathcal{S}^- \cup \mathcal{S},$$

respectively, where $\partial\mathcal{R}(t) = \mathcal{S}^+(t) \cup \mathcal{S}^-(t)$, as drawn in Figure 4.11. All of these boundaries have piecewise continuously differentiable parametrizations.

4. *The functions Ψ, ρ, $\mathbf{\Pi}$, ϖ, and $\mathbf{v}$ are continuously differentiable in $\mathcal{R}^-$ and $\mathcal{R}^+$, and they have well-defined one-sided limits*

$$\Psi(\mathbf{x}_S+,t) = \lim_{\mathbf{x}\to\mathbf{x}_S+} \Psi(\mathbf{x},t) = \Psi^+(\mathbf{x}_S,t),$$
$$\Psi(\mathbf{x}_S-,t) = \lim_{\mathbf{x}\to\mathbf{x}_S-} \Psi(\mathbf{x},t) = \Psi^-(\mathbf{x}_S,t),$$

and so forth. Figure 4.12 illustrates a path along which one might compute such a limit.

5. *The differences in all of these one-sided limits are continuous functions of position $\mathbf{x}_S$ along the surface $\mathcal{S}(t)$.*

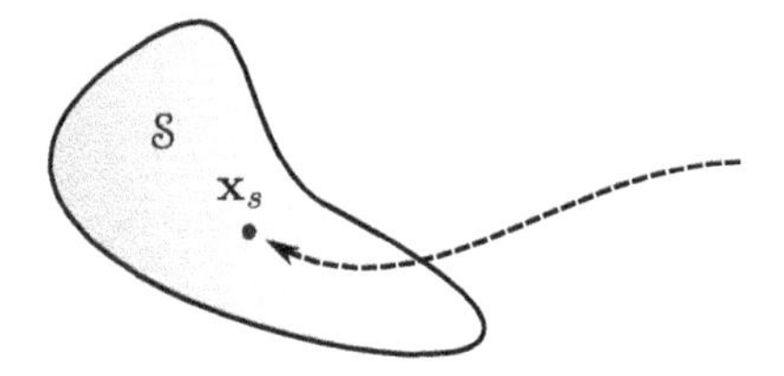

Figure 4.12. A path along which $\mathbf{x} \to \mathbf{x}_S+$.

DEFINITION. *The* **jump** *in a function $f(\mathbf{x},t)$ at $\mathbf{x}_S \in \mathcal{S}(t)$ is*

$$[\![f(\mathbf{x}_S,t)]\!] = f(\mathbf{x}_S+,t) - f(\mathbf{x}_S-,t).$$

This concept extends in a straightforward way to vector- and tensor-valued functions.

EXERCISE 116 *Assuming that f and g possesses at worst jump discontinuities, prove the following identities:*

1. $[\![fg]\!] = [\![f]\!]\, g^+ + f^-[\![g]\!] = [\![f]\!] g^- + f^+[\![g]\!]$.
2. *If f is continuous, then* $[\![fg]\!] = f\,[\![g]\!]$.
3. $[\![f]\!]\,[\![g]\!] = [\![fg]\!] - [\![f]\!]\, g^- - f^-[\![g]\!]$.

Since the velocity $\mathbf{v}_S$ of the singular surface is well defined, it has no jump, so

$$[\![\mathbf{v}_S \cdot \mathbf{n}]\!] = 0. \tag{4.6.3}$$

To derive the jump condition from the generic balance law requires a tool analogous to the Reynolds transport theorem. Recall from Equation (2.4.11) that if $\mathcal{R}(t)$ is a material region and f is continuously differentiable, then

$$\frac{d}{dt}\int_{\mathcal{R}(t)} f\, dv = \int_{\mathcal{R}(t)} \frac{\partial f}{\partial t}\, dv + \int_{\partial\mathcal{R}(t)} f\mathbf{v}\cdot\mathbf{n}\, d\sigma.$$

We wish to apply such a theorem to the regions $\mathcal{R}^+(t)$ and $\mathcal{R}^-(t)$ lying on either side of the singular surface $\mathcal{S}(t)$. Unfortunately, these regions are not generally material regions, since $\mathcal{S}(t)$ is not necessarily a material surface. We must extend the Reynolds transport theorem to accommodate this more general type of region.

THEOREM 4.6.1 (EXTENDED REYNOLDS TRANSPORT THEOREM). *If $f(\mathbf{x}, t)$ is continuously differentiable in the material region $\mathcal{R}(t)$ except for jump discontinuities on a singular surface $\mathcal{S}(t)$, then*

$$\frac{d}{dt}\int_{\mathcal{R}(t)} f\,dv = \int_{\mathcal{R}(t)} \frac{\partial f}{\partial t}\,dv + \int_{\partial\mathcal{R}(t)} f\mathbf{v}\cdot\mathbf{n}\,d\sigma - \int_{\mathcal{S}(t)} [\![f]\!]\,\mathbf{v}_S\cdot\mathbf{n}\,d\sigma. \quad (4.6.4)$$

PROOF: To circumvent the difficulty that $\mathcal{R}^+(t)$ and $\mathcal{R}^-(t)$ may not be material regions, define regions $\mathcal{V}^+(t)$ and $\mathcal{V}^-(t)$ such that, at the instant t,

1. The unit normal vector field for $\mathcal{S}$ points toward $\mathcal{V}^+$.
2. $\mathcal{V}^+$ occupies the same spatial points as $\mathcal{R}^+ \cup \mathcal{S}$, and $\mathcal{V}^-$ occupies the same spatial points as $\mathcal{R}^- \cup \mathcal{S}$. Thus $\partial\mathcal{V}^\pm = \mathcal{S}^\pm \cup \mathcal{S}$.
3. $\mathcal{V}^+$ has a velocity field $\mathbf{v}^+(\mathbf{x}, t)$ that satisfies the following conditions on $\partial\mathcal{V}^\pm$.

$$\mathbf{v}^+(\mathbf{x}, t) = \begin{cases} \mathbf{v}(\mathbf{x}, t), & \text{if} \quad \mathbf{x} \in \mathcal{S}^+, \\ (\mathbf{v}_S\cdot\mathbf{n})\mathbf{n}, & \text{if} \quad \mathbf{x} \in \mathcal{S}. \end{cases}$$

$\mathcal{V}^-$ has a velocity field $\mathbf{v}^-(\mathbf{x}, t)$ defined analogously.

Then $\mathcal{V}^+(t)$ and $\mathcal{V}^-(t)$ are material regions. Since $\mathcal{V}^+ \cup \mathcal{V}^- = \mathcal{R}$ at time t,

$$\frac{d}{dt}\int_{\mathcal{R}(t)} f\,dv = \frac{d}{dt}\int_{\mathcal{V}^+(t)} f\,dv + \frac{d}{dt}\int_{\mathcal{V}^-(t)} f\,dv.$$

Apply the Reynolds transport theorem to $\mathcal{V}^+$ and $\mathcal{V}^-$:

$$\frac{d}{dt}\int_{\mathcal{V}^+} f\,dv = \int_{\mathcal{V}^+} \frac{\partial f}{\partial t}\,dv + \int_{\mathcal{S}^+} f\mathbf{v}^+\cdot\mathbf{n}\,d\sigma + \int_{\mathcal{S}} f(\mathbf{x}_{S}+, t)\mathbf{v}^+\cdot(-\mathbf{n})\,d\sigma,$$

$$\frac{d}{dt}\int_{\mathcal{V}^-} f\,dv = \int_{\mathcal{V}^-} \frac{\partial f}{\partial t}\,dv + \int_{\mathcal{S}^-} f\mathbf{v}^-\cdot\mathbf{n}\,d\sigma + \int_{\mathcal{S}} f(\mathbf{x}_{S}-, t)\mathbf{v}^-\cdot\mathbf{n}\,d\sigma.$$

At time t we may substitute $\mathbf{v}^+ = \mathbf{v}^- = \mathbf{v}$ in the integrals over $\mathcal{S}^+$ and $\mathcal{S}^-$ and $\mathbf{v}^+ = \mathbf{v}^- = \mathbf{v}_S$ in the integrals over $\mathcal{S}$. Adding the resulting equations gives Equation (4.6.4). ■

The divergence theorem (see Appendix B) has an analogous extension.

THEOREM 4.6.2 (EXTENDED DIVERGENCE THEOREM). *For a vector field* $\mathbf{a}(\mathbf{x}, t)$ *subject to the hypotheses of the previous theorem,*

$$\int_{\mathcal{R}(t)} \operatorname{div} \mathbf{a} \, dv + \int_{\mathcal{S}(t)} [\![\mathbf{a}]\!] \cdot \mathbf{n} \, d\sigma = \int_{\partial \mathcal{R}(t)} \mathbf{a} \cdot \mathbf{n} \, d\sigma.$$

EXERCISE 117 *Prove Theorem 4.6.2 for the geometry in Figure 4.11.*

4.6.2 Localization

We use both the extended Reynolds transport theorem and the extended divergence theorem to localize the global balance equation in the vicinity of a singular surface $\mathcal{S}(t)$.

THEOREM 4.6.3 (GENERIC JUMP DISCONTINUITY). *At a singular surface* $\mathcal{S}(t)$, *the generic global balance equation (4.6.1) reduces to the following condition:*

$$[\![\rho \Psi (\mathbf{v}_S - \mathbf{v}) + \mathbf{\Pi}]\!] \cdot \mathbf{n} = 0. \tag{4.6.5}$$

Equation (4.6.5) is the generic **jump condition.**

PROOF: By the extended Reynolds transport theorem, the first integral in Equation (4.6.1) reduces as follows.

$$\begin{aligned} \frac{d}{dt} \int_{\chi(\mathcal{P}, t)} \rho \Psi \, dv = & \int_{\chi(\mathcal{P}, t)} \frac{\partial}{\partial t} (\rho \Psi) \, dv + \int_{\partial \chi(\mathcal{P}, t)} \rho \Psi \mathbf{v} \cdot \mathbf{n} \, d\sigma \\ & - \int_{\mathcal{S}(t)} [\![\rho \Psi]\!] \, \mathbf{v} \cdot \mathbf{n} \, d\sigma. \end{aligned}$$

Applying the extended divergence theorem to the integral over $\partial \chi(\mathcal{P}, t)$ yields

$$\begin{aligned} \frac{d}{dt} \int_{\chi(\mathcal{P}, t)} \rho \Psi \, dv = & \int_{\chi(\mathcal{P}, t)} \left[\frac{\partial}{\partial t} (\rho \Psi) + \operatorname{div} (\rho \Psi \mathbf{v}) \right] dv \\ & - \int_{\mathcal{S}(t)} [\![\rho \Psi (\mathbf{v}_S - \mathbf{v})]\!] \cdot \mathbf{n} \, d\sigma. \end{aligned}$$

The extended divergence theorem also applies to the flux integral in Equation (4.6.1):

$$\int_{\partial \chi(\mathcal{P}, t)} \mathbf{\Pi} \cdot \mathbf{n} \, d\sigma = \int_{\chi(\mathcal{P}, t)} \operatorname{div} \mathbf{\Pi} \, dv + \int_{\mathcal{S}(t)} [\![\mathbf{\Pi}]\!] \cdot \mathbf{n} \, d\sigma.$$

Combining these results gives

$$\int_{\chi(\mathcal{P},t)} \left[\rho \frac{D\Psi}{Dt} + \Psi \left(\frac{D\rho}{Dt} + \rho \operatorname{div} \mathbf{v} \right) - \operatorname{div} \mathbf{\Pi} - \rho \varpi \right] dv$$
$$- \int_{\mathcal{S}(t)} [\![\rho\Psi(\mathbf{v}_S - \mathbf{v}) + \mathbf{\Pi}]\!] \cdot \mathbf{n} \, d\sigma = 0. \qquad (4.6.6)$$

To analyze the integrals in this equation near the singular surface $\mathcal{S}(t)$, fix t, and take for the material region $\chi(\mathcal{P},t)$ a layer around $\mathcal{S}(t)$, as shown in Figure 4.13, such that all particles in the region lie within distance δ of $\mathcal{S}(t)$. As $\delta \to 0$, the volume $\chi(\mathcal{P},t)$ vanishes. Hence so does the integral in Equation (4.6.6). However, the surface integral remains unchanged, leaving

$$\int_{\mathcal{S}(t)} [\![\rho\Psi(\mathbf{v}_S - \mathbf{v}) + \mathbf{\Pi}]\!] \cdot \mathbf{n} \, d\sigma = 0.$$

Since the extent of $\mathcal{S}(t)$ is arbitrary and the integrand is continuous,

$$[\![\rho \, \Psi \, (\mathbf{v}_S - \mathbf{v}) + \mathbf{\Pi}]\!] \cdot \mathbf{n} = 0, \qquad (4.6.7)$$

by the reasoning used in the Dubois-Reymond localization principle. ■

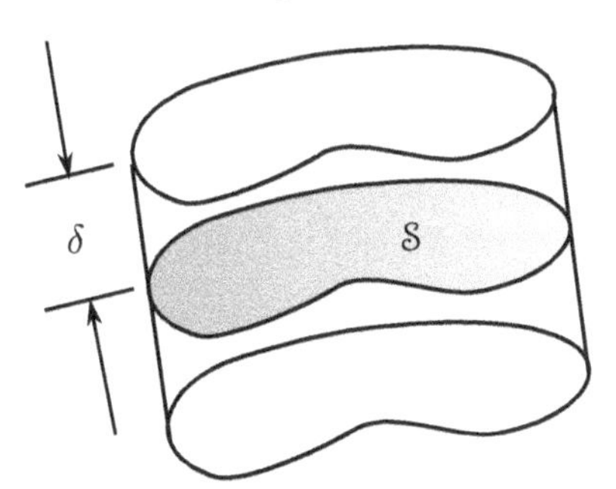

Figure 4.13. Layer-shaped material region of thickness 2δ around the singular surface $\mathcal{S}$.

Substituting concrete values for the generic quantities Ψ, $\mathbf{\Pi}$, and ϖ gives specific jump conditions corresponding to the balance laws. For example, taking $\Psi = 1$, $\mathbf{\Pi} = \mathbf{0}$, and $\varpi = 0$ yields

$$[\![\rho(\mathbf{v}_S - \mathbf{v}) \cdot \mathbf{n}]\!] = 0 \qquad \text{mass.} \qquad (4.6.8)$$

Taking $\Psi = \mathbf{v}$, $\mathbf{\Pi} = \mathsf{T}$, and $\varpi = \mathbf{b}$ gives

$$[\![\rho\mathbf{v}(\mathbf{v}_S - \mathbf{v}) \cdot \mathbf{n}]\!] + [\![\mathsf{T}\mathbf{n}]\!] = \mathbf{0} \qquad \text{momentum.} \qquad (4.6.9)$$

Finally, taking $\Psi = \varepsilon + \frac{1}{2}\mathbf{v} \cdot \mathbf{v}$, $\mathbf{\Pi} = \mathsf{T}\mathbf{v} - \mathbf{q}$, and $\varpi = \mathbf{b} \cdot \mathbf{v} + r$ produces

$$[\![\rho(\varepsilon + \tfrac{1}{2}\mathbf{v} \cdot \mathbf{v})(\mathbf{v}_S - \mathbf{v}) \cdot \mathbf{n}]\!] + [\![(\mathsf{T}\mathbf{v} - \mathbf{q}) \cdot \mathbf{n}]\!] = 0 \qquad \text{energy.}$$

Singular surfaces occur in many applications to fluid mechanics. We close this section with a glimpse of these physics.

DEFINITION. *A singular surface* $\mathcal{S}(t)$ *is a* **contact discontinuity** *if* $(\mathbf{v}_S - \mathbf{v})\cdot\mathbf{n} = 0$ *everywhere on* $\mathcal{S}(t)$. *It is a* **shock** *if* $[\![\mathbf{v}\cdot\mathbf{n}]\!] \neq 0$.

EXERCISE 118 *For isochoric motion,*

$$\frac{d}{dt}\int_{\boldsymbol{\chi}(\mathcal{P},t)} 1\, dv = 0,$$

for every part $\mathcal{P}$ *of the body. Show that in this case, if a singular surface is present, then* $[\![\mathbf{v}\cdot\mathbf{n}]\!] = 0$. *Thus in isochoric motion shocks do not exist.*

EXERCISE 119 *If* $[\![\mathbf{v}\cdot\mathbf{n}]\!] = 0$ *but* $[\![\mathbf{v}\times\mathbf{n}]\!] \neq 0$, *then the jump in* $\mathbf{v}$ *across* $\mathcal{S}(t)$ *is purely tangential. We call such a discontinuity a* **vortex sheet**. *Prove that a vortex sheet is a contact discontinuity.*

The following theorem plays a crucial role in shock dynamics.

THEOREM 4.6.4 (RANKINE–HUGONIOT CONDITION). *Denote* $V = (\mathbf{v}-\mathbf{v}_S)\cdot\mathbf{n}$ *and* $T = (\mathsf{T}\mathbf{n})\cdot\mathbf{n}$. *Across a shock, the following relationship holds:*

$$[\![T]\!] = -V^+V^-[\![\rho]\!]. \tag{4.6.10}$$

PROOF: The definition of V reduces the jump condition for mass to $[\![\rho V]\!] = 0$, that is,

$$\rho^+V^+ = \rho^-V^-. \tag{4.6.11}$$

Taking the inner product of the momentum jump condition (4.6.9) with $\mathbf{n}$ yields

$$[\![T]\!] = [\![\rho V v_n]\!],$$

where $v_n = -\mathbf{v}\cdot\mathbf{n}$. But Equation (4.6.3) implies that $[\![v_n]\!] = -[\![V]\!]$, and, by Exercise 120, this identity and the mass jump condition (4.6.8) imply that

$$[\![\rho V v_n]\!] = [\![\rho V^2]\!]. \tag{4.6.12}$$

It follows that $[\![T]\!] = [\![\rho V^2]\!]$, that is,

$$[\![T]\!] = \rho^+(V^+)^2 - \rho^-(V^-)^2.$$

Substituting the identities $V^+ = \rho^-V^-/\rho^+$ and $V^- = \rho^+V^+/\rho^-$, derivable from the mass jump condition (4.6.11), yields Equation (4.6.10). ■

EXERCISE 120 *Verify the identity (4.6.12).*

4.6.3 Summary

Global balance laws have the generic form

$$\frac{d}{dt}\int_{\chi(\mathcal{P},t)} \rho\,\Psi\,dv = \int_{\partial\chi(\mathcal{P},t)} \mathbf{\Pi}\cdot\mathbf{n}\,d\sigma + \int_{\chi(\mathcal{P},t)} \rho\,\varpi\,dv,$$

where the symbols have the following interpretations:

$\Psi =$ amount per unit mass of the balanced quantity

$\mathbf{\Pi} =$ influx through $\partial\chi(\mathcal{P},t)$

$\varpi =$ external supply of the balanced quantity.

The quantity $\mathbf{\Pi}$ is one tensor order higher than Ψ and ϖ. Table 4.1 summarizes the assignments of physically meaningful functions to the generic quantities Ψ, $\mathbf{\Pi}$, and ϖ.

Table 4.1. Assignments of Physical Values to Generic Quantities in Balance Laws

Balance Law	Ψ	$\mathbf{\Pi}$	ϖ
Mass	1	$\mathbf{0}$	0
Momentum	$\mathbf{v}$	T	$\mathbf{b}$
Angular Momentum	$\mathbf{x}\times\mathbf{v}$	$\mathbf{x}\times\mathsf{T}$	$\mathbf{x}\times\mathbf{b}$
Energy	$\varepsilon+\frac{1}{2}\mathbf{v}\cdot\mathbf{v}$	$\mathsf{T}\mathbf{v}-\mathbf{q}$	$r+\mathbf{v}\cdot\mathbf{b}$

The standard localization argument fails when any of the integrands is discontinuous. But if the discontinuities are jump discontinuities on a smooth, orientable surface with outward unit normal vector field $\mathbf{n}$, then the generic balance law reduces to a generic jump condition,

$$[\![\rho\,\Psi\,(\mathbf{v}_S-\mathbf{v})+\mathbf{\Pi}]\!]\cdot\mathbf{n}=0.$$

For example, mass, momentum, and energy obey the jump conditions

$$[\![\rho(\mathbf{v}_S-\mathbf{v})\cdot\mathbf{n}]\!]=0,$$

$$[\![\rho\mathbf{v}(\mathbf{v}_S-\mathbf{v})\cdot\mathbf{n}]\!]+[\![\mathsf{T}\mathbf{n}]\!]=\mathbf{0},$$

and

$$[\![\rho(\varepsilon+\tfrac{1}{2}\mathbf{v}\cdot\mathbf{v})(\mathbf{v}_S-\mathbf{v})\cdot\mathbf{n}]\!]+[\![(\mathsf{T}\mathbf{v}-\mathbf{q})\cdot\mathbf{n}]\!]=0,$$

respectively.

CHAPTER 5

CONSTITUTIVE RELATIONS: EXAMPLES OF MATHEMATICAL MODELS

The balance laws introduced in Chapter 4 furnish basic integral, differential, and jump equations used to model continua. Yet, since they hold for all continua, they provide no means for distinguishing among different materials. The mass balance for air flowing over a mountain range is the same as that for water in an estuary and rock in the earth's crust. To accommodate responses of different materials to forces and motion, we adopt **constitutive relations**, which give functional relationships among quantities that appear in the balance laws. Typically, constitutive relations treat stress T, specific internal energy ε, heat flux $\mathbf{q}$, and specific entropy η as functions of quantities derivable from the deformation $\boldsymbol{\chi}_\kappa$ or the temperature θ.

Among the most familiar constitutive relations are those that relate two quantities:

- Stress is proportional to strain.
- Heat flux is proportional to the temperature gradient.
- Internal energy is proportional to temperature.

Simple linear examples such as these hold only for certain materials and, typically, only within restricted ranges of conditions, with coefficients of proportionality that vary from one material to another. Nature exhibits much richer behavior. The behavior is not arbitrary, however. Constitutive relations must respect several physical principles.

This chapter examines several constitutive relations commonly appearing in applications, deducing some of the classic results of heat transfer and fluid and solid mechanics as consequences. Chapter 6 reviews the axiomatic framework of constitutive theory, probing the restrictions that more general constitutive relations must obey.

We distinguish constitutive relations from **internal constraints**. An internal constraint is a statement about the allowable geometries of motion. The following definitions give two important examples.

DEFINITION. *A material is* **rigid** *if it can sustain only rigid motions. It is* **incompressible** *if it can sustain only isochoric motions.*

Mathematically, a material is rigid if it satisfies the internal constraint $\mathsf{F} = \mathsf{I}$. It is incompressible if it obeys the internal constraint $\det \mathsf{F} = 1$ (see Exercise 48) or, equivalently, $\operatorname{div} \mathbf{v} = 0$ (see Exercise 94). In contrast, constitutive relations give functional relationships among variables that appear in the kinematics or in the balance laws themselves. The examples in this chapter—taken from heat transfer, fluid mechanics, and solid mechanics—help to clarify this distinction.

The vignettes in Sections 5.1, 5.3 and 5.4 are naïve, in the sense that they do not examine the formal consistency of the constitutive relations, nor do they refer to precise definitions of such terms as "fluid" and "solid." Chapter 6 explores these issues in more depth.

5.1 HEAT TRANSFER

Consider a rigid solid body in an inertial frame of reference, and let the temperature θ vary as a function of spatial position $\mathbf{x}$ and time t. In this setting we can adopt a frame of reference with respect to which the body is stationary. In this frame the mass, momentum, and angular momentum balance laws are uninformative. The local thermal energy balance for the body is Equation (4.4.9):

$$\rho \frac{D\varepsilon}{Dt} = \operatorname{tr}(\mathsf{TD}) - \operatorname{div} \mathbf{q} + \rho r.$$

EXERCISE 121 *For a body consisting of a rigid material, show that the thermal energy balance reduces to*

$$\rho \frac{D\varepsilon}{Dt} = -\operatorname{div} \mathbf{q} + \rho r. \tag{5.1.1}$$

Consider the following constitutive relations:

$$\rho = \rho(\theta), \quad \varepsilon = \varepsilon(\theta), \quad r = 0, \quad \mathbf{q} = -k(\theta) \operatorname{grad} \theta. \tag{5.1.2}$$

The last relationship in Equations (5.1.2) is **Fourier's law**; it states that heat flows down the temperature gradient from regions of high temperature to regions of low temperature. The coefficient $k(\theta)$ is the **thermal conductivity**, subject to the constraint that $k(\theta) \geqslant k_{\min} > 0$ for some constant $k_{\min}$. The functional dependencies postulated in Equations (5.1.2) relate the mass density ρ, the specific internal energy ε, and the heat flux $\mathbf{q}$, which we treated simply as functions of $(\mathbf{x}, t)$ in Chapter 4, to other functions of $(\mathbf{x}, t)$, namely, the temperature θ and its gradient. Thus the constitutive relations assert that the spatial and temporal dependencies of ρ, ε, and $\mathbf{q}$ are implicit, through these functions' explicit dependence on θ and $\operatorname{grad} \theta$.

This observation suggests a view of constitutive relations as equations needed to close the system. Equation (5.1.1) furnishes a single differential equation in the unknowns $\rho, \varepsilon, \mathbf{q}$, and r, which comprise six scalar functions of $(\mathbf{x}, t)$. We regard the functional dependencies in Equations (5.1.2) as being known or measurable, as opposed to being unknowns to be solved for in the partial differential equation (5.1.1).

Substituting the constitutive relations into the reduced thermal energy balance (5.1.1) yields

$$\rho(\theta)\varepsilon'(\theta)\frac{D\theta}{Dt} - \operatorname{div}\left[k(\theta) \operatorname{grad} \theta\right] = 0,$$

where we use the chain rule to obtain a partial differential equation in which temperature $\theta(\mathbf{x}, t)$ is the principal unknown. The quantity $\varepsilon'(\theta)$ is the **specific heat** of the material; it obeys the constraint $\varepsilon'(\theta) \geqslant \varepsilon'_{\min} > 0$.

An important special case arises when ρ, ε', and k are positive constants, independent of temperature. In this case, the thermal energy balance further simplifies:

$$\frac{\partial \theta}{\partial t} - K\,\Delta\theta = 0. \tag{5.1.3}$$

Here $K = k/(\rho\varepsilon')$ is a positive coefficient called the **thermal diffusivity**, and $\Delta = \operatorname{div}(\operatorname{grad})$ stands for the Laplace operator. Equation (5.1.3) is the **heat equation**, a classic partial differential equation of applied mathematics.

5.1.1 Properties of the Heat Equation

One can gain insight into the physics of heat conduction by examining how the one-dimensional analog of Equation (5.1.3),

$$\frac{\partial \theta}{\partial t} - K\frac{\partial^2 \theta}{\partial x^2} = 0, \tag{5.1.4}$$

propagates spatial and temporal variations in temperature. Consider a temperature distribution $\theta(x,t)$ having Fourier expansion

$$\theta(x,t) = \sum_{\beta=-\infty}^{\infty} \upsilon_\beta(t)\exp(i\beta x),$$

with time-dependent coefficients $\upsilon_\beta(t)$. (Here $i^2 = -1$.) Since Equation (5.1.4) is linear, it suffices to examine how it propagates a typical mode $\upsilon_\beta(t)\exp(i\beta x)$, with mode number β, in the Fourier expansion. Substituting the mode into Equation (5.1.4) reveals how $\upsilon_\beta(t)$ varies in time.

EXERCISE 122 *Show that*

$$\upsilon_\beta'(t) + K\beta^2\upsilon_\beta(t) = 0.$$

This ordinary differential equation has solution $\upsilon_\beta(t) = \upsilon_\beta(0)\exp(-K\beta^2 t)$. Since $\beta^2 \geqslant 0$ and $K > 0$, the heat equation damps all Fourier modes as $t \to \infty$, except that associated with the Fourier mode number $\beta = 0$. This mode $\upsilon_0(t) = \upsilon_0(0)$ corresponds to a uniform, constant, steady-state temperature distribution.

More generally, the heat equation tends to damp any deviation from steady-state solutions, which satisfy $\partial\theta/\partial t = 0$. For this reason we associate the positivity of the thermal diffusivity K with the physical stability of heat conduction, in a sense to be made more precise shortly. An interesting consequence of this stability is that heat conduction establishes a preferred direction for time, a concept that the next three exercises explore.

EXERCISE 123 *Consider the problem*

$$\begin{aligned} \frac{\partial \theta}{\partial t} - K\Delta u &= 0, && \mathbf{x} \in \mathbb{E}, \quad t < t_f, \\ \theta(\mathbf{x}, t_f) &= 1, && \mathbf{x} \in \mathbb{E}, \end{aligned}$$

where $K > 0$ is constant. The goal here is to find $\theta(\mathbf{x},t)$ for prior times $t < t_f$. Show that the change of variables $\tau = t_f - t$ converts this **final-value problem** *into* *an initial-value problem for the* **backward heat equation**,

$$\frac{\partial u}{\partial \tau} + K\Delta u = 0, \qquad \mathbf{x} \in \mathbb{E}, \quad \tau > 0, \tag{5.1.5}$$
$$u(\mathbf{x}, 0) = 1, \qquad \mathbf{x} \in \mathbb{E},$$

where $u(\mathbf{x}, \tau) = \theta(\mathbf{x}, t)$.

EXERCISE 124 *Clearly* $u(\mathbf{x}, \tau) = 1$ *is a solution to the initial-value problem (5.1.5). Show that the function*

$$u_n(\mathbf{x}, t) = \frac{1}{n} e^{3Kn^2\tau} \sin(x_1 + x_2 + x_3)$$

is a solution to the perturbed problem

$$\frac{\partial u_n}{\partial \tau} + K\Delta u_n = 0, \qquad \mathbf{x} \in \mathbb{E}, \quad \tau > 0,$$
$$u_n(\mathbf{x}, 0) = 1 + \frac{1}{n} \sin[n(x_1 + x_2 + x_3)], \qquad \mathbf{x} \in \mathbb{E},$$

for any positive integer n.

In Exercise 124, a perturbation to the initial condition that is arbitrarily small in magnitude yields a solution whose magnitude grows without bound as τ increases (as we look further back into history), in sharp contrast to the solution to the unperturbed problem.

For problems that look forward in time, we expect natural phenomena to be far more stable than this against small physical disturbances and measurement errors. This expectation motivates the following definition:

DEFINITION. *An initial-value problem is* **well posed** *if it has three properties:*

1. **Existence:** *a solution exists.*
2. **Uniqueness:** *the solution is unique, and*
3. **Stability:** *the solution depends continuously on the initial data.*

EXERCISE 125 *In the context of initial-value problems, continuous dependence on the initial data means that for every number* $\epsilon > 0$ *and for every value of* $\tau > 0$, *there exists a number* $\delta > 0$ *such that whenever* $|u(\mathbf{x}, 0) - u_n(\mathbf{x}, 0)| < \delta$ *for all values of* $\mathbf{x}$, $|u(\mathbf{x}, \tau) - u_n(\mathbf{x}, \tau)| < \epsilon$ *for all values of* $\mathbf{x}$. *Here* u_n *denotes the solution to an initial-value problem with a perturbed initial condition. Show that the initial-value problem (5.1.5) does not satisfy this condition.*

The result of Exercise 125 has two interpretations. First, it is generally impossible to determine earlier temperature distributions given the temperature distribution at a later time. Second, for heat-flow models based on Equation (5.1.3) to be physically meaningful, K cannot be negative. Section 6.5 explores a different avenue leading to this second conclusion.

EXERCISE 126 *Consider steady-state, constant-coefficient Fourier heat conduction when a bounded, integrable heat supply $r(\mathbf{x})$ is present. Show that the equation governing the temperature is the* **Poisson equation**,

$$-\Delta\theta = f(\mathbf{x}), \qquad f(\mathbf{x}) = \frac{\rho r(\mathbf{x})}{K}. \tag{5.1.6}$$

When $r(\mathbf{x}) = 0$, Equation (5.1.6) reduces to the **Laplace equation**, $\Delta\theta = 0$, which governs unforced, steady-state heat flow, among many other physical phenomena. The Poisson and Laplace equations figure so prominently in applied mathematics that we devote the next section to their properties.

5.1.2 Summary

Constitutive relations supplement the balance laws discussed in Chapter 4. One way to view constitutive relations is as a set of functional relationships that distinguish among the various responses exhibited by different materials. An example is Fourier's law. This constitutive relation gives heat flux as a function of temperature and temperature gradient: $\mathbf{q} = -k(\theta)\,\mathrm{grad}\,\theta$, with $k(\theta) \geqslant k_{\min} > 0$. Combining this expression with constitutive relations for the mass density $\rho = \rho(\theta)$ and specific internal energy $\varepsilon = \varepsilon(\theta)$ and assuming a rigid material in which external heat supplies are absent yield the following version of the energy balance:

$$\frac{\partial\theta}{\partial t} - K\,\Delta\theta = 0,$$

where $K = k(\theta)/[\rho\varepsilon'(\theta)]$. When K is a positive constant, this equation is the heat equation, one of the classic partial differential equations of applied mathematics. Simple Fourier analysis shows that the heat equation tends to damp initial departures from steady-state solutions, establishing a preferred direction for time.

5.2 POTENTIAL THEORY

5.2.1 Motivation

As shown in Exercise 79, when a body's motion is isochoric and irrotational, the velocity's scalar potential Φ satisfies the Laplace equation,

$$-\Delta\Phi = -\operatorname{div}(\operatorname{grad}\Phi) = 0.$$

In heat transfer problems, as shown in Exercise 126, the simplest model of steady-state heat flow yields the Poisson equation,

$$-\Delta\theta = f,$$

where we typically regard the forcing function f as known and θ as unknown. From a mathematical perspective, the first equation is a special case of the second. These two equations arise frequently in applied mathematics and have been subjects of intensive study for over two centuries. The resulting field of **potential theory** is by now a rich mathematical province, with results that bear significantly on the analysis of fluids, heat, gravitation, electrostatics, and other physical phenomena. For this reason, it is useful to consider properties of the equations that are independent of the particular physics.

This optional section, containing material not required in subsequent sections (except for a brief mention in Section 7.3), briefly reviews a few qualitative results of this type. Specifically, we examine the most basic facts about

- boundary-value problems commonly found in applications,
- conditions that guarantee unique solutions,
- standard qualitative results on the behavior of the solutions, such as continuous dependence of the solution on boundary data and the mean value theorem for harmonic functions.

Since the Laplace equation amounts to a special case of the Poisson equation, we develop many of the results for the latter, which we write as

$$-\Delta u = f. \tag{5.2.1}$$

For deeper analysis, including significant theorems on the existence of solutions, refer to [30].

Throughout this section, we seek classical solutions to the Laplace and Poisson equations having the form $u(\mathbf{x})$, defined on a **normal region** $\mathcal{R} \subset \mathbb{E}$. Any such region is nonempty, open, and bounded, and it has an orientable boundary $\partial\mathcal{R}$ with outward unit normal vector field $\mathbf{n}(\mathbf{x})$. We use the shorthand $\overline{\mathcal{R}}$ to denote the set $\mathcal{R} \cup \partial\mathcal{R}$. For further details about the nature of $\mathcal{R}$ and the integral theorems used in this section, see Appendix B. To qualify as a classical solution to Equation (5.2.1), $u(\mathbf{x})$ must be twice continuously differentiable in $\mathcal{R}$ and continuously differentiable on $\overline{\mathcal{R}}$, and it must satisfy the differential equation.

EXERCISE 127 *Show that* $u(\mathbf{x}) = 1/\|\mathbf{x}\|$ *satisfies the Laplace equation at every* $\mathbf{x}$ *outside a neighborhood of* $\mathbf{0}$.

EXERCISE 128 *Find an infinite family of solutions to the Laplace equation, each valid on all of* $\mathbb{E}$.

5.2.2 Boundary Conditions

As Exercise 128 shows, the Laplace equation has many solutions. The same is true for the Poisson equation. To specify a particular solution requires additional information about its behavior on the boundary $\partial\mathcal{R}$. While many types of boundary conditions are possible, three occur most frequently: Dirichlet conditions, Neumann conditions, and Robin conditions.

Dirichlet conditions prescribe the value of $u(\mathbf{x})$ on $\partial\mathcal{R}$:

$$\begin{aligned} -\Delta u(\mathbf{x}) &= f(\mathbf{x}), & \mathbf{x} &\in \mathcal{R} \\ u(\mathbf{x}) &= g(\mathbf{x}), & \mathbf{x} &\in \partial\mathcal{R}, \end{aligned} \tag{5.2.2}$$

where f and g are known functions. This type of boundary-value problem is arguably the simplest of the three.

Neumann conditions prescribe the total flux of u across the boundary $\partial\mathcal{R}$. For example, in the context of heat transfer, where u represents temperature, the outward component of the heat flux across the boundary $\partial\mathcal{R}$ is $\mathbf{q} \cdot \mathbf{n}$, where $\mathbf{q}$ stands for the heat flux vector. Following the derivation in Section 5.1, the constant-coefficient version of Fourier's law gives $\mathbf{q} = -K \operatorname{grad} u$, where the thermal diffusivity K is a positive constant. Therefore, the Neumann problem becomes

$$\begin{aligned} -\Delta u(\mathbf{x}) &= f(\mathbf{x}), & \mathbf{x} &\in \mathcal{R} \\ -K \operatorname{grad} u(\mathbf{x}) \cdot \mathbf{n}(\mathbf{x}) &= g(\mathbf{x}), & \mathbf{x} &\in \partial\mathcal{R}, \end{aligned} \tag{5.2.3}$$

where again f and g are known functions.

Again in the context of heat transfer, **Robin conditions** prescribe the heat radiation rate across $\partial\mathcal{R}$. If the exterior of $\overline{\mathcal{R}}$ stays at a constant temperature u_e, then **Newton's law of cooling**—justified largely by observation rather than theory—postulates that the heat flux across the surface $\partial\mathcal{R}$ is proportional to the temperature difference between $\partial\mathcal{R}$ and the exterior:

$$-K \operatorname{grad} u(\mathbf{x}) \cdot \mathbf{n} = -k(\mathbf{x})\,[u_e - u(\mathbf{x})],$$

where k denotes a positive function of position. Rearranging this equation and dividing through by the thermal diffusivity yield the following boundary-value problem:

$$\begin{aligned} -\Delta u(\mathbf{x}) &= f(\mathbf{x}), && \mathbf{x} \in \mathcal{R} \\ h(\mathbf{x})\,u(\mathbf{x}) + \operatorname{grad} u(\mathbf{x}) \cdot \mathbf{n}(\mathbf{x}) &= g(\mathbf{x}), && \mathbf{x} \in \partial\mathcal{R}, \end{aligned} \tag{5.2.4}$$

where $h(\mathbf{x}) = k(\mathbf{x})/K > 0$ for all $\mathbf{x} \in \partial\mathcal{R}$.

In some geometries, with certain boundary conditions, problems involving the three-dimensional Poisson equation reduce to problems in two space dimensions. As an example, consider a region of the form $\mathcal{R} = \mathcal{D} \times [0, b]$, where $\mathcal{D}$ is a nonempty, connected, open set in the (x_1, x_2)-plane and b is the constant distance between the parallel bounding planes $x_3 = 0$ and $x_3 = b$, as illustrated in Figure 5.1. In this setting we define the **vertical average** of an integrable function $\Psi(\mathbf{x})$ by

$$\tilde{\Psi}(x_1, x_2) = \frac{1}{b} \int_0^b \Psi(\mathbf{x})\, dx_3.$$

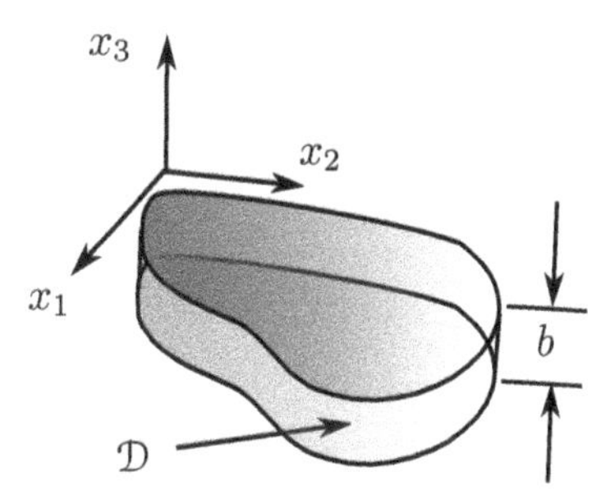

Figure 5.1. A region of the form $\mathcal{D} \times [0, b]$ in three-space with Cartesian coordinates.

EXERCISE 129 *Suppose that* $-\Delta u = f$ *in a region* $\mathcal{D} \times [0, b]$ *of this form and that the Neumann condition* $\partial u/\partial x_3 = 0$ *holds on the planar surfaces* $x_3 = 0$ *and* $x_3 = b$. *Show that* $\tilde{u}$ *obeys the two-dimensional Poisson equation,*

$$-\Delta_2 \tilde{u}(x_1, x_2) = \tilde{f}, \qquad (x_1, x_2) \in \mathcal{D},$$

where

$$\Delta_2 = \frac{\partial^2}{\partial x_1^2} + \frac{\partial^2}{\partial x_2^2}.$$

EXERCISE 130 *Show that*

$$\tilde{u}(x_1, x_2) = \log\left(\frac{1}{\sqrt{x_1^2 + x_2^2}}\right)$$

is a solution to the two-dimensional Laplace equation $-\Delta_2 \tilde{u} = 0$ *in any region* $\mathcal{D}$ *of the* (x_1, x_2)*-plane that excludes a neighborhood of the origin.*

5.2.3 Uniqueness of Solutions to the Poisson Equation

A well-posed boundary-value problem must possess a unique solution. The following theorem establishes uniqueness of solutions for the Robin problem (5.2.4). A related but simpler argument establishes uniqueness for the Dirichlet problem (5.2.2). As Exercise 132 shows, no such uniqueness guarantee is available for the Neumann problem (5.2.3).

THEOREM 5.2.1 (UNIQUENESS IN THE ROBIN PROBLEM). *There exists at most one classical solution to the Robin problem (5.2.4).*

PROOF: If u_1 and u_2 are both classical solutions to problem (5.2.4), then by linearity the function $w = u_1 - u_2$ solves the following boundary-value problem for the Laplace equation:

$$\begin{aligned} \Delta w = 0 &\quad \text{on} \quad \mathcal{R} \\ hw + \operatorname{grad} w \cdot \mathbf{n} = 0 &\quad \text{on} \quad \partial\mathcal{R}, \end{aligned}$$

where $h > 0$. We show that $w = 0$ everywhere on $\overline{\mathcal{R}}$. Applying Green's identity (B.2.1) with $f = g = w$ yields

$$\int_{\mathcal{R}} (w\,\Delta w + \|\operatorname{grad} w\|^2)\,dv = \int_{\partial\mathcal{R}} w \operatorname{grad} w \cdot \mathbf{n}\,d\sigma = -\int_{\partial\mathcal{R}} h\,w^2\,d\sigma,$$

the last step following from the Robin boundary condition for w. The first term in the integral on the left vanishes by the Laplace equation, and the second term is nonnegative. However, the right side is nonpositive. It follows that the integrand on the left vanishes, that is, $\operatorname{grad} w = \mathbf{0}$ on $\mathcal{R}$. Hence, w is constant on $\mathcal{R}$ and, by continuity, on $\overline{\mathcal{R}}$. This constant must be 0, since $0 = hw + \operatorname{grad} w \cdot \mathbf{n} = hw$ on $\partial\mathcal{R}$. Therefore, $w = 0$ everywhere on $\overline{\mathcal{R}}$. ■

EXERCISE 131 *Prove an analogous uniqueness result for the Dirichlet problem (5.2.2).*

EXERCISE 132 *Show that the Neumann problem (5.2.3) does not enjoy an analogous uniqueness result: given one solution, construct another.*

EXERCISE 133 *Why is the following identity a necessary condition for the existence of a solution to the Neumann problem?*

$$\int_{\mathcal{R}} f \, dv = \int_{\partial\mathcal{R}} g \, d\sigma.$$

5.2.4 Maximum Principle

The Laplace equation obeys a maximum principle that constrains the magnitudes of solution values in the interior $\mathcal{R}$ in terms of the values on the boundary $\partial\mathcal{R}$. From a physical point of view, this principle is an important qualitative property in its own right. Mathematically, it leads to a guarantee that certain boundary-value problems are well posed.

The maximum principle rests on two facts from advanced calculus. First, any continuous function $u\colon \overline{\mathcal{R}} \to \mathbb{R}$ attains its minimum and maximum values, respectivel, at some points $\mathbf{x}_m, \mathbf{x}_M \in \overline{\mathcal{R}}$. Second, if u is twice continuously differentiable in $\mathcal{R}$ and the points $\mathbf{x}_m$ and $\mathbf{x}_M$ lie in $\mathcal{R}$, then by the second-derivative test $\Delta u(\mathbf{x}_m) \geqslant 0$ and $\Delta u(\mathbf{x}_M) \leqslant 0$.

DEFINITION. *A function $u\colon \overline{\mathcal{R}} \to \mathbb{R}$ that is twice continuously differentiable on $\mathcal{R}$ and continuously differentiable on $\overline{\mathcal{R}}$ is*

- **subharmonic** *if $-\Delta u(\mathbf{x}) \leqslant 0$ for every $\mathbf{x} \in \mathcal{R}$,*
- **harmonic** *if $-\Delta u(\mathbf{x}) = 0$ for every $\mathbf{x} \in \mathcal{R}$,*
- **superharmonic** *if $-\Delta u(\mathbf{x}) \geqslant 0$ for every $\mathbf{x} \in \mathcal{R}$.*

Thus a harmonic function is both subharmonic and superharmonic.

For the remainder of this section we assume that the function u satisfies the continuity conditions required in this definition. In this case the maximum principle follows from three lemmas.

LEMMA 5.2.2 (FIRST MAXIMUM PRINCIPLE LEMMA).

1. *If* $-\Delta u(\mathbf{x}) < 0$ *for every* $\mathbf{x} \in \mathcal{R}$ *and* $u(\mathbf{x}) \leqslant 0$ *for every* $\mathbf{x} \in \partial\mathcal{R}$*, then* $u(\mathbf{x}) \leqslant 0$ *for every* $\mathbf{x} \in \mathcal{R}$*.*

2. *If* $-\Delta u(\mathbf{x}) > 0$ *for every* $\mathbf{x} \in \mathcal{R}$ *and* $u(\mathbf{x}) \geqslant 0$ *for every* $\mathbf{x} \in \partial\mathcal{R}$*, then* $u(\mathbf{x}) \geqslant 0$ *for every* $\mathbf{x} \in \mathcal{R}$*.*

EXERCISE 134 *In Lemma 5.2.2, show that assertion 2 follows from assertion 1.*

EXERCISE 135 *Prove Lemma 5.2.2. (Assume* $u(\mathbf{x}) > 0$ *for some* $\mathbf{x}$*, and observe, as in the proof of Lemma 5.2.3, that* u *must attain its maximum value at some point* $\mathbf{x}_M \in \mathcal{R}$*. Examine* $-\Delta u(\mathbf{x}_M)$ *to reach a contradiction.)*

The next lemma extends Lemma 5.2.2 to subharmonic and superharmonic functions by relaxing the strict inequalities on $-\Delta u$. The key tactic is the construction of a new function, related to u, that satisfies the hypotheses of Lemma 5.2.2.

LEMMA 5.2.3 (SECOND MAXIMUM PRINCIPLE LEMMA).

1. *If* $-\Delta u(\mathbf{x}) \leqslant 0$ *for every* $\mathbf{x} \in \mathcal{R}$ *and* $u(\mathbf{x}) \leqslant 0$ *for every* $\mathbf{x} \in \partial\mathcal{R}$*, then* $u(\mathbf{x}) \leqslant 0$ *for every* $\mathbf{x} \in \mathcal{R}$*.*

2. *If* $-\Delta u(\mathbf{x}) \geqslant 0$ *for every* $\mathbf{x} \in \mathcal{R}$ *and* $u(\mathbf{x}) \geqslant 0$ *for every* $\mathbf{x} \in \partial\mathcal{R}$*, then* $u(\mathbf{x}) \geqslant 0$ *for every* $\mathbf{x} \in \mathcal{R}$*.*

PROOF: As in Lemma 5.2.2, the second assertion follows from the first, which we prove by contradiction. Assume that $u(\mathbf{z}) > 0$ for some $\mathbf{z} \in \mathcal{R}$. Since u is continuous on $\overline{\mathcal{R}}$, it attains its maximum value at some point $\mathbf{x}_M \in \overline{\mathcal{R}}$, and $u(\mathbf{x}_M) \geqslant u(\mathbf{z}) > 0$. But $u \leqslant 0$ on $\partial\mathcal{R}$, so $\mathbf{x}_M \in \mathcal{R}$. Consider the function u_ϵ defined by

$$u_\epsilon(\mathbf{x}) = u(\mathbf{x}) + \epsilon\|\mathbf{x} - \mathbf{x}_M\|^2,$$

where ϵ is an arbitrary positive number. Clearly $u_\epsilon(\mathbf{x}_M) = u(\mathbf{x}_M)$. Also, since $-\Delta u \leqslant 0$ on $\mathcal{R}$,

$$-\Delta u_\epsilon(\mathbf{x}) = \Delta u(\mathbf{x}) - 2\epsilon < 0, \tag{5.2.5}$$

for every $\mathbf{x} \in \mathcal{R}$. The fact that $\overline{\mathcal{R}}$ is bounded implies that there is a distance D such that for every $\mathbf{x} \in \partial\mathcal{R}$, $\|\mathbf{x} - \mathbf{x}_M\| \leqslant D$. It follows that for every $\mathbf{x} \in \partial\mathcal{R}$,

$$u_\epsilon(\mathbf{x}) \leqslant u(\mathbf{x}) + \epsilon D^2 \leqslant \epsilon D^2,$$

since $u \leqslant 0$ on $\partial\mathcal{R}$. Pick $\epsilon < u(\mathbf{x}_M)/D^2$. Then

$$u_\epsilon(\mathbf{x}) < u(\mathbf{x}_M) = u_\epsilon(\mathbf{x}_M)$$

for every $\mathbf{x} \in \partial\mathcal{R}$. Therefore, u_ϵ does not attain its maximum value on $\partial\mathcal{R}$; instead, it attains it at some point $\mathbf{y} \in \mathcal{R}$. At this point the second derivative test guarantees $-\Delta u_\epsilon(\mathbf{y}) \geqslant 0$, contradicting the inequality (5.2.5). ■

LEMMA 5.2.4 (THIRD MAXIMUM PRINCIPLE LEMMA). *Denote by M and m, respectively, the maximum and minimum values of u on $\partial\mathcal{R}$.*

1. *If $-\Delta u(\mathbf{x} \leqslant 0$ for every $\mathbf{x} \in \mathcal{R}$, then $u(\mathbf{x}) \leqslant M$ for every $\mathbf{x} \in \mathcal{R}$.*
2. *If $-\Delta u(\mathbf{x}) \geqslant 0$ for every $\mathbf{x} \in \mathcal{R}$, then $u(\mathbf{x}) \geqslant m$ for every $\mathbf{x} \in \mathcal{R}$.*

In other words, a subharmonic function attains its maximum value on the boundary $\partial\mathcal{R}$, and a superharmonic function attains its minimum value on $\partial\mathcal{R}$.

EXERCISE 136 *Prove Lemma 5.2.4 by examining the functions defined by $u(\mathbf{x}) - M$ and $u(\mathbf{x}) - m$.*

With Lemma 5.2.4 in hand, it is a straightforward exercise to establish the maximum principle.

THEOREM 5.2.5 (MAXIMUM PRINCIPLE FOR HARMONIC FUNCTIONS). *If $-\Delta u = 0$ in $\mathcal{R}$, then u attains its maximum and minimum values on $\partial\mathcal{R}$.*

EXERCISE 137 *Prove Theorem 5.2.5.*

The maximum principle has many uses, one of which is to guarantee that certain boundary-value problems involving the Poisson equation are well posed. Consider, for example, the Dirichlet problem (5.2.2). Having shown that any solution is unique, we can now establish continuous dependence of the solution on the boundary data:

COROLLARY 5.2.6 (STABILITY OF THE DIRICHLET PROBLEM). *Let u be the solution to the Dirichlet problem (5.2.2), and suppose that u_p is the solution to a perturbed problem*

$$\begin{aligned} -\Delta u_p(\mathbf{x}) &= f(\mathbf{x}), & \mathbf{x} &\in \mathcal{R} \\ u_p(\mathbf{x}) &= g_p(\mathbf{x}), & \mathbf{x} &\in \partial\mathcal{R}, \end{aligned}$$

where g_p is the perturbed boundary condition. If $|g_p(\mathbf{x}) - g(\mathbf{x})| < \epsilon$ for every $\mathbf{x} \in \partial\mathcal{R}$, then $|u_p(\mathbf{x}) - u(\mathbf{x})| < \epsilon$ for every $\mathbf{x} \in \mathcal{R}$.

In other words, small differences in the boundary data yield small differences in the solution.

PROOF: By linearity, the function $w = u_p - u$ is harmonic, so it attains its maximum and minimum values on $\partial\mathcal{R}$. But on the boundary $|w(\mathbf{x})| = |g_p(\mathbf{x}) - g(\mathbf{x})| < \epsilon$. ∎

5.2.5 Mean Value Property

We close this section with a classic result that characterizes harmonic functions in terms of average values. In analyzing this property, we fix a point $\mathbf{x}_0$ and use the following notation for the open ball of radius R centered at $\mathbf{x}_0$:

$$\mathcal{D}_R = \Big\{\mathbf{x} \in \mathbb{E} \;\Big|\; \|\mathbf{x} - \mathbf{x}_0\| < R\Big\}.$$

EXERCISE 138 *When $\mathbf{x}_0 = \mathbf{0}$, the surface of the ball $\mathcal{D}_R$ is the level set of the function $\|\mathbf{x}\|^2$ associated with the value R^2. Show that the unit normal vector field pointing outward from this surface is $\mathbf{n}(\mathbf{x}) = \mathbf{x}/\|\mathbf{x}\|$.*

THEOREM 5.2.7 (MEAN VALUE THEOREM FOR HARMONIC FUNCTIONS). *If $-\Delta u = 0$ in $\mathcal{D}_R$, then*

$$u(\mathbf{x}_0) = \frac{1}{4\pi R^2} \int_{\partial\mathcal{D}_R} u \, d\sigma. \tag{5.2.6}$$

The expression on the right side of Equation (5.2.6) represents the average value of u over the surface $\partial\mathcal{D}_R$, which has area $4\pi R^2$.

PROOF: Since the Laplace operator Δ is invariant under the change of variables $\mathbf{y} = \mathbf{x} - \mathbf{x}_0$, it suffices to establish the theorem for the case $\mathbf{x}_0 = \mathbf{0}$. Consider the open region $\mathcal{R}$ obtained from $\mathcal{D}_R$ by deleting a smaller, concentric closed ball $\mathcal{D}_\delta \cup \partial\mathcal{D}_\delta$, where $0 < \delta < R$, as drawn in Figure 5.2. We apply Green's identity (B.2.1),

$$\int_{\mathcal{R}} (f\Delta g - g\Delta f)\, dv = \int_{\partial\mathcal{R}} (g \operatorname{grad} f - f \operatorname{grad} g) \cdot \mathbf{n}\, d\sigma, \tag{5.2.7}$$

to the functions $f(\mathbf{x}) = u(\mathbf{x})$ and $g(\mathbf{x}) = 1/\|\mathbf{x}\|$. Since $\Delta u = 0 = \Delta(1/\|\mathbf{x}\|)$ in $\mathcal{R}$ (see Exercise 127), the left side of Equation (5.2.7) vanishes. Also, by Exercise 138,

$$\operatorname{grad} g(\mathbf{x}) \cdot \mathbf{n}(\mathbf{x}) = \begin{cases} -1/R^2, & \text{if } \mathbf{x} \in \partial\mathcal{D}_R \\ 1/\delta^2, & \text{if } \mathbf{x} \in \partial\mathcal{D}_\delta. \end{cases}$$

Therefore, the identity (5.2.7) yields the following equation:

$$0 = \int_{\partial \mathfrak{D}_R} \frac{1}{\|\mathbf{x}\|} \operatorname{grad} u \cdot \mathbf{n}\, d\sigma - \int_{\partial \mathfrak{D}_\delta} \frac{1}{\|\mathbf{x}\|} \operatorname{grad} u \cdot \mathbf{n}\, d\sigma$$
$$+ \frac{1}{R^2} \int_{\partial \mathfrak{D}_R} u\, d\sigma - \frac{1}{\delta^2} \int_{\partial \mathfrak{D}_\delta} u\, d\sigma.$$

The first two integrals vanish: by the divergence theorem and the hypothesis that u is harmonic,

$$\int_{\partial \mathfrak{D}_R} \frac{1}{\|\mathbf{x}\|} \operatorname{grad} u \cdot \mathbf{n}\, d\sigma = \frac{1}{R} \int_{\mathfrak{D}_R} \Delta u\, dv = 0,$$

and similarly for the second integral. Multiplying the surviving terms by $(4\pi)^{-1}$ yields

$$\frac{1}{4\pi R^2} \int_{\partial \mathfrak{D}_R} u\, d\sigma = \frac{1}{4\pi \delta^2} \int_{\partial \mathfrak{D}_\delta} u\, d\sigma. \tag{5.2.8}$$

Because u is continuous, the right side tends to $u(\mathbf{0})$ as $\delta \to 0$ (see Exercise 139), completing the proof. ■

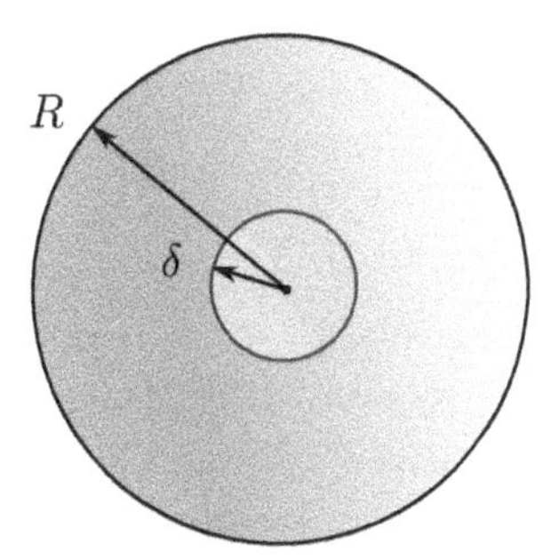

Figure 5.2. Deleted ball used in the proof of the mean value theorem for harmonic functions.

EXERCISE 139 *Prove that the right side of Equation (5.2.8) tends to* $u(\mathbf{0})$ *as* $\delta \to 0$.

5.2.6 Summary

Potential theory focuses on the properties of the Laplace and Poisson equations, which appear in virtually every branch of applied mathematics. In continuum mechanics these equations arise most often in steady-state problems or in settings where a vector field can be represented in terms of the gradient of a scalar potential. Several qualitative properties of solutions to these equations promote insight into the physics that they govern.

Three classic boundary-value problems, involving Dirichlet, Neumann, and Robin boundary conditions, frequently appear in applications. It is possible to establish uniqueness theorems for Dirichlet and Robin problems involving the Poisson equation, but solutions to Neumann problems are unique only up to an arbitrary constant. In addition, the Laplace equation obeys a maximum principle that ensures that the solutions to the Poisson equation depend continuously on the boundary data, a key requirement for boundary-value problems to be well posed. Finally, the Laplace equation possesses an averaging property known as the mean value theorem for harmonic functions.

5.3 FLUID MECHANICS

This section briefly explores five topics in fluid mechanics: (1) ideal fluids, which are the simplest models of fluid behavior; (2) the effects of rotation on fluid motion; (3) the acoustic equations, which describe how sound waves propagate; (4) Navier–Stokes fluids, which include viscous effects; and (5) Stokes flows, which model a limiting case of Navier–Stokes flows in which viscous effects dominate the physics.

5.3.1 Ideal Fluids

The simplest model of fluids involves two assumptions. The first is the internal constraint that the material is incompressible. Hence, $\det \mathsf{F} = 1$ (see Exercise 48), and by Exercise 94 the mass balance reduces to $\operatorname{div} \mathbf{v} = 0$. By Equation (4.1.3), incompressibility also implies that the mass density in any configuration equals that in the reference configuration, $\rho = \rho_{\mathcal{K}}$.

The second assumption is a constitutive relation: we assume that there are no shear stresses, and the normal stresses are independent of direction:

$$\mathsf{T} = -p\,\mathsf{I}. \tag{5.3.1}$$

Here p is a positive quantity called the **(mechanical) pressure**, which for ideal fluids we treat as a function of density ρ. Tensors such as that in Equation (5.3.1) that are multiples of the identity tensor are called **isotropic** tensors. They exhibit no preferred directions, in the sense that any nonzero vector is an eigenvector. This stress tensor (5.3.1) is clearly symmetric and, hence, is consistent with the angular momentum balance.

EXERCISE 140 *Show that the form (5.3.1) is invariant under coordinate rotations. Thus an isotropic tensor is a multiple of* I *in any Cartesian coordinate system.*

Substituting these assumptions into Cauchy's first law of motion (4.2.7) yields

$$\rho \frac{D\mathbf{v}}{Dt} = -\operatorname{grad} p + \rho \mathbf{b}.$$

Assume further that the body force is conservative, so that $\mathbf{b} = -\operatorname{grad}\Phi$ for some time-independent scalar potential Φ. This assumption is reasonable for the gravitational field near the earth's surface. In this case, the momentum balance reduces to the following equation:

$$\frac{D\mathbf{v}}{Dt} = -\operatorname{grad}\left(\frac{p}{\rho} + \Phi\right). \tag{5.3.2}$$

In words, the acceleration is a gradient field. By Kelvin's theorem (see Section 3.4), the next theorem follows immediately.

THEOREM 5.3.1 (CIRCULATION OF IDEAL FLUIDS). *The motion of an ideal fluid under a conservative body force is circulation-preserving.*

In particular, if such a flow is irrotational at any particular time t_0, it is irrotational for all t. This result highlights a shortcoming of the ideal fluid as a model of real fluids: vorticity can never develop in a flow that exhibits none initially. Later in this section we examine models of fluid flow that permit the development of vorticity.

Ideal fluids are the subject of one of the most basic and well-known theorems in fluid mechanics.

THEOREM 5.3.2 (BERNOULLI'S THEOREM). *The following equations characterize steady, irrotational motion of an ideal fluid under a conservative body force:*

$$\operatorname{div}\mathbf{v} = 0, \qquad \operatorname{curl}\mathbf{v} = \mathbf{0}, \qquad \frac{\|\mathbf{v}\|^2}{2} + \frac{p}{\rho} + \Phi = \text{Constant}.$$

This theorem explains why pressure is high where the velocity is small and vice versa and why gas in a pipeline loses pressure as it ascends a mountain range.

PROOF: The equation $\operatorname{div}\mathbf{v} = 0$ is simply the mass balance equation, and the equation $\operatorname{curl}\mathbf{v} = \mathbf{0}$ characterizes irrotational motion. To prove the third equation, consider the momentum balance (5.3.2):

$$\frac{\partial \mathbf{v}}{\partial t} + (\mathbf{v}\cdot\operatorname{grad})\mathbf{v} = -\operatorname{grad}\left(\frac{p}{\rho} + \Phi\right).$$

Since the motion is steady, $\partial\mathbf{v}/\partial t = \mathbf{0}$. Also, $(\mathbf{v}\cdot\operatorname{grad})\mathbf{v} = \frac{1}{2}\operatorname{grad}(\mathbf{v}\cdot\mathbf{v})$. Making these substitutions yields

$$\operatorname{grad}\left(\frac{\|\mathbf{v}\|^2}{2} + \frac{p}{\rho} + \Phi\right) = \mathbf{0}.$$

The desired conclusion follows. ■

5.3.2 An Ideal Fluid in a Rotating Frame of Reference

A modification of Equation (5.3.2) models the motion of an ideal fluid in a rotating frame of reference, such as a frame fixed to the Earth. As discussed in Section 4.2, in such a frame the momentum balance equation has additional terms, owing to the fact that the acceleration is not objective. If we restrict attention to the rotating Earth, two of these additional terms vanish. Specifically, when the origin of the rotating frame of reference coincides with the origin in an inertial frame and the rotation rate $\boldsymbol{\omega}$ is constant, then the translational inertia and the Euler effect identified in Equation (4.2.11) vanish, and Equation (5.3.2) takes the form

$$\frac{D\mathbf{v}}{Dt} = -\operatorname{grad}\left(\frac{p}{\rho} + \Phi\right) + 2\boldsymbol{\omega} \times \mathbf{v} - \boldsymbol{\omega} \times (\boldsymbol{\omega} \times \mathbf{x}). \tag{5.3.3}$$

The last two terms on the right represent the Coriolis effect and centrifugal force, respectively.

EXERCISE 141 *Using Exercises 20 and 21, show that*

$$-\boldsymbol{\omega} \times (\boldsymbol{\omega} \times \mathbf{x}) = \frac{1}{2} \operatorname{grad} \|\boldsymbol{\omega} \times \mathbf{x}\|^2.$$

The result of this exercise suggests treating the centrifugal effect as a pressure-like effect by including it in an effective potential in Equation (5.3.3):

$$\frac{D\mathbf{v}}{Dt} = -\operatorname{grad} \Phi_{\text{eff}} + 2\boldsymbol{\omega} \times \mathbf{v}, \tag{5.3.4}$$

where

$$\Phi_{\text{eff}} = \frac{p}{\rho} + \Phi - \frac{1}{2}\|\boldsymbol{\omega} \times \mathbf{x}\|^2.$$

Writing the momentum balance equation in this way prompts a question: when does the Coriolis effect $2\boldsymbol{\omega} \times \mathbf{v}$ exert a significant influence on the motion? To gauge the contributions of the different terms in Equation (5.3.4), we must filter out any effects on magnitudes attributable solely to our choices of measurement units. For this purpose, let us cast the partial differential equation in terms of unit-free independent variables appropriate to the scales present in the problem. Let L be a characteristic length associated with the geometry, and let V be a representative velocity. Define dimensionless space and time variables as follows:

$\boldsymbol{\xi} = \mathbf{x}/L$	dimensionless position
$\tau = Vt/L$	dimensionless time.

EXERCISE 142 *Show that* $\text{grad}_{\mathbf{x}} = L^{-1}\,\text{grad}_{\xi}$ *and* $\partial/\partial t = VL^{-1}\partial/\partial\tau$. *Use these results to rewrite Equation (5.3.4) as follows:*

$$\frac{\partial\mathbf{v}}{\partial\tau} + \left(\frac{\mathbf{v}}{V}\cdot\text{grad}_{\xi}\right)\mathbf{v} = -\frac{1}{V}\text{grad}\,\Phi_{\text{eff}} + \frac{2}{\text{Ro}}\mathbf{e}_{\omega}\times\mathbf{v}. \tag{5.3.5}$$

Here $\mathbf{e}_{\omega} = \boldsymbol{\omega}/\|\boldsymbol{\omega}\|$ *is a unit-length vector, and* $\text{Ro} = V/(L\|\boldsymbol{\omega}\|)$ *is a dimensionless scalar called the* **Rossby number**.

The Rossby number measures the significance of the Coriolis effect in specific problems. Roughly speaking, when $\text{Ro} > 10^2$, inertial effects associated with the terms

$$\frac{\partial\mathbf{v}}{\partial\tau} + \frac{\mathbf{v}}{V}\cdot\text{grad}_{\xi}\mathbf{v}$$

greatly dominate the Coriolis effect. When $\text{Ro} \simeq 1$, Coriolis effects are comparable in magnitude to inertial effects. When $\text{Ro} < 10^{-2}$, the Coriolis effect dominates.

EXERCISE 143 *Calculate Rossby numbers for the following physical settings on the Earth's surface:*

1. *Air flow in the jet stream, a current in the Earth's atmosphere that is typically around 200 km wide, with speeds commonly ranging around 100 km/hr*

2. *Water flow near a bathtub drain having a diameter of 5 cm, assuming an average initial velocity of magnitude around* 10^{-2} *mm/s (nearly calm).*

5.3.3 Acoustics

Small disturbances in fluid density can propagate through a fluid in the form of waves traveling with a common speed. Our ears and brains perceive these disturbances—at least those within a particular range of frequencies—as sound. The equations that describe the propagation of sound waves arise from the mass and momentum balance equations, together with a simple constitutive relation and some modeling assumptions about the magnitudes of the disturbances.

Begin with the mass and momentum balances,

$$\frac{D\rho}{Dt} + \rho\,\text{div}\,\mathbf{v} = 0$$

$$\rho\frac{D\mathbf{v}}{Dt} - \text{div}\,\mathsf{T} - \rho\mathbf{b} = \mathbf{0},$$

which we regard as differential equations for the unknowns ρ and $\mathbf{v}$. We neglect the influence of body forces $\mathbf{b}$ and adopt the constitutive relation $\mathsf{T} = -p(\rho)\mathsf{I}$ for an ideal fluid. For purposes of the model we assume that p is twice continuously differentiable and that the functions ρ, p, and p' are positive, bounded above, and bounded away from 0. The momentum balance reduces to the following equation:

$$\rho\frac{D\mathbf{v}}{Dt} + \operatorname{grad} p(\rho) = \mathbf{0}.$$

In the classic acoustic model, fluid density disturbances amount to small departures from a constant equilibrium value ρ_0. More precisely, we assume that $\rho(\mathbf{x}, t) = \rho_0 + \rho_1(\mathbf{x}, t)$ and $\operatorname{grad} \rho(\mathbf{x}, t) = \operatorname{grad} \rho_1(\mathbf{x}, t)$, with

$$\begin{aligned} |\rho_1(\mathbf{x}, t)| &< \epsilon \\ \|\operatorname{grad} \rho_1(\mathbf{x}, t)\| &< \epsilon, \end{aligned}$$

for some small parameter $\epsilon > 0$. We assume that departures of the velocity from the equilibrium value $\mathbf{v} = \mathbf{0}$ are similarly small:

$$\begin{aligned} \|\mathbf{v}(\mathbf{x}, t)\| &< \epsilon, \\ \|[\operatorname{grad} \mathbf{v}(\mathbf{x}, t)]\mathbf{a}\| &< \epsilon\|\mathbf{a}\|, \end{aligned} \tag{5.3.6}$$

for every vector $\mathbf{a} \in \mathbb{E}$.

EXERCISE 144 *Show that the inequality (5.3.6) guarantees that* $|\operatorname{div} \mathbf{v}| < 3\epsilon$.

Expanding the mass balance using the decomposition of ρ yields

$$\frac{\partial \rho}{\partial t} + \mathbf{v} \cdot \operatorname{grad} \rho_1 + \rho_0 \operatorname{div} \mathbf{v} + \rho_1 \operatorname{div} \mathbf{v} = 0.$$

But by the Cauchy–Schwarz inequality $|\mathbf{v} \cdot \operatorname{grad} \rho_1| \leqslant \|\mathbf{v}\|\,\|\operatorname{grad} \rho_1\| < \epsilon^2$, and by Exercise 144 $|\rho_1 \operatorname{div} \mathbf{v}| < 3\epsilon^2$. If ϵ is small, ϵ^2 is much smaller. Neglecting terms proportional to ϵ^2 yields the following reduced mass balance:

$$\frac{\partial \rho}{\partial t} + \rho_0 \operatorname{div} \mathbf{v} = 0. \tag{5.3.7}$$

We consider this equation to be **linearized** in the sense that it contains no products of the unknown density ρ and unknown velocity $\mathbf{v}$.

To treat the momentum balance similarly, begin by expanding it as follows:

$$\frac{\partial \mathbf{v}}{\partial t} + (\mathbf{v} \cdot \operatorname{grad})\mathbf{v} + n(\rho) \operatorname{grad} \rho = \mathbf{0},$$

where $n(\rho) = p'(\rho)/\rho$ is a positive, bounded function. To gauge the effects of small density disturbances on the function n, note that by the mean value theorem there exists a density value ζ between ρ_0 and ρ such that $n(\rho) = n(\rho_0) + n'(\zeta)\rho_1$. Thus

$$\frac{\partial \mathbf{v}}{\partial t} + (\mathbf{v} \cdot \operatorname{grad})\mathbf{v} + n(\rho_0) \operatorname{grad} \rho + n'(\zeta)\rho_1 \operatorname{grad} \rho = \mathbf{0}.$$

In this equation, $\|(\mathbf{v}\cdot\operatorname{grad})\mathbf{v}\| < \epsilon^2$, and $|n'(\zeta)|\,\|\rho_1 \operatorname{grad}\rho_1\| < |n'(\zeta)|\epsilon^2$. Again neglecting terms proportional to ϵ^2 yields a linearized momentum balance,

$$\frac{\partial \mathbf{v}}{\partial t} + \frac{p'(\rho_0)}{\rho_0}\operatorname{grad}\rho = \mathbf{0}. \tag{5.3.8}$$

Equations (5.3.7) and (5.3.8) constitute the **acoustic equations**.

EXERCISE 145 *Assuming that ρ and $\mathbf{v}$ are twice differentiable as functions of $(\mathbf{x}, t)$, use the acoustic equations to derive the three-dimensional* **wave equation**,

$$\frac{\partial^2 \rho}{\partial t^2} - c^2 \Delta\rho = 0, \tag{5.3.9}$$

where $c^2 = p'(\rho_0)$, and $\Delta = \operatorname{div}(\operatorname{grad})$ denotes the Laplace operator.

The quantity $c = \sqrt{p'(\rho)}$ is the **sound speed**. To justify this nomenclature, it is useful to examine particular types of solution to the wave equation.

EXERCISE 146 *Let $\mathbf{k} \in \mathbb{E}$ be a fixed vector having unit length. Show that any sufficiently differentiable function $\rho(\mathbf{x}, t) = f(\mathbf{k}\cdot\mathbf{x} \pm ct)$ is a solution to Equation (5.3.9).*

A solution of this form is a **plane wave**. The terminology arises from the following geometric observation:

EXERCISE 147 *Let $\mathcal{S}$ be any plane in $\mathbb{E}$ that is perpendicular to the unit vector $\mathbf{k}$. Show that $\mathbf{x}\cdot\mathbf{k}$ remains constant as $\mathbf{x}$ ranges over $\mathcal{S}$. Hence, at each time t, $f(\mathbf{x}\cdot\mathbf{k} \pm ct)$ has the same value at all points in $\mathcal{S}$.*

For plane waves, the time dependence of the argument $\mathbf{x}\cdot\mathbf{k} \pm ct$ indicates that planes of constant density undergo translation parallel to $\mathbf{k}$ with speed $\pm c$. This translation is even easier to grasp in fewer space dimensions:

EXERCISE 148 *Consider the one-dimensional wave equation*

$$\frac{\partial^2 \rho}{\partial t^2} - c^2\frac{\partial^2 \rho}{\partial x^2} = 0.$$

Let f be any sufficiently differentiable function of a scalar variable, and show that $\rho(x, t) = [f(x + ct) - f(x - ct)]/2$ is a solution. Sketch the graph of such a function $f(x)$ that vanishes outside some finite-length interval of the x-axis, then sketch the solution components $f(x + ct)/2$ and $f(x - ct)/2$ for $t > 0$.

5.3.4 Incompressible Newtonian Fluids

Anyone who has watched vortices form in the wake of a boat or dust devils swirl above a prairie knows that many fluid motions are not circulation-preserving, and therefore not all fluids are ideal. We now examine a more sophisticated model of fluids: one that supports shear stresses. Again, we assume that the fluid is incompressible, so that $\operatorname{div} \mathbf{v} = 0$ and $\rho = \rho_\kappa$, which we assume to be uniform in space. For the stress tensor, we adopt a constitutive relation in which stress depends linearly on the stretching tensor $\mathsf{D} = \frac{1}{2}[\operatorname{grad} \mathbf{v} + (\operatorname{grad} \mathrm{v})^\top)]$:

$$\mathsf{T} = -p\mathsf{I} + \lambda(\operatorname{tr} \mathsf{D})\mathsf{I} + 2\mu\mathsf{D}. \tag{5.3.10}$$

We examine this relationship more closely in Section 6.5. The parameters λ and μ are positive constants called the **coefficients of viscosity**. Fluids obeying this constitutive law are called **Newtonian fluids**. Equation (5.3.10) respects the angular momentum balance, since $\mathsf{T} = \mathsf{T}^\top$.

Since the fluid is incompressible, $\operatorname{tr} \mathsf{D} = \operatorname{div} \mathbf{v} = 0$, and therefore the constitutive relation for stress reduces to $\mathsf{T} = -p\mathsf{I} + 2\mu\mathsf{D}$.

EXERCISE 149 *Show that in this case,* $\operatorname{div} \mathsf{T} = -\operatorname{grad} p + \mu\Delta\mathbf{v}$, *where* $\Delta\mathbf{v} = \Delta v_i\, \mathbf{e}_i$, *and* Δ *denotes the Laplace operator.*

It follows from these observations that the momentum balance (4.2.7) for incompressible, Newtonian fluids takes the form

$$\frac{\partial \mathbf{v}}{\partial t} + (\mathbf{v} \cdot \operatorname{grad})\mathbf{v} = -\frac{1}{\rho_\kappa} \operatorname{grad} p + \nu\Delta\mathbf{v} + \mathbf{b}. \tag{5.3.11}$$

Here, $\nu = \mu/\rho_\kappa$ is the **kinematic viscosity**. Equation (5.3.11) is the **incompressible Navier–Stokes equation**. The **inertial term** $(\mathbf{v} \cdot \operatorname{grad})\mathbf{v}$ has a reputation as one of the most troublesome nonlinearities in fluid mechanics.

The following theorem demonstrates that in the presence of nonzero viscosity, fluid motions are not generally circulation-preserving.

THEOREM 5.3.3 (CIRCULATION IN A NEWTONIAN FLUID). *For an incompressible, Newtonian fluid under a conservative body force,*

$$\frac{dC_\gamma}{dt} = \int_\gamma \nu\, \Delta\mathbf{v} \cdot d\mathbf{x},$$

where $\gamma \colon [a, b] \to \mathbb{E}$ *is any smooth, closed arc in the fluid.*

PROOF: Recall from Section 4.4 that a conservative body force has the form $\mathbf{b} = -\operatorname{grad}\Phi$ for some scalar potential $\Phi(\mathbf{x}, t)$. With this representation, the incompressible Navier–Stokes equation becomes

$$\frac{D\mathbf{v}}{Dt} = \nu\,\Delta\mathbf{v} - \operatorname{grad}\left(\frac{p}{\rho_\kappa} + \Phi\right).$$

By Theorem 3.4.2,

$$\begin{aligned}\frac{dC_\gamma}{dt} &= \int_\gamma \frac{D\mathbf{v}}{Dt}\cdot d\mathbf{x}\\ &= \int_\gamma \nu\,\Delta\mathbf{v}\cdot d\mathbf{x} - \int_\gamma \operatorname{grad}\left(\frac{p}{\rho_\kappa} + \Phi\right)\cdot d\mathbf{x}.\end{aligned}$$

The last integral vanishes since the arc γ is closed:

$$\int_\gamma \operatorname{grad}\left(\frac{p}{\rho_\kappa} + \Phi\right)\cdot d\mathbf{x} = \left[\frac{p(\gamma(s))}{\rho_\kappa} + \Phi(\gamma(s))\right]\Bigg|_a^b = 0.$$

This completes the proof. ■

5.3.5 Stokes Flow

In many applications to slow flow, the viscous effects in the Navier–Stokes equation—far from being negligible as in ideal fluids—dominate other effects, and the momentum balance simplifies. In settings like these, deriving the governing equations requires an analysis of scales in addition to the adoption of constitutive equations. This subsection examines slow flow of an incompressible Newtonian fluid around a fixed sphere of radius R suspended in a viscous fluid. Figure 5.3 shows the geometry. We assume that the kinematic viscosity ν is constant, neglect body forces, and assume that the fluid velocity is uniform far from the sphere: $\mathbf{v} \to \mathbf{v}_\infty = v_\infty\mathbf{e}_1$ as $\|\mathbf{x}\| \to \infty$.

The idea is to treat the inertial effects $D\mathbf{v}/Dt$ in the momentum balance as being small. To make this reasoning precise we must assess the relative magnitudes of the terms in the Navier–Stokes equation,

$$\frac{\partial\mathbf{v}}{\partial t} + (\mathbf{v}\cdot\operatorname{grad})\mathbf{v} = -\frac{1}{\rho_\kappa}\operatorname{grad} p + \nu\,\Delta\mathbf{v}. \tag{5.3.12}$$

This task is delicate. As in the rotating fluids examined earlier in this section, the choice of units can have a deceptive influence. To eliminate this subjective effect, we

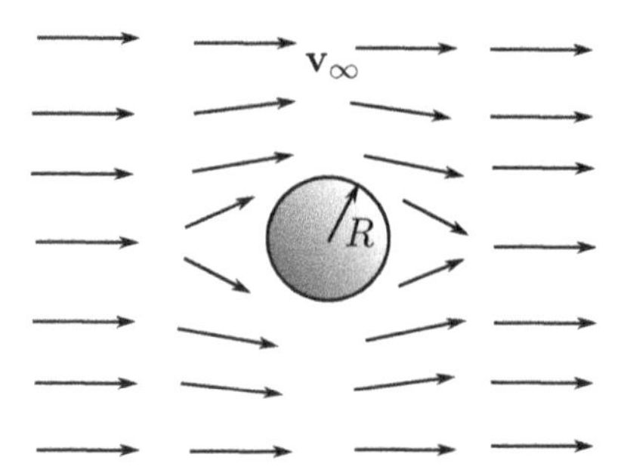

Figure 5.3. Geometry of Stokes flow around a sphere S.

cast the partial differential equation in unit-free length, time, velocity, and pressure variables using the characteristic length R, far-field fluid speed v_∞, and density ρ_κ, obtaining the following dimensionless variables:

$$\boldsymbol{\xi} = \frac{\mathbf{x}}{R}, \quad \tau = \frac{v_\infty t}{R}, \quad \mathbf{v}^* = \frac{\mathbf{v}}{v_\infty}, \quad p^* = \frac{p}{\rho_\kappa v_\infty^2}. \tag{5.3.13}$$

EXERCISE 150 *Verify that $\rho_\kappa^2 v_\infty$ and p have the same physical dimensions.*

EXERCISE 151 *Show that, in the change of variables defined in Equation (5.3.13), the derivative operators in the Navier–Stokes equation transform as follows:*

$$\frac{\partial}{\partial t} = \frac{v_\infty}{R}\frac{\partial}{\partial \tau}, \quad \operatorname{grad} = \frac{1}{R}\operatorname{grad}_\xi, \quad \Delta = \frac{1}{R^2}\Delta_\xi,$$

where, with respect to the basis $\{\mathbf{e}_1, \mathbf{e}_2, \mathbf{e}_3\}$,

$$\operatorname{grad}_\xi = \left(\frac{\partial}{\partial \xi_1}, \frac{\partial}{\partial \xi_2}, \frac{\partial}{\partial \xi_3}\right), \qquad \Delta_\xi = \operatorname{div}_\xi(\operatorname{grad}_\xi).$$

EXERCISE 152 *Show that the Navier–Stokes equation (5.3.12) is equivalent to the following nondimensional form:*

$$\frac{\partial \mathbf{v}^*}{\partial \tau} + (\mathbf{v}^* \cdot \operatorname{grad}_\xi)\mathbf{v}^* = -\operatorname{grad}_\xi p^* + \frac{1}{\mathrm{Re}}\Delta_\xi \mathbf{v}^*, \tag{5.3.14}$$

where $\mathrm{Re} = v_\infty R/\nu$ *is a dimensionless constant.*

The constant Re is the **Reynolds number** of the flow. It characterizes, without reference to specific physical units, the relative influence of the nonlinear inertial terms

$$\frac{D\mathbf{v}^*}{D\tau} = \frac{\partial \mathbf{v}^*}{\partial \tau} + (\mathbf{v}^* \cdot \operatorname{grad}_\xi)\mathbf{v}^*$$

and the viscous term $(1/\text{Re})\Delta_\xi \mathbf{v}^*$ in the Navier–Stokes equation. When Re is much larger than 1, for example when the far-field fluid speed v_∞ is large, the inertial terms dominate. In slow flows, when Re is much smaller than 1, the inertial terms are negligible, and the viscous term balances the pressure-gradient term:

$$\frac{1}{\text{Re}}\Delta_\xi \mathbf{v}^* = \text{grad}_\xi p^*.$$

Using this reasoning, we pose the following boundary-value problem for slow flow around a sphere of radius R. For $\|\mathbf{x}\| > R$,

$$\mu\Delta\mathbf{v} = \text{grad}\, p, \qquad \text{div}\,\mathbf{v} = 0, \tag{5.3.15}$$

subject to the boundary conditions

$$\begin{aligned} \mathbf{v}(\mathbf{x}) &= \mathbf{0}, & \|\mathbf{x}\| &= R, \\ \mathbf{v} &\to \mathbf{v}_\infty, & \|\mathbf{x}\| &\to \infty. \end{aligned}$$

Equations (5.3.15) are the **Stokes equations** for slow, viscous flow. These equations often serve as models for flows in which $\text{Re} < 10^{-1}$. The only input data to this boundary-value problem are the far-field fluid speed v_∞, the radius R of the sphere, and the fluid's dynamic viscosity μ.

Historically, the motivation for studying this problem was to compute the drag force $\mathbf{F}_{\text{drag}}$ exerted by the fluid on the sphere. Stokes solved this problem exactly, via intricate calculations, obtaining $\mathbf{F}_{\text{drag}} = F_{\text{drag}}\mathbf{e}_1$, where

$$F_{\text{drag}} = 6\pi\mu R v_\infty. \tag{5.3.16}$$

We call this expression the **Stokes drag** on the sphere. See [32, Section 20] for a derivation.

It is possible to derive a qualitatively similar result through much simpler reasoning, based on **dimensional analysis**. Because the viscosity μ, far-field speed v_∞, and radius R are the only input data appearing in the boundary-value problem, the drag force F_{drag} must be a function of only these parameters. Let us examine the conditions that must hold if this function takes the form of a power-law relationship:

$$F_{\text{drag}} = F_{\text{drag}}(\mu, v_\infty, R) = \mu^{\alpha_1} v_\infty^{\alpha_2} R^{\alpha_3},$$

where α_1, α_2, and α_3 are parameters to be determined. Since both sides of this equation have the same physical dimension,

$$\begin{aligned} [\text{MT}^{-2}\text{L}] &= [\text{MT}^{-1}\text{L}^{-1}]^{\alpha_1}[\text{LT}^{-1}]^{\alpha_2}[\text{L}]^{\alpha_3} \\ &= [\text{M}^{\alpha_1}\text{T}^{-\alpha_1-\alpha_2}\text{L}^{-\alpha_1+\alpha_2+\alpha_3}]. \end{aligned}$$

Matching powers yields the linear system

$$\begin{bmatrix} 1 & 0 & 0 \\ -1 & -1 & 0 \\ -1 & 1 & 1 \end{bmatrix} \begin{bmatrix} \alpha_1 \\ \alpha_2 \\ \alpha_3 \end{bmatrix} = \begin{bmatrix} 1 \\ -2 \\ 1 \end{bmatrix},$$

with solution $\alpha_1 = \alpha_2 = \alpha_3 = 1$. Therefore, if a power-law expression for the drag force holds,

$$F_{\text{drag}} = C\mu R v_\infty, \tag{5.3.17}$$

where C is a dimensionless constant. This form is consistent with Stokes's exact solution (5.3.16) for a sphere.

EXERCISE 153 *If the sphere drops through the fluid under the influence of gravity, the net body force on the sphere is* $\frac{4}{3}\pi R^3 g(\rho_S - \rho_L)$, *where* g *is the magnitude of the gravitational acceleration,* ρ_S *is the density of the sphere, and* ρ_F *is the density of the fluid. Assume that all of these quantities are constant, and balance the net body force with the Stokes drag to find the steady-state—or terminal—velocity of the sphere.*

The result of Exercise 153 furnishes a method for determining the dynamic viscosity μ experimentally.

EXERCISE 154 *Based on the results of Section 5.1, one model for cooking an initially cold, spherical turkey in an oven at fixed temperature* θ_e *is the heat equation with initial condition* $\theta(\mathbf{x}, 0) = \theta_0$ *and boundary condition* $\theta(\mathbf{x}, t) = \theta_e$ *for* $\|\mathbf{x}\| = R$. *Recasting the equation in terms of the dimensionless temperature* $u(\mathbf{x}, t) = \theta(\mathbf{x}, t)/\theta_e$ *yields the initial–boundary–value problem*

$$\begin{aligned} \frac{\partial u}{\partial t} - K\Delta u = 0, &\qquad \|\mathbf{x}\| < R,\ t > 0; \\ u(\mathbf{x}, 0) = u_0, &\qquad \|\mathbf{x}\| < R; \\ u(\mathbf{x}, t) = 1, &\qquad \|\mathbf{x}\| = R. \end{aligned}$$

Thus the time t_{done} *required for the center of the turkey to reach a prescribed dimensionless temperature* u_{done} *depends only on the parameters* K, R, u_0, *and* u_{done}. *Use dimensional analysis to deduce that this time is proportional to* $\text{mass}^{2/3}$.

Dimensional analysis yields much deeper and sometimes surprisingly powerful results. For more thorough introductions to this subject, see [27] and [4].

5.3.6 Summary

Fluids exhibit an enormous range of behaviors. The simplest model is the ideal fluid, for which stress obeys the constitutive relation $\mathsf{T} = -p\,\mathsf{I}$, where $p = p(\rho) > 0$ denotes mechanical pressure. Using this law in the momentum balance and assuming that the specific body force is a potential field $\mathbf{b} = -\text{grad}\,\Phi$ yield two classic theorems for fluids in a conservative body force:

- The motion of an ideal fluid is circulation-preserving.
- Steady, irrotational motions of an ideal fluid obey Bernoulli's law, $\frac{1}{2}\|\mathbf{v}\| + \rho/\rho_\kappa + \Phi = \text{constant}$.

The form of the momentum balance changes when we work in a noninertial frame of reference. For an ideal fluid in a frame attached to the Earth, the momentum balance includes additional terms accounting for the Coriolis effect and the centrifugal force. The dimensionless Rossby number gauges the significance of the Coriolis effect.

A special case of the equations of motion for an ideal fluid arises when fluid density and velocity disturbances are small. In this case, it is common to model the fluid motions using versions of the mass and momentum balance linearized about a constant density ρ_0. This model leads to the three-dimensional wave equation:

$$\frac{\partial^2 \rho}{\partial t^2} - c^2 \Delta \rho = 0.$$

Wave-like solutions to this equation travel with the sound speed $c = \sqrt{p'(\rho)}$.

Newtonian fluids obey a more complicated constitutive relation for stress:

$$\mathsf{T} = -p\mathsf{I} + \lambda(\text{tr}\,\mathsf{D})\mathsf{I} + 2\mu\mathsf{D},$$

where λ and μ denote the positive coefficients of viscosity. The momentum balance in this case reduces to the Navier–Stokes equation,

$$\frac{\partial \mathbf{v}}{\partial t} + (\mathbf{v} \cdot \text{grad})\mathbf{v} = -\frac{1}{\rho_\kappa}\,\text{grad}\,p + \nu \Delta \mathbf{v} + \mathbf{b},$$

where $\nu = \mu/\rho_\kappa$. The fact that flows obeying this equation are not generally circulation-preserving squares with empirical observations that vorticity can develop in flows that are initially irrotational.

A classic problem for viscous fluids examines slow flows, where we neglect inertial terms in the Navier–Stokes equation to arrive at the steady-state equations for Stokes flows:

$$\mu \Delta \mathbf{v} = \text{grad}\,p, \qquad \text{div}\,\mathbf{v} = 0.$$

A standard problem for this equation, dating from the nineteenth century, calculates the drag force on a rigid sphere embedded in a steady velocity field that is uniform in the far field. Dimensional analysis corroborates that the drag force is proportional to the far-field fluid speed, a result known as Stokes drag.

5.4 SOLID MECHANICS

This section reviews the most common constitutive relation for stress in an elastic solid. We assume that the material is isothermal and that it undergoes only small displacement gradients, so that the infinitesimal strain tensor E developed in Section 3.2 gives a reasonable approximation to the strain.

The momentum balance appropriate for this application is Equation (4.2.13), derived in Section 4.2 for the reference configuration:

$$\rho \frac{\partial^2 \mathbf{u}}{\partial t^2} = \operatorname{Div} \mathsf{T}_\kappa + \rho \mathbf{b}.$$

Here $\mathbf{u}(\mathbf{x}, t)$ denotes the displacement, illustrated in Figure 2.4, and $\rho = \rho_\kappa$ denotes the density in the reference configuration. As a constitutive relation for the first Piola–Kirchhoff stress T_k, we adopt **Hooke's law**,

$$\mathsf{T}_\kappa = \lambda(\operatorname{tr} \mathsf{E})\mathsf{I} + 2\mu \mathsf{E}. \tag{5.4.1}$$

Section 6.5 illustrates a derivation.

In contrast to the constitutive law (5.3.10) for Newtonian fluids, in which stress depends linearly on the strain rate, the stress in Equation (5.4.1) depends linearly on the strain itself. The coefficients λ and μ are positive scalars called the **Lamé parameters**. Substituting this constitutive relation into the momentum balance gives the following equation of motion for linearly elastic solids, known as **Navier's equation**:

$$\rho \frac{\partial^2 \mathbf{u}}{\partial t^2} = \mu \operatorname{Div}(\operatorname{Grad} \mathbf{u}) + (\lambda + \mu) \operatorname{Grad}(\operatorname{Div} \mathbf{u}) + \rho \mathbf{b}. \tag{5.4.2}$$

EXERCISE 155 *Invert the constitutive relation (5.4.1) to get*

$$\mathsf{E} = \frac{1}{2\mu} \left[\mathsf{T}_\kappa - \frac{\lambda}{2\mu + 3\lambda} (\operatorname{tr} \mathsf{T}_\kappa) \mathsf{I} \right]. \tag{5.4.3}$$

5.4.1 Static Displacements

The Lamé parameters λ and μ completely characterize the solid's linear stress-strain responses. A few simple types of static strain, analyzed with respect to the orthonormal basis $\{\mathbf{e}_1, \mathbf{e}_2, \mathbf{e}_3\}$, illustrate how these parameters, and others derived from them, quantify these responses.

Consider first a simple shear. From Equation (2.1.2),

$$\mathbf{u}(\mathbf{X}) = \mathbf{x}(\mathbf{X}, t) - \mathbf{X} = \alpha\, \mathbf{e}_1 \otimes \mathbf{e}_2\, \mathbf{X} = \alpha X_2 \mathbf{e}_1.$$

Figure 5.4 illustrates this displacement. For the time being, we treat α as a nonzero constant. The infintesimal strain tensor in this case is

$$\mathsf{E} = \frac{1}{2}\begin{bmatrix} 0 & \alpha & 0 \\ \alpha & 0 & 0 \\ 0 & 0 & 0 \end{bmatrix} = \frac{\alpha}{2}(\mathbf{e}_1 \otimes \mathbf{e}_2 + \mathbf{e}_2 \otimes \mathbf{e}_1),$$

and the corresponding stress tensor is, by Equation (5.4.1),

$$\mathsf{T}_\kappa = \begin{bmatrix} 0 & \mu\alpha & 0 \\ \mu\alpha & 0 & 0 \\ 0 & 0 & 0 \end{bmatrix} = 2\mu\mathsf{E}.$$

By comparing E with T_κ, we see that shearing strain amplifies the stress by a factor 2μ. For this reason, we call the Lamè parameter μ the **shear modulus**.

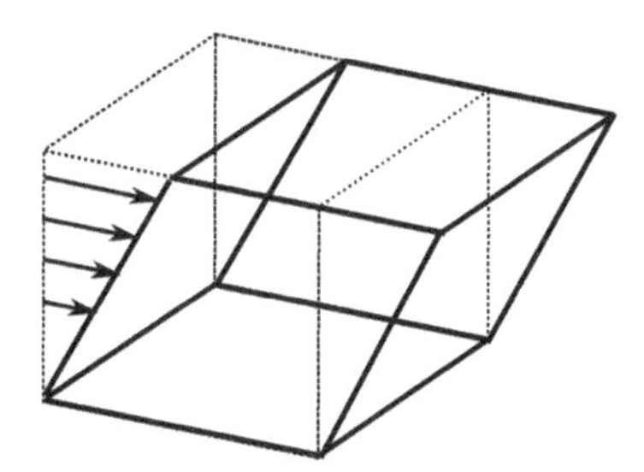

Figure 5.4. Pure shear applied to a cubic material volume.

Next consider a uniform extension, $\mathbf{u}(\mathbf{X}) = \alpha\mathbf{X}$, as illustrated in Figure 5.5. In this case, the strain and stress tensors are

$$\mathsf{E} = \alpha\mathsf{I}, \qquad \mathsf{T}_\kappa = 3\underbrace{\left(\lambda + \frac{2}{3}\mu\right)}_{\kappa}\alpha\mathsf{I} = (2\mu + 3\lambda)\,\alpha\,\mathsf{I}.$$

The factor $\kappa = \lambda + \frac{2}{3}\mu$ measures the response of the stress to the compressive or tensile strains imposed by the extension; we call it the **modulus of compression**.

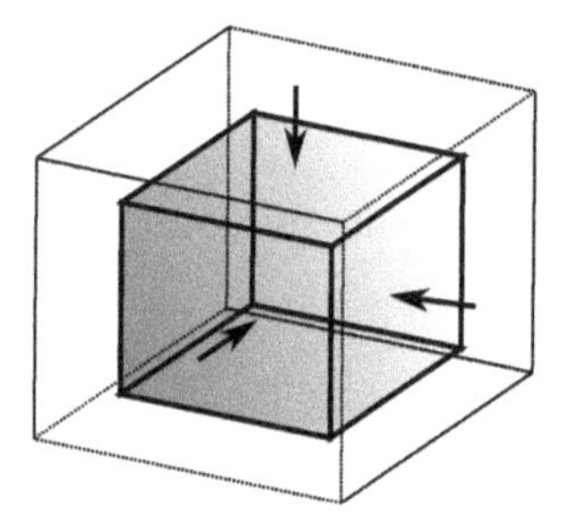

Figure 5.5. Uniform extension (with $0 < \alpha < 1$) applied to a cubic material volume.

Finally, consider tension in the X_1-direction, as illustrated in Figure 5.6. The stress tensor has the form

$$\mathsf{T}_\kappa = \begin{bmatrix} T & 0 & 0 \\ 0 & 0 & 0 \\ 0 & 0 & 0 \end{bmatrix}.$$

From Equation (5.4.3) the corresponding infinitesimal strain is

$$\mathsf{E} = \begin{bmatrix} T/E & 0 & 0 \\ 0 & -\nu(T/E) & 0 \\ 0 & 0 & -\nu(T/E) \end{bmatrix}.$$

Here the parameter

$$E = \frac{\mu(2\mu + 3\lambda)}{\mu + \lambda},$$

which measures the ratio of stress to longitudinal strain, is **Young's modulus**, while

$$\nu = \frac{\lambda}{2(\mu + \lambda)},$$

measuring the ratio of transverse contraction to longitudinal strain, is **Poisson's ratio**. Figure 5.6 illustrates the effects of longitudinal strain coupled with transverse contraction.

EXERCISE 156 *Show that, in terms of the engineering parameters E and ν, Navier's equation (5.4.2) takes the form*

$$\rho \frac{\partial^2 \mathbf{u}}{\partial t^2} = \frac{E}{2(1+\nu)} \left[\operatorname{Div}(\operatorname{Grad} \mathbf{u}) + \frac{1}{1-2\nu} \operatorname{Grad}(\operatorname{Div} \mathbf{u}) \right] + \rho \mathbf{b}. \tag{5.4.4}$$

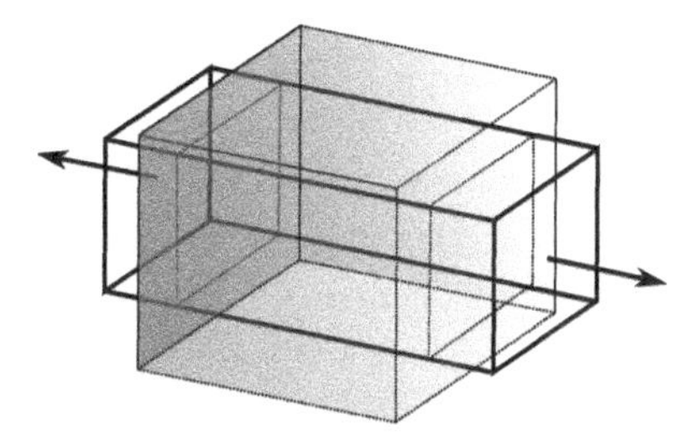

Figure 5.6. Effects of tension in the X_1-direction on a cubic material volume.

EXERCISE 157 *Using index notation, prove the identity*

$$\Delta_\kappa \mathbf{u} = \mathrm{Grad}\,(\mathrm{Div}\,\mathbf{u}) - \mathrm{Curl}\,(\mathrm{Curl}\,\mathbf{u}). \tag{5.4.5}$$

Here $\Delta_\kappa = \mathrm{Div}\,(\mathrm{Grad})$ *denotes the Laplace operator in referential coordinates, and* $\Delta_\kappa \mathbf{u}$ *is the vector field with components* $(\Delta_\kappa \mathbf{u}) \cdot \mathbf{e}_i = \Delta_\kappa(\mathbf{u} \cdot \mathbf{e}_i) = \Delta_\kappa u_i$ *with respect to the orthonormal basis* $\{\mathbf{e}_1, \mathbf{e}_2, \mathbf{e}_3\}$.

EXERCISE 158 *Use the identity (5.4.5), together with the assumption that body forces are negligible, to derive the following form of Navier's equation:*

$$\rho \frac{\partial^2 \mathbf{u}}{\partial t^2} = (\lambda + 2\mu)\,\mathrm{Grad}\,(\mathrm{Div}\,\mathbf{u}) - \mu\,\mathrm{Curl}\,(\mathrm{Curl}\,\mathbf{u}). \tag{5.4.6}$$

5.4.2 Elastic Waves

Although so far we have considered only static strains, Equation (5.4.2) clearly admits time-dependent solutions as well. An important class of motions in many applications consists of waves. The remainder of this section examines wave solutions to Navier's equation, under the assumption that body forces are negligible.

The question of what waves Equation (5.4.2) supports is complicated in general, but it is illuminating to explore a simple case. Assume that the Lamé parameters λ and μ are constant. We ask what types of **plane waves** the equation

$$\rho \frac{\partial^2 \mathbf{u}}{\partial t^2} = \mu\,\mathrm{Div}\,(\mathrm{Grad}\,\mathbf{u}) + (\lambda + \mu)\,\mathrm{Grad}\,(\mathrm{Div}\,\mathbf{u}) \tag{5.4.7}$$

supports. Paralleling the discussion of acoustics in Section 5.3, waves of this type are superpositions of displacements having the form

$$\mathbf{u}(\mathbf{X}, t) = \mathbf{a} \exp[i(\mathbf{k} \cdot \mathbf{X} - ct)], \tag{5.4.8}$$

where $i^2 = -1$. Thus at any time t the displacement is constant in planes in which $\mathbf{k} \cdot \mathbf{X}$ is constant. Here c is the speed of the wave, and the constant unit vector $\mathbf{k}$ gives the direction in which the wave propagates. The vector $\mathbf{a}$ is the **amplitude vector**. The wave is **longitudinal** when $\mathbf{a} = \|\mathbf{a}\|\mathbf{k}$, in which case the displacements are parallel to the direction of propagation; it is **transverse** if $\mathbf{a} \cdot \mathbf{k} = 0$, in which case the displacements are perpendicular to the direction of propagation. Figure 5.7 illustrates longitudinal and transverse waves schematically.

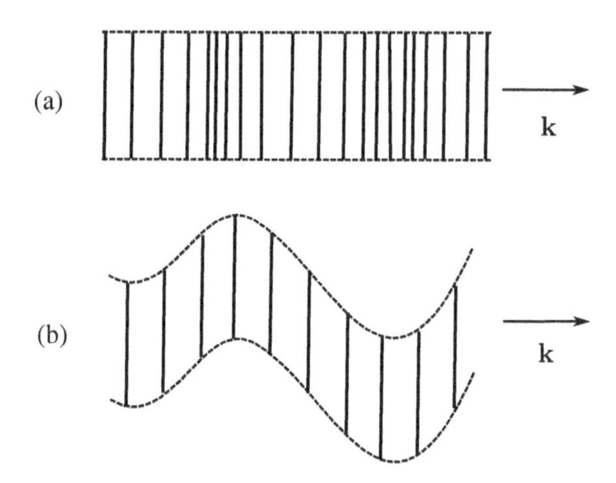

Figure 5.7. (a) Schematic picture of a longitudinal wave. (b) Schematic picture of a transverse wave.

To see what types of plane waves satisfy Equation (5.4.7), substitute the generic form (5.4.8) into the differential equation and examine the constraints that result. We find that

$$\begin{aligned}
\operatorname{Grad} \mathbf{u}(\mathbf{X}, t) &= i(\mathbf{k} \otimes \mathbf{a}) \exp[i(\mathbf{k} \cdot \mathbf{X} - ct)], \\
\operatorname{Div} \mathbf{u}(\mathbf{X}, t) &= i(\mathbf{k} \cdot \mathbf{a}) \exp[i(\mathbf{k} \cdot \mathbf{X} - ct)], \\
\frac{\partial^2 \mathbf{u}}{\partial t^2} &= -c^2 \mathbf{a} \exp[i(\mathbf{k} \cdot \mathbf{X} - ct)].
\end{aligned}$$

It follows that

$$\begin{aligned}
\operatorname{Div} (\operatorname{Grad} \mathbf{u})(\mathbf{X}, t) &= -\mathbf{a} \exp[i(\mathbf{k} \cdot \mathbf{X} - ct)], \\
\operatorname{Grad} (\operatorname{Div} \mathbf{u})(\mathbf{X}, t) &= - \underbrace{\mathbf{k}(\mathbf{k} \cdot \mathbf{a})}_{(\mathbf{k} \otimes \mathbf{k})\mathbf{a}} \exp[i(\mathbf{k} \cdot \mathbf{X} - ct)].
\end{aligned}$$

Substituting these expressions into Equation (5.4.7) and dividing through by the nonzero exponential factor and the mass density ρ yield

$$c^2\mathbf{a} = \underbrace{\frac{1}{\rho}[\mu\mathsf{I} + (\lambda+\mu)(\mathbf{k}\otimes\mathbf{k})]}_{\mathsf{A}(\mathbf{k})}\,\mathbf{a}.$$

The linear transformation A identified in this equation is the **acoustic tensor**. We rewrite it in a more useful form as

$$\mathsf{A}(\mathbf{k}) = \left(\frac{\lambda+2\mu}{\rho}\right)(\mathbf{k}\otimes\mathbf{k}) + \frac{\mu}{\rho}[\mathsf{I} - (\mathbf{k}\otimes\mathbf{k})]. \tag{5.4.9}$$

These calculations establish the following sufficient condition for the plane wave (5.4.8) to be a solution to the momentum balance equation (5.4.7) for a linearly elastic solid:

$$\mathsf{A}(\mathbf{k})\mathbf{a} = c^2\mathbf{a}.$$

In words, the material supports the plane wave (5.4.8) provided $(c^2, \mathbf{a})$ is an eigenpair of the tensor $\mathsf{A}(\mathbf{k})$. The representation (5.4.9) makes the identification of such pairs especially easy.

EXERCISE 159 *Show that*

(*a*) *for longitudinal waves* ($\mathbf{a} = \|\mathbf{a}\|\mathbf{k}$), $\mathsf{A}(\mathbf{k})\mathbf{a} = [(\lambda+2\mu)/\rho]\,\mathbf{a}$;

((*b*) *for transverse waves* ($\mathbf{a}\cdot\mathbf{k} = 0$), $\mathsf{A}(\mathbf{k})\mathbf{a} = (\mu/\rho)\,\mathbf{a}$.

The representation (5.4.9) is the spectral decomposition of A. It expresses A as the sum of linear transformations, each of which maps a subspace of $\mathbb{E}$ into itself by mapping vectors in the subspace onto eigenvalue multiples of themselves.

The results of Exercise 159 show that Equation (5.4.7) supports two types of waves. One type is longitudinal and travels with wavespeed

$$c_P = \sqrt{\frac{\lambda+2\mu}{\rho}}.$$

In applications to seismic waves in the earth's crust these waves are called **compressional waves** or **P-waves**. The other type of wave is transverse and travels with the slower wavespeed

$$c_S = \sqrt{\frac{\mu}{\rho}}.$$

In seismic applications these are **shear waves** or **S-waves**.

EXERCISE 160 *Prove the* **Helmholtz decomposition**: *if the vector field* $\mathbf{u}$ *is continuously differentiable and vanishes outside a bounded region, then there are twice continuously differentiable functions* Φ *(a scalar field) and* $\boldsymbol{\Psi}$ *(a vector field) such that*

$$\mathbf{u} = \operatorname{Grad}\Phi + \operatorname{Curl}\boldsymbol{\Psi}. \tag{5.4.10}$$

Φ *is the* **scalar potential**, *and* $\boldsymbol{\Psi}$ *is the* **vector potential**. *Observe that* $\operatorname{grad}\Phi$ *is irrotational and* $\operatorname{curl}\boldsymbol{\Psi}$ *is solenoidal.*

EXERCISE 161 *Assume* ρ, λ, *and* μ *are constants, and substitute the Helmholtz decomposition (5.4.10) into the constant-coefficient Equation (5.4.6). Show that, to solve for* $\mathbf{u}$, *it suffices to find a scalar field* Φ *and a vector field* $\boldsymbol{\Psi}$ *that satisfy the following versions of the classic* **wave equation**,

$$\frac{\partial^2\Phi}{\partial t^2} = c_1^2\Delta_\kappa\Phi,$$
$$\frac{\partial^2\boldsymbol{\Psi}}{\partial t^2} = c_2^2\Delta_\kappa\boldsymbol{\Psi},$$

where $\Delta_\kappa = \operatorname{Div}(\operatorname{Grad})$ *stands for the Laplace operator with respect to referential coordinates. Compare the wave speeds* c_1 *and* c_2 *with those derived for P- and S-waves.*

5.4.3 Summary

The most common model for elastic solids arises through substitution of Hooke's constitutive relation for stress

$$\mathsf{T}_\kappa = \lambda(\operatorname{tr}\mathsf{E})\mathsf{I} + 2\mu\mathsf{E}$$

into the momentum balance derived for the reference configuration,

$$\rho\frac{\partial^2\mathbf{u}}{\partial t^2} = \operatorname{Div}\mathsf{T}_\kappa + \rho\mathbf{b}.$$

Here, λ and μ are positive coefficients called the Lamé parameters. The resulting momentum balance law is Navier's equation,

$$\rho\frac{\partial^2\mathbf{u}}{\partial t^2} = \mu\operatorname{Div}(\operatorname{Grad}\mathbf{u}) + (\lambda+\mu)\operatorname{Grad}(\operatorname{Div}\mathbf{u}) + \rho\mathbf{b},$$

which governs the displacement $\mathbf{u}$ from the solid's reference configuration.

It is possible to relate the Lamé parameters to quantities measurable in static displacement tests. The following is a summary of three such tests and the engineering parameters that characterize the material responses in terms of the infinitesimal strain tensor E:

1. Applying pure shear of the form

$$\mathsf{E} = \frac{\alpha}{2}(\mathbf{e}_1 \otimes \mathbf{e}_2 + \mathbf{e}_2 \otimes \mathbf{e}_1)$$

   to the solid yields the shear modulus, $\mu = T_{12}/(2E_{12})$.

2. Applying uniform extension or compression yields the modulus of compression, $\kappa = \lambda + \frac{2}{3}\mu$.

3. Applying longitudinal tension yields Young's modulus,

$$E = \frac{\mu(2\mu + 3\lambda)}{\mu + \lambda},$$

   which measures the ratio of longitudinal stress to longitudinal strain, and Poisson's ratio,

$$\nu = \frac{\lambda}{2(\mu + \lambda)},$$

   which is the ratio of transverse contraction to longitudinal strain.

Navier's equation has the form of a wave equation. When body forces are negligible, it supports two types of plane waves: longitudinal waves, traveling with speed $c_P = \sqrt{(\lambda + 2\mu)/\rho}$, and transverse waves, traveling with the slower speed $c_S = \sqrt{\mu/\rho}$. In applications involving seismic waves in the earth's crust, longitudinal plane waves are called P-waves, and transverse plane waves are called S-waves.

CHAPTER 6

CONSTITUTIVE THEORY

This chapter focuses on conceptual aspects of constitutive theory. The development of constitutive relations is far from a capricious enterprise. To be acceptable on physical grounds, constitutive functions must respect certain principles. These principles take the form of axioms, which place restrictions on which variables may appear in various functional dependencies and frequently on the structure of these dependencies. Here we take a first look at the theory underlying constitutive functions, with reference to five **constitutive axioms**:

1. Determinism
2. Equipresence
3. Objectivity
4. Symmetry
5. Admissibility.

Continuum Mechanics: The Birthplace of Mathematical Models, First Edition. M.B. Allen.

As a platform for illustrating these principles, we examine their application to a specific type of material, the thermoviscous fluid. In Section 6.5 we briefly examine another type of material, the thermoelastic solid. Advanced texts, including several listed in the bibliography, discuss many additional types of continua.

6.1 CONCEPTUAL SETTING

The balance equations derived in Chapter 4 provide eight scalar equations:

$$\frac{D\rho}{Dt} + \rho \operatorname{div} \mathbf{v} = 0 \qquad \text{1 equation}$$

$$\rho \frac{D\mathbf{v}}{Dt} - \operatorname{div} \mathsf{T} - \rho \mathbf{b} = \mathbf{0} \qquad \text{3 equations}$$

$$\mathsf{T} = \mathsf{T}^\top \qquad \text{3 equations}$$

$$\rho \frac{D\varepsilon}{Dt} - \operatorname{tr}(\mathsf{TD}) + \operatorname{div} \mathbf{q} - \rho r = 0 \qquad \underline{\text{1 equation}}$$

$$\text{8 equations.}$$

The vector momentum balance counts as three scalar equations. The symmetry of the stress tensor, which is equivalent to the balance of angular momentum, imposes three additional independent constraints on the off-diagonal entries of T. The entropy inequality

$$\rho \frac{D\eta}{Dt} \geqslant -\operatorname{div}\left(\frac{\mathbf{q}}{\theta}\right) + \frac{\rho r}{\theta},$$

developed in Section 4.5, imposes yet another constraint on the system, as we explore later, but it does not count as an equation *per se*.

6.1.1 The Need to Close the System

Appearing in this system are the variables ρ, $\mathbf{x}$, T, $\mathbf{b}$, ε, $\mathbf{q}$, r, and, in the entropy balance, η and θ. In all, the balance laws govern 23 scalar variables. Eight scalar differential equations clearly do not suffice, even were we to include boundary and initial conditions sufficient to yield well-posed problems. Thus, in addition to providing a vehicle for particularizing the balance laws to specific materials, constitutive theory furnishes a means for mathematically closing the system of governing equations.

We regard the eight scalar balance equations listed above as serving two purposes: (1) determining the five scalar components of the **principal unknowns** ρ, $\mathbf{x}$, and θ

and (2) guaranteeing the symmetry of T. Thus from now on we consider the stress tensor to have only six independent entries. In this view, 15 scalar unknowns remain. Of these, the body force $\mathbf{b}$ and the heat supply r account for four scalar unknowns. In many applications the physics determines these quantities; for example, the body force is often a known gravitational or electromagnetic field, and the heat supply is often a known rate of radiative heat transfer. In this common scenario, there remain the 11 scalar unknowns needed to determine T, ε, $\mathbf{q}$, and η.

To determine these remaining unknowns, we impose additional equations that relate T, ε, $\mathbf{q}$, and η to the principal unknowns determined by the balance laws. We seek these equations in the form of **constitutive functions**,

$$
\begin{aligned}
\mathsf{T} &= \mathsf{T}(\rho, \mathbf{x}, \theta, \text{and their derivatives and history}),\\
\varepsilon &= \varepsilon(\rho, \mathbf{x}, \theta, \text{and their derivatives and history}),\\
\mathbf{q} &= \mathbf{q}(\rho, \mathbf{x}, \theta, \text{and their derivatives and history}),\\
\eta &= \eta(\rho, \mathbf{x}, \theta, \text{and their derivatives and history}).
\end{aligned}
$$

The derivatives can be with respect to spatial variables or time or both. The choice of which independent variables to include in the list is a matter for the modeler to determine. Once this choice has been made, each assignment of values to the variables

$$
\big(\rho(\mathbf{x},t), \mathbf{x}(\mathbf{X},t), \theta(\mathbf{x},t), \text{derivatives of } \rho,\ \mathbf{x}, \text{ and } \theta, \text{ and history}\big)
$$

determines the variables T, ε, $\mathbf{q}$, and η uniquely.

The remaining sections of this chapter examine constraints imposed on constitutive relations by the five constitutive axioms listed above. In the literature of continuum mechanics one sometimes finds reference to other axioms of constitutive theory. In particular, some authors impose an **axiom of just setting**, requiring that constitutive functions yield well-posed problems in the sense defined in Section 5.1. Still, the set listed above constitutes a core of widely agreed-upon restrictions.

These axioms allow for a wide range of behaviors, including nonlinear dependencies, history dependence, and dependencies that are nonlocal in space. These exotic cases offer an important window into some of the most intriguing material behaviors of interest in modern science and technology, but their analysis is often delicate. In this text we restrict attention to a more elementary class of materials, in which the constitutive functions T, ε, $\mathbf{q}$, and η depend at most on derivatives of ρ, $\mathbf{x}$, and θ through first order in time and space. Thus, if we use the symbol G to stand for a **generic constitutive function**

$$
G = (\mathsf{T}, \varepsilon, \mathbf{q}, \theta), \tag{6.1.1}
$$

then the constitutive functions of interest depend at most on the following variables:

$$G = G\left(\rho, \operatorname{grad}\rho, \frac{D\rho}{Dt}, \frac{D(\operatorname{grad}\rho)}{Dt}, \mathbf{x}, \mathsf{F}, \mathbf{v}, \mathsf{L}, \theta, \operatorname{grad}\theta, \frac{D\theta}{Dt}, \frac{D(\operatorname{grad}\theta)}{Dt}\right).$$

(Recall that $\mathsf{F} = \operatorname{Grad}\mathbf{x}$ and $\mathsf{L} = \operatorname{grad}\mathbf{v}$ and that we can relate grad to Grad via the chain rule once we know $\mathbf{x}(\mathbf{X}, t)$.)

Even with this restriction, the possibilities are myriad. To focus the discussion in the remaining sections we explore the implications of the constitutive axioms on a particular material, the **thermoviscous fluid**. The generic constitutive function for this material is

$$G = G\left(\mathbf{x}, \mathsf{F}, \mathbf{v}, \mathsf{L}, \theta, \operatorname{grad}\theta\right).$$

This modeling assumption does not explicitly include functional dependencies on the density ρ; however, we show that such dependencies arise in the development. Examples where ρ is an explicit dependent variable appear in Chapter 7. The idea in the present chapter is to subject this generic function to a logical filter, in the form of the constitutive axioms, to determine the allowable forms for the functions T, ε, $\mathbf{q}$, and η.

6.1.2 Summary

From a conceptual point of view, constitutive relations are necessary to close the systems of equations derived from the balance laws, which yield eight independent scalar equations for 23 scalar unknowns. We regard the balance laws as governing the principal unknowns $\rho(\mathbf{x}, t)$, $\mathbf{x}(\mathbf{X}, t)$, and $\theta(\mathbf{x}, t)$ and guaranteeing the symmetry of the stress tensor $\mathsf{T}(\mathbf{x}, t)$. In many problems the physics naturally yield expressions for the specific body force $\mathbf{b}$ and the external energy supply r. Therefore, closing the system requires functional relations for the remaining variables T, ε, $\mathbf{q}$, and η in terms of the principal unknowns, their derivatives, and possibly their history. Five constitutive axioms impose restrictions on these functional relations. We apply these axioms to the model case of a thermoviscous fluid, for which the constitutive relations take the *a priori* form

$$G = G\left(\mathbf{x}, \mathsf{F}, \mathbf{v}, \mathsf{L}, \theta, \operatorname{grad}\theta\right).$$

Here G denotes an array of functions $(\mathsf{T}, \varepsilon, \mathbf{q}, \eta)$ sought for stress, specific internal energy, heat flux, and specific entropy, respectively.

6.2 DETERMINISM AND EQUIPRESENCE

The most general constitutive relations allow the values of the constitutive functions T, ε, $\mathbf{q}$, and η at a specific particle $\mathbf{X}$ and time t to depend on a range of particles and times. To indicate this possibility, we write

$$G = G\big(\rho(\mathbf{x}(\mathbf{X}^*, t^*), t^*), \mathbf{x}(\mathbf{X}^*, t^*), \mathsf{F}(\mathbf{X}^*, t^*), \mathbf{v}(\mathbf{x}(\mathbf{X}^*, t^*), t^*), \ldots\big), \qquad (6.2.1)$$

where G is the generic constitutive function introduced in Equation (6.1.1) and the notation $(\mathbf{X}^*, t^*)$ signifies the range of allowable values of $(\mathbf{X}, t)$.

6.2.1 Determinism

The first axiom of constitutive theory restricts these allowable values.

AXIOM (DETERMINISM). *In the constitutive functions (6.2.1) valid at* $(\mathbf{X}, t)$, *the variable* $\mathbf{X}^*$ *ranges over a subset of* $\mathcal{B}$, *and* t^* *ranges over a subset of* $(-\infty, t]$.

The subsets can be improper, that is, all of $\mathcal{B}$ or $(-\infty, t]$. This axiom asserts that the only values of the independent variables that can affect the behavior at $(\mathbf{X}, t)$ are those at particles of $\mathbf{X}^*$ that belong to the body and at times up to and including the present time t.

For the remainder of this chapter we assume that the material is **local**, that is,

$$G = G\big(\rho(\mathbf{x}(\mathbf{X}, t), t), \mathbf{x}(\mathbf{X}, t), \mathsf{F}(\mathbf{X}, t), \mathbf{v}(\mathbf{x}(\mathbf{X}, t), t), \ldots\big).$$

Nonlocal theories are possible. These may account for history dependence, in which we allow $t^* \in (-\infty, t]$, and regional dependence, in which values of constitutive variables in some neighborhood of $\mathbf{X}$ may affect the behavior at $\mathbf{X}$.

6.2.2 Equipresence

The second axiom of constitutive theory requires that, *a priori*, we allow similar functional dependencies in all of the constitutive variables.

AXIOM (EQUIPRESENCE). *An independent variable that appears in one constitutive relation must appear in all of them, in principle, except where its presence contradicts other constitutive axioms.*

One may reasonably argue that this axiom amounts to a prescription against methodological caprice, not an axiom. Appearance in principle hardly precludes the physically realistic possibility that some material properties may depend weakly or not at all on some of the independent variables. For example, in a particular application the stress may depend on L only, while the heat flux for the same material may depend only on $\operatorname{grad}\theta$. Nevertheless, in filtering the logical possibilities the axiom of equipresence dictates that we start with the same argument list for the function T as for $\mathbf{q}$ and all other constitutive variables, to avoid missing any possible dependencies.

As a consequence of the axioms of determinism and equipresence and the assumption that the material is local, we seek constitutive relations for thermoviscous fluids having the form

$$\begin{aligned}
\mathsf{T}(\mathbf{x},t) &= \mathsf{T}(\mathbf{x},\mathsf{F},\mathbf{v},\mathsf{L},\theta,\ \operatorname{grad}\theta),\\
\varepsilon(\mathbf{x},t) &= \varepsilon(\mathbf{x},\mathsf{F},\mathbf{v},\mathsf{L},\theta,\ \operatorname{grad}\theta),\\
\mathbf{q}(\mathbf{x},t) &= \mathbf{q}(\mathbf{x},\mathsf{F},\mathbf{v},\mathsf{L},\theta,\ \operatorname{grad}\theta),\\
\eta(\mathbf{x},t) &= \eta(\mathbf{x},\mathsf{F},\mathbf{v},\mathsf{L},\theta,\ \operatorname{grad}\theta),
\end{aligned}$$

where

$$\begin{aligned}
\mathbf{x} &= \mathbf{x}(\mathbf{X},t),\\
\mathsf{F} &= \mathsf{F}(\mathbf{X},t),\\
\mathbf{v} &= \mathbf{v}(\mathbf{x}(\mathbf{X},t),t),\\
\mathsf{L} &= \mathsf{L}(\mathbf{x}(\mathbf{X},t),t),\\
\theta &= \theta(\mathbf{x}(\mathbf{X},t),t),\\
\operatorname{grad}\theta &= \operatorname{grad}\theta(\mathbf{x}(\mathbf{X},t),t).
\end{aligned}$$

Generically, we write these relations as

$$G(\mathbf{x},t) = G(\mathbf{x},\mathsf{F},\mathbf{v},\mathsf{L},\theta,\ \operatorname{grad}\theta),$$

leaving understood the local dependence of the arguments on the right side.

6.2.3 Summary

The axiom of determinism demands that any constitutive function G depend only on values of the arguments associated with particles belonging to the body and times up to and including the present. For the thermoviscous fluid examined in this chapter as a model, we assume that the material is local, in the sense that every constitutive function depends only on values of the arguments

$$(\mathbf{x},\mathsf{F},\mathbf{v},\mathsf{L},\theta,\ \operatorname{grad}\theta)$$

at the specific particle $\mathbf{X}$ and time t under examination. The axiom of equipresence insists that, *a priori*, we allow all of the constitutive functions $\mathsf{T}, \varepsilon, \mathbf{q}, \eta$ to depend on these arguments, ruling out specific dependencies only if they contradict the balance laws, entropy inequality, or other constitutive axioms.

6.3 OBJECTIVITY

The third axiom of constitutive theory requires that constitutive relations be valid for all observers.

AXIOM (OBJECTIVITY.) *The functional forms of constitutive relations must be invariant under changes in frame of reference.*

Recall from Section 3.5 that a frame of reference is a choice $(\boldsymbol{f}, t_0)$ of a one-to-one correspondence $\boldsymbol{f}$ between the Euclidean point space $\mathbb{X}$ in which the objects of interest reside and the space $\mathbb{E}$ of position vectors, together with a choice t_0 of temporal instant assigned to the origin of the time axis $\mathbb{R}$.

As introduced in Section 3.5, a change in frame of reference takes the form of an observer transformation

$$\hat{\mathbf{y}} = \mathsf{Q}(t)\mathbf{y} + \mathbf{c}(t), \qquad \hat{t} = t - b, \tag{6.3.1}$$

where the functions $\mathsf{Q}\colon \mathbb{R} \to \mathrm{O}(\mathbb{E})$ and $\mathbf{c}\colon \mathbb{R} \to \mathbb{E}$ are continuously differentiable and $b \in \mathbb{R}$. Here we use $\mathbf{y}$ to denote a generic position vector, reserving $\mathbf{x}$ to denote the deformation $\mathbf{x}(\mathbf{X}, t)$. Physically, this transformation represents a rigid motion of the hatted observer with respect to the original observer, together with a translation in time that leaves time intervals unchanged in length. In applying the axiom of objectivity, it is common to examine specific changes in frame of reference—that is, specific choices of $\mathsf{Q}(t)$ and $\mathbf{c}(t)$—defined by values of $\mathsf{Q}, \mathsf{Q}', \mathbf{c}, \mathbf{c}'$ at some instant t.

EXERCISE 162 *Show that* $\mathsf{Q}'(t)$ *is skew at any instant* t *for which* $\mathsf{Q}(t) = \mathsf{I}$.

For thermoviscous fluids, the axiom of objectivity requires that the generic constitutive function $G = (\mathsf{T}, \varepsilon, \mathbf{q}, \eta)$ obey an identity of the form

$$G(\mathbf{x}, \mathsf{F}, \mathbf{v}, \mathsf{L}, \theta, \operatorname{grad}\theta) = G(\hat{\mathbf{x}}, \hat{\mathsf{F}}, \hat{\mathbf{v}}, \hat{\mathsf{L}}, \hat{\theta}, \operatorname{grad}\hat{\theta}), \tag{6.3.2}$$

the hatted quantities being those measured in the transformed frame. In imposing Equation (6.3.2), we recall the polar decomposition $\mathsf{F} = \mathsf{RU}$ of the deformation gradient and the decomposition $\mathsf{L} = \mathsf{D} + \mathsf{W}$ of the velocity gradient into symmetric and skew parts. Of special interest is the dependence of the generic constitutive function G on the variables $\mathbf{x}(\mathbf{X},t)$, $\mathsf{F}(\mathbf{X},t)$, $\mathbf{v}(\mathbf{x}(\mathbf{X},t),t)$, and $\mathsf{L}(\mathbf{x}(\mathbf{X},t),t)$, shown in Section 3.5 to be nonobjective. Using the transformation rules for $\mathbf{x}$, F, $\mathbf{v}$, and L derived in Section 3.5, we have

$$\begin{aligned} &G(\mathbf{x},\mathsf{F},\mathbf{v},\mathsf{L},\theta,\operatorname{grad}\theta) \\ &= G(\mathsf{Q}\mathbf{x}+\mathbf{c},\mathsf{QRU},\mathsf{Q}\mathbf{v}+\mathsf{Q}'\mathbf{x}+\mathbf{c}',\mathsf{QDQ}^\top+\mathsf{QWQ}^\top+\mathsf{Q}'\mathsf{Q}^\top,\theta,\operatorname{grad}\theta), \end{aligned} \tag{6.3.3}$$

for any continuously differentiable, orthogonal tensor-valued function $\mathsf{Q}(t)$ and any continuously differentiable vector-valued function $\mathbf{c}(t)$.

6.3.1 Reducing Functional Dependencies

We examine two principles used to apply the axiom. The first is crafty: let $\boldsymbol{\Xi}(\mathbf{X},t)$ denote any argument of the generic constitutive function G, and let $\boldsymbol{\xi}$ be an arbitrary but fixed value of that argument. Suppose that for every space-time point $(\mathbf{X},t)$, there exists an observer transformation (6.3.1) such that

$$G(\ldots,\boldsymbol{\Xi}(\mathbf{X},t),\ldots) = G(\ldots,\boldsymbol{\xi},\ldots),$$

leaving the other arguments of G unchanged. Then the function G does not depend on the independent variable $\boldsymbol{\Xi}$. The following theorem exploits this principle.

THEOREM 6.3.1 (INDEPENDENCE FROM POSITION AND VELOCITY). *For thermoviscous fluids, each of the constitutive functions* T, ε, $\mathbf{q}$, *and* η *is independent of* $\mathbf{x}$ *and* $\mathbf{v}$.

PROOF: Given a space-time point $(\mathbf{X},t)$, choose an arbitrary vector $\boldsymbol{\xi} \in \mathbb{E}$. In examining values of the generic constitutive function G at $(\mathbf{X},t)$, first define an observer transformation (6.3.1) such that at time t,

$$\mathsf{Q}(t) = \mathsf{I}, \qquad \mathbf{c}(t) = \boldsymbol{\xi} - \mathbf{x}(\mathbf{X},t),$$
$$\mathsf{Q}'(t) = 0, \qquad \mathbf{c}'(t) = \mathbf{0}.$$

Figure 6.1(a) illustrates this observer transformation. In this case, $\hat{\mathbf{x}} = \mathsf{Q}\mathbf{x}+\mathbf{c} = \boldsymbol{\xi}$; $\hat{\mathbf{v}} = \mathsf{Q}\mathbf{v}+\mathsf{Q}'\mathbf{x}+\mathbf{c}' = \mathbf{v}$; and the other arguments remain unchanged. The invariance requirement (6.3.3) now yields

$$G(\mathbf{x},\mathsf{RU},\mathbf{v},\mathsf{D}+\mathsf{W},\theta,\operatorname{grad}\theta) = G(\boldsymbol{\xi},\mathsf{RU},\mathbf{v},\mathsf{D}+\mathsf{W},\theta,\operatorname{grad}\theta). \tag{6.3.4}$$

In other words, given an arbitrary vector $\boldsymbol{\xi}$, there exists an observer transformation for which Equation (6.3.4) holds at $(\mathbf{X},t)$. Therefore, G is independent of $\mathbf{x}$.

To prove that G is independent of $\mathbf{v}$, choose

$$\begin{aligned} \mathsf{Q}(t) &= \mathsf{I}, & \mathbf{c}(t) &= \mathbf{0}, \\ \mathsf{Q}'(t) &= 0, & \mathbf{c}'(t) &= \boldsymbol{\xi} - \mathbf{v}(\mathbf{x}(\mathbf{X},t),t). \end{aligned}$$

Figure 6.1(b) shows this observer transformation. At time t, $\hat{\mathbf{x}} = \mathsf{Q}\mathbf{x} + \mathbf{c} = \mathbf{x}$; $\hat{\mathbf{v}} = \mathsf{Q}\mathbf{v} + \mathsf{Q}'\mathbf{x} + \mathbf{c}' = \boldsymbol{\xi}$; and the other arguments remain unchanged. This time the invariance of G implies that

$$G(\mathsf{RU}, \mathbf{v}, \mathsf{D} + \mathsf{W}, \theta, \operatorname{grad}\theta) = G(\mathsf{RU}, \boldsymbol{\xi}, \mathsf{D} + \mathsf{W}, \theta, \operatorname{grad}\theta).$$

Therefore G is independent of $\mathbf{v}$ as well. ■

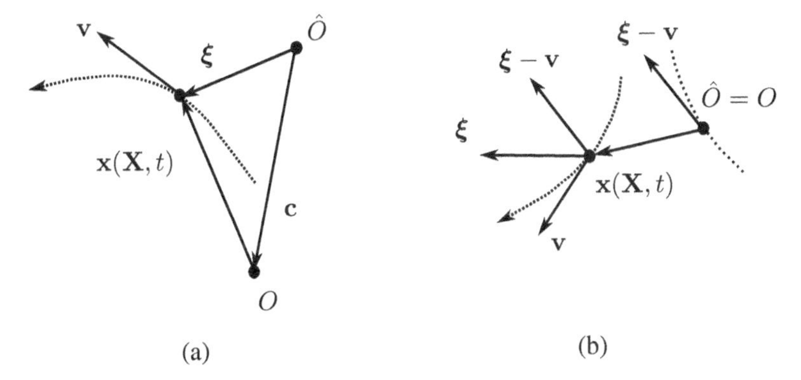

Figure 6.1. Observer transformations used to establish that constitutive functions for thermoviscous fluids do not depend explicitly on $\mathbf{x}$ or $\mathbf{v}$. Part (a) shows the transformation used for $\mathbf{x}$; part (b) shows the transformation used for $\mathbf{v}$. In each case, O and $\hat{O}$ represent origins in the original and transformed frames of reference, respectively.

The invariance condition (6.3.3) thus reduces, in the case of thermoviscous fluids, to the form

$$G(\mathsf{RU}, \mathsf{D} + \mathsf{W}, \theta, \operatorname{grad}\theta) = G(\mathsf{QRU}, \mathsf{Q}(\mathsf{D} + \mathsf{W})\mathsf{Q}^\top + \mathsf{Q}'\mathsf{Q}^\top, \theta, \operatorname{grad}\theta). \quad (6.3.5)$$

The second principle is more straightforward. As before, let $\Xi(\mathbf{X},t)$ denote the value of any one of the independent variables. If G is independent of Ξ in one frame of reference, it must be independent of Ξ in them all. The next theorem illustrates this principle.

THEOREM 6.3.2 (INDEPENDENCE FROM SPIN). *No constitutive function G depends on the spin tensor* W.

PROOF: Select any observer transformation (6.3.1) such that, at time t,

$$\mathsf{Q}(t) = \mathsf{I}, \qquad \mathbf{c}(t) = \mathbf{0},$$
$$\mathsf{Q}'(t) = -\mathsf{W}(t), \qquad \mathbf{c}'(t) = \mathbf{0}.$$

Such a choice is consistent with the result of Exercise 162, and it yields $\mathsf{QRU} = \mathsf{RU}$ and $\mathsf{QDQ}^\top + \mathsf{QWQ}^\top + \mathsf{Q}'\mathsf{Q} = \mathsf{D}$. For this observer transformation, at time t, the relation (6.3.5) reduces to the following:

$$G(\mathsf{RU}, \mathsf{D} + \mathsf{W}, \theta, \operatorname{grad}\theta) = G(\mathsf{RU}, \mathsf{D}, \theta, \operatorname{grad}\theta).$$

Since G is independent of W in one frame of reference, it is independent of W in them all. ■

At this point, we have reduced the constitutive dependencies to

$$G = G(\mathbf{x}, \mathbf{v}, \mathsf{F}, \mathsf{L}, \theta, \operatorname{grad}\theta) = G(\mathsf{RU}, \mathsf{D}, \theta, \operatorname{grad}\theta).$$

Still further reduction is possible, as the following exercise shows.

EXERCISE 163 *Show that* $G = G(\mathsf{U}, \mathsf{D}, \theta, \operatorname{grad}\theta)$.

No logical difference in functional dependency arises when we replace U in this result with the right Cauchy–Green tensor $\mathsf{C} = \mathsf{U}^2$. By doing so, we see that constitutive functions for thermoviscous fluids exhibit the reduced dependencies indicated by the equation

$$G = G(\mathsf{C}, \mathsf{D}, \theta, \operatorname{grad}\theta). \tag{6.3.6}$$

The axiom of objectivity enables several additional types of reasoning, not covered here, about the functional dependencies allowed in constitutive relationships. For a deeper exploration, refer to [35, Chapter 3].

6.3.2 Summary

The axiom of objectivity requires that each constitutive function G be form-invariant under any observer transformation

$$\hat{\mathbf{y}} = \mathsf{Q}(t)\mathbf{y} + \mathbf{c}(t), \qquad \hat{t} = t - b.$$

For thermoviscous fluids, this requirement takes the following form:

$$G(\mathbf{x}, \mathsf{F}, \mathbf{v}, \mathsf{L}, \theta, \operatorname{grad}\theta) = G(\hat{\mathbf{x}}, \hat{\mathsf{F}}, \hat{\mathbf{v}}, \hat{\mathsf{L}}, \hat{\theta}, \operatorname{grad}\hat{\theta}).$$

We apply this axiom by examining specific observer transformations to rule out certain types of dependencies. These arguments reduce the constitutive functions for thermoviscous fluids to the following generic form:

$$G = G(\mathsf{C}, \mathsf{D}, \theta, \operatorname{grad}\theta),$$

where C denotes the right Cauchy–Green tensor and D is the symmetric part of the velocity gradient.

6.4 SYMMETRY

The axiom of symmetry is more subtle than the three axioms discussed so far, and its statement at first seems more obscure.

AXIOM (SYMMETRY). *Constitutive relations must be consistent with the known symmetries of the material.*

To make sense of this mandate, we must clarify what we mean by the symmetries of a material. The idea is to ask how the mathematical representations of constitutive responses (T, ε, $\mathbf{q}$, η) depend on the choice of reference configuration $\boldsymbol{\kappa}\colon \mathcal{B} \to \mathbb{E}$. The symmetries of a material are essentially those changes in reference configuration that yield equivalent descriptions of the material's constitutive responses.

6.4.1 Changes in Reference Configuration

To pin down this idea, we examine in more detail how changes in the choice of reference configuration affect the description of a body's responses. First, consider two reference configurations, $\boldsymbol{\kappa}$ and $\boldsymbol{\lambda}$, assigning referential coordinates to a point X as follows:

$$\mathbf{X} = \boldsymbol{\kappa}(X), \qquad \mathbf{Y} = \boldsymbol{\lambda}(X).$$

By the invertibility of configurations,

$$\mathbf{Y} = \boldsymbol{\lambda}(\boldsymbol{\kappa}^{-1}(\mathbf{X})) = \boldsymbol{\psi}(\mathbf{X}),$$

where the mapping

$$\boldsymbol{\psi} = \boldsymbol{\lambda} \circ \boldsymbol{\kappa}^{-1}\colon \boldsymbol{\kappa}(\mathcal{B}) \to \boldsymbol{\lambda}(\mathcal{B})$$

has derivative, or **Jacobian transformation**, $\mathsf{J} = \operatorname{Grad}_{\mathbf{X}}\boldsymbol{\psi}$. Figure 6.2 illustrates the relationships among these mappings.

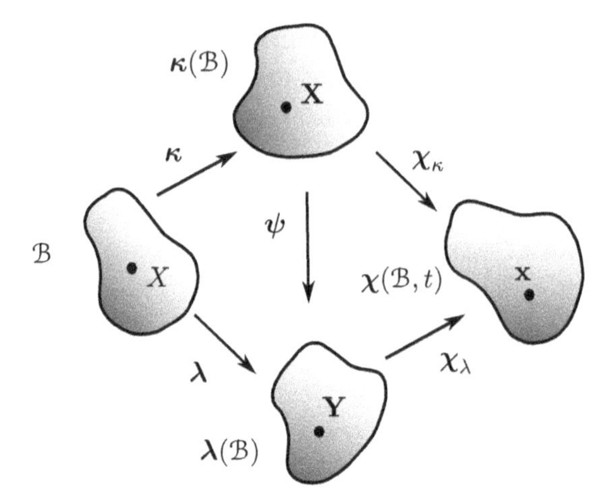

Figure 6.2. Schematic diagram of the relationships among mappings associated with two different reference configurations $\boldsymbol{\kappa}$ and $\boldsymbol{\lambda}$.

Suppose that the body undergoes a motion $\chi \colon \mathcal{B} \times \mathbb{R} \to \mathbb{E}$. With respect to the two reference configurations $\boldsymbol{\kappa}$ and $\boldsymbol{\lambda}$, the motion χ defines two deformations, namely,

$$\chi_\kappa \colon \boldsymbol{\kappa}(\mathcal{B}) \times \mathbb{R} \to \chi_\kappa(\boldsymbol{\kappa}(\mathcal{B}), t) \subset \mathbb{E},$$

$$\chi_\kappa(\mathbf{X}, t) = \chi(\boldsymbol{\kappa}^{-1}(\mathbf{X}), t),$$

and

$$\chi_\lambda \colon \boldsymbol{\lambda}(\mathcal{B}) \times \mathbb{R} \to \chi_\lambda(\boldsymbol{\lambda}(\mathcal{B}), t) \subset \mathbb{E},$$

$$\chi_\lambda(\mathbf{Y}, t) = \chi(\boldsymbol{\lambda}^{-1}(\mathbf{Y}), t).$$

These deformations furnish two possibly distinct representations of the same motion, related by the equation

$$\chi_\kappa(\mathbf{X}, t) = \chi_\lambda(\mathbf{Y}, t) = \chi_\lambda(\psi(\mathbf{X}), t).$$

The two deformations give rise to deformation gradients $\mathsf{F}_\kappa(\mathbf{X}, t)$ and $\mathsf{F}_\lambda(\mathbf{Y}, t)$, respectively, and to other quantities derived from the deformations.

Clearly, the choice of reference configuration can affect the description of a body's motion in terms of referential coordinates. Until now we have largely suppressed this dependence, assuming that the choice of reference configuration remains fixed for any particular application. In this section, we recognize the dependence explicitly using the symbol G_κ to refer to a material's responses $G = (\mathsf{T}, \varepsilon, \mathbf{q}, \eta)$ with respect to $\boldsymbol{\kappa}$. We ask which changes in reference configuration, from $\boldsymbol{\kappa}$ to $\boldsymbol{\lambda}$, yield identical values of G_κ and G_λ, after accounting for the configuration-related changes in the arguments of these two functions.

As Figure 6.2 indicates, the mapping $\boldsymbol{\psi} = \boldsymbol{\lambda} \circ \boldsymbol{\kappa}^{-1}$ effects the change in reference configuration. In examining which changes in reference configuration leave the values of constitutive functions invariant, we restrict attention to mappings $\boldsymbol{\psi}$ that leave the mass density invariant, using a criterion developed in the following exercise.

EXERCISE 164 *Show that the mapping $\boldsymbol{\psi}$ leaves the mass density invariant if and only if* $|\det \mathsf{J}| = 1$.

In addition we require consistency with the restriction, imposed in Section 2.4, that any configuration have the same orientation as the motion. Thus we insist that $\boldsymbol{\kappa}$ and $\boldsymbol{\lambda}$ have the same orientation. The condition

$$\mathsf{J} \in \mathrm{SL}(\mathbb{E}) = \left\{ \mathsf{A} \in \mathrm{L}(\mathbb{E}) \;\middle|\; \det \mathsf{A} = 1 \right\} \tag{6.4.1}$$

respects both constraints: the change in reference configuration must not affect the mass density, and it must not change orientation.

EXERCISE 165 *Show that* $\mathrm{SL}(\mathbb{E})$, *as defined in the condition (6.4.1), is a group under composition of mappings. We call it the* **special linear group**. *Observe that*

$$\mathrm{O}^+(\mathbb{E}) \subset \left\{ \begin{matrix} \mathrm{SL}(\mathbb{E}) \\ \mathrm{O}(\mathbb{E}) \end{matrix} \right\} \subset \mathrm{GL}(\mathbb{E}),$$

where $\mathrm{GL}(\mathbb{E})$ *denotes the group of tensors in* $\mathrm{L}(\mathbb{E})$ *having nonzero determinant.*

With this background, we turn next to the question of how the arguments of a generic constitutive function G change under a change in reference configuration. For a thermoviscous fluid, previous sections have reduced these arguments to $\mathsf{C}(\mathbf{X}, t)$, $\mathsf{D}(\mathbf{x}, t)$, $\theta(\mathbf{x}, t)$, and $\operatorname{grad} \theta(\mathbf{x}, t)$.

Let us begin with the arguments D, θ, and $\operatorname{grad} \theta$, for which the answers are simple. Observe that $\mathsf{D}_\kappa = \mathsf{D}_\lambda$, since we compute

$$\mathsf{L} = \operatorname{grad} \mathbf{v} = \frac{\partial v_j}{\partial x_k} \mathbf{e}_j \otimes \mathbf{e}_k,$$

and hence its symmetric part D, without reference to $\boldsymbol{\kappa}$ or $\boldsymbol{\lambda}$. For similar reasons, θ and $\operatorname{grad} \theta$ also remain unchanged under a change in reference configuration.

The remaining argument of G is the symmetric right Cauchy–Green tensor $\mathsf{C}(\mathbf{X}, t)$. To see how this tensor transforms under a change in reference configurations, consider first the deformation gradient. With respect to $\boldsymbol{\kappa}$, this function is

$$\begin{aligned}
\mathsf{F}_\kappa(\mathbf{X}, t) &= \operatorname{Grad} \chi_\kappa(\mathbf{X}, t) \\
&= \operatorname{Grad}(\chi_\lambda \circ \psi)(\mathbf{X}, t) \\
&= \operatorname{Grad} \chi_\lambda(\mathbf{Y}, t) \operatorname{Grad} \psi(\mathbf{X}, t) \\
&= \mathsf{F}_\lambda(\mathbf{Y}, t)\, \mathsf{J}(\mathbf{X}, t) = \mathsf{R}_\lambda(\mathbf{Y}, t)\, \mathsf{U}_\lambda(\mathbf{Y}, t)\, \mathsf{J}(\mathbf{X}, t).
\end{aligned}$$

Here we have used the chain rule and the polar decomposition of F_λ. The resulting identity $\mathsf{F}_\kappa = \mathsf{R}_\lambda \mathsf{U}_\lambda \mathsf{J}$ leads to the desired result for C:

EXERCISE 166 *Show that the right Cauchy–Green tensors with respect to $\boldsymbol{\kappa}$ and $\boldsymbol{\lambda}$ stand in the relationship $\mathsf{C}_\kappa = \mathsf{J}^\top \mathsf{C}_\lambda \mathsf{J}$.*

This reasoning establishes the following result.

THEOREM 6.4.1 (CHANGE OF REFERENCE CONFIGURATION IN A THERMOVISCOUS FLUID). *For a thermoviscous fluid, if $G_\lambda(\mathsf{C}, \mathsf{D}, \theta, \operatorname{grad}\theta)$ gives a constitutive response with respect to $\boldsymbol{\lambda}$, then $G_\kappa(\mathsf{J}^\top\mathsf{C}\mathsf{J}, \mathsf{D}, \theta, \operatorname{grad}\theta)$ gives the corresponding response with respect to $\boldsymbol{\kappa}$. In symbols,*

$$G_\lambda(\mathsf{C}, \mathsf{D}, \theta, \operatorname{grad}\theta) = G_\kappa(\mathsf{J}^\top\mathsf{C}\mathsf{J}, \mathsf{D}, \theta, \operatorname{grad}\theta). \tag{6.4.2}$$

6.4.2 Symmetry Groups

We now have the tools needed to tackle the central question: for what changes $\psi\colon \boldsymbol{\kappa}(\mathcal{B}) \to \boldsymbol{\lambda}(\mathcal{B})$ in reference configuration are the thermoviscous fluid's constitutive responses at X with respect to $\boldsymbol{\kappa}$ indistinguishable from those with respect to $\boldsymbol{\lambda}$? More precisely, for what Jacobian transformations $\mathsf{J} \in \mathrm{L}(\mathbb{E})$ does $G_\lambda(\cdot) = G_\kappa(\cdot)$ for all values of the arguments $(\mathsf{C}, \mathsf{D}, \theta, \operatorname{grad}\theta)$? By Equation (6.4.2), this condition is equivalent to the identity

$$G_\kappa(\mathsf{J}^\top\mathsf{C}\mathsf{J}, \mathsf{D}, \theta, \operatorname{grad}\theta) = G_\kappa(\mathsf{C}, \mathsf{D}, \theta, \operatorname{grad}\theta). \tag{6.4.3}$$

In posing the question, we allow the tensors C to range over $\mathrm{Sym}(\mathbb{E})$, in light of the fact that the right Cauchy–Green tensor is symmetric; see Equation (3.1.11).

By this observation and the condition (6.4.1), for a given material the answer to the central question is some set of linear transformations having determinant 1. The following definition accommodates the possibility that this set can vary from one particle X to another.

DEFINITION. *Given a reference configuration* $\boldsymbol{\kappa}$ *and a particle* $X \in \mathcal{B}$,

$$\Gamma_{\kappa}(X) = \Big\{ \mathsf{J} \in \mathrm{SL}(\mathbb{E}) \;\Big|\; \text{Equation (6.4.3) holds for all } \mathsf{C} \in \mathrm{Sym}(\mathbb{E}) \Big\}.$$

In words, $\Gamma_{\kappa}(X)$ consists of the unit-determinant transformations on reference configurations that leave constitutive responses at X invariant. The nature of this set depends on the material under investigation, as well, perhaps, as on the particle X. And even at a fixed particle X the set may be different for different constitutive functions. In any case, $\Gamma_{\kappa}(X)$ has an interesting mathematical structure:

THEOREM 6.4.2 (THE GROUP OF SYMMETRIES). $\Gamma_{\kappa}(X)$ *is a group under composition of mappings.*

PROOF: $\Gamma_{\kappa}(X)$ inherits associativity from $\mathrm{SL}(\mathbb{E})$. Also, it contains the identity mapping I. To prove that $\Gamma_{\kappa}(X)$ is closed, consider two transformations $\mathsf{J}_1, \mathsf{J}_2 \in \Gamma_{\kappa}(X)$. Since $\mathsf{J}_1^{\top}\mathsf{C}\mathsf{J}_1$ is symmetric,

$$\begin{aligned} G_{\kappa}((\mathsf{J}_1\mathsf{J}_2)^{\top}\mathsf{C}(\mathsf{J}_1\mathsf{J}_2), \ldots) &= G_{\kappa}(\mathsf{J}_2^{\top}(\mathsf{J}_1^{\top}\mathsf{C}\mathsf{J}_1)\mathsf{J}_2, \ldots) \\ &= G_{\kappa}(\mathsf{J}_1^{\top}\mathsf{C}\mathsf{J}_1, \ldots) = G_{\kappa}(\mathsf{C}, \ldots). \end{aligned}$$

Therefore, $\mathsf{J}_1\mathsf{J}_2 \in \Gamma_{\kappa}(X)$. To show that every element of $\Gamma_{\kappa}(X)$ has an inverse in $\Gamma_{\kappa}(X)$, note that $\Gamma_{\kappa}(X) \subset \mathrm{SL}(\mathbb{E})$, so every $\mathsf{J} \in \Gamma_{\kappa}(X)$ possesses an inverse J^{-1} with $\det \mathsf{J}^{-1} = 1$. It remains only to show that $\mathsf{J}^{-1} \in \Gamma_{\kappa}(X)$. For this task, pick any symmetric transformation C, and define $\mathsf{C}_J = \mathsf{J}^{-\top}\mathsf{C}\mathsf{J}^{-1}$, which is also symmetric. Then by the fact that $\mathsf{J} \in \Gamma_{\kappa}(X)$,

$$G_{\kappa}(\mathsf{J}^{-\top}\mathsf{C}\mathsf{J}^{-1}, \ldots) = G_{\kappa}(\mathsf{C}_J, \ldots) = G_{\kappa}(\mathsf{J}^{\top}\mathsf{C}_J\mathsf{J}, \ldots) = G_{\kappa}(\mathsf{C}, \ldots).$$

This completes the proof. ∎

We call $\Gamma_{\kappa}(X)$ the **symmetry group** for the material at the particle X. We use it to characterize the symmetries of a material, as detailed shortly. The **symmetries** of the material at X are the elements of $\Gamma_{\kappa}(X)$. The physical interpretation is as follows: if $\mathsf{J} \in \Gamma_{\kappa}(X)$, then changes from $\boldsymbol{\kappa}$ in the reference configuration that have derivative J cannot be detected by measuring the responses governed by the constitutive function G.

An exercise in elementary group theory shows how $\Gamma_{\kappa}(X)$ changes under an arbitrary change in reference configuration whose Jacobian transformation belongs to $\mathrm{SL}(\mathbb{E})$:

THEOREM 6.4.3 (NOLL'S THEOREM). *If* $\mathsf{K} \in \mathrm{SL}(\mathbb{E})$ *is the derivative of the mapping from reference configuration* $\boldsymbol{\kappa}$ *to reference configuration* $\boldsymbol{\lambda}$*, then*

$$\Gamma_\lambda(X) = \mathsf{K}\,\Gamma_\kappa(X)\,\mathsf{K}^{-1}. \tag{6.4.4}$$

In other words, $\mathsf{J} \in \Gamma_\kappa(X)$ if and only if $\mathsf{KJK}^{-1} \in \Gamma_\lambda(X)$. Algebraically, Equation (6.4.4) states that $\Gamma_\lambda(X)$ is a **conjugate group** of $\Gamma_\kappa(X)$. Noll's theorem shows how the symmetry groups associated with two different reference configurations are related, even in cases when changing between the reference configurations does not leave the material's responses unchanged.

EXERCISE 167 *Prove Theorem 6.4.3. The key relationships are*

1. $G_\lambda(\mathsf{C}_\lambda, \ldots) = G_\kappa(\mathsf{C}_\kappa, \ldots)$, *and*
2. $\mathsf{C}_\lambda = \mathsf{K}^{-\top}\mathsf{C}_\kappa\mathsf{K}^{-1}$.

Figure 6.3 illustrates.

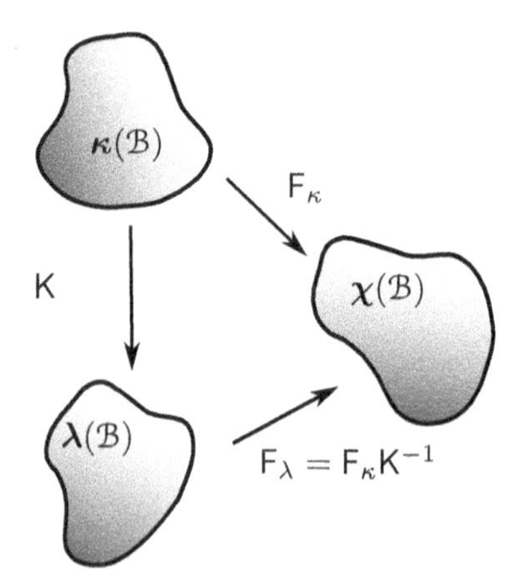

Figure 6.3. Mappings involved in a change of reference configuration, including the deformation gradients F_κ and F_λ that give rise to corresponding right Cauchy–Green tensors C_κ and C_λ.

Another group-theoretic observation relates more directly to the discussion that follows. The trivial group $\{\mathsf{I}\}$ is the smallest possible symmetry group, in the sense that for any symmetry group $\Gamma_\kappa(X)$ the following inclusions automatically hold:

$$\{\mathsf{I}\} \subset \Gamma_\kappa(X) \subset \mathrm{SL}(\mathbb{E}).$$

Between these two bounds on $\Gamma_\kappa(X)$ there is a remarkable variety of groups. For example, there are finite groups, such as those that contain only rotations by integer fractions of 2π about certain axes, and infinite groups, such as those whose elements depend on a continuously varying parameter such as angle of rotation.

6.4.3 Classification of Materials

Noll [41] proposed using the detailed structure of symmetry groups for the stress T to classify materials in a manner that is more formal than what we have done so far.

DEFINITION. *Let $\Gamma_\kappa(X)$ be the symmetry group for the stress T at X.*

1. *A material is* **isotropic** *at X if there exists a reference configuration $\boldsymbol{\kappa}$ such that $\mathrm{O}^+(\mathbb{E}) \subset \Gamma_\kappa(X)$. In words, there is a reference configuration with respect to which proper rotations have no effect on the constitutive function for stress. Isotropic materials exhibit no preferred directions.*

2. *A material is a* **fluid** *at X if there exists a reference configuration $\boldsymbol{\kappa}$ with respect to which $\Gamma_\kappa(X) = \mathrm{SL}(\mathbb{E})$. In words, the material is isotropic at X, and no density-preserving change in reference configuration has any effect on the constitutive function T. Less formally, a fluid is a material for which there is no preferred reference configuration.*

3. *A material is a* **solid** *at X if there exists a reference configuration $\boldsymbol{\kappa}$ with respect to which $\Gamma_\kappa(X) \subset \mathrm{O}^+(\mathbb{E})$. (Note that $\mathrm{O}^+(\mathbb{E})$ is a proper subgroup of $\mathrm{SL}(\mathbb{E})$.) In words, changes in reference configuration other than rigid rotations affect the constitutive function T. If $\Gamma_\kappa(X)$ is a proper subgroup of $\mathrm{O}^+(\mathbb{E})$, the stress function T is invariant only under some restricted set of rigid rotations.*

This classification deserves three comments. First, related to the notion of an isotropic material is a corresponding concept for constitutive functions:

DEFINITION: *A function of scalar, vector, or tensor arguments is* **isotropic** *if its action is invariant under rigid-body rotations and reflections.*

The precise formulation of this condition depends on the tensor order of the function's values as well as on the tensor order of its arguments. The general principle is that the function's output must rotate or reflect in concert with rotations or reflections applied to its arguments, in a manner that eliminates directional dependence. As discussed in Section 3.5, a rotation or reflection $\mathsf{Q} \in \mathrm{O}(\mathbb{E})$ transforms a scalar a, a vector a, and a tensor A into a, Qa, and $\mathsf{QAQ}^\top$, respectively. Table 6.1 summarizes

Table 6.1. Transformation Properties of Isotropic Functions

Scalar Argument	Vector Argument	Tensor Argument
$\varphi(a) = \varphi(a)$	$\varphi(\mathsf{Q}\mathbf{a}) = \varphi(\mathbf{a})$	$\varphi(\mathsf{QAQ}^\top) = \varphi(\mathsf{A})$
$\boldsymbol{\varphi}(a) = \mathsf{Q}\boldsymbol{\varphi}(a)$	$\boldsymbol{\varphi}(\mathsf{Q}\mathbf{a}) = \mathsf{Q}\boldsymbol{\varphi}(\mathbf{a})$	$\boldsymbol{\varphi}(\mathsf{QAQ}^\top) = \mathsf{Q}\boldsymbol{\varphi}(\mathsf{A})$
$\Phi(a) = \mathsf{Q}\Phi(a)\mathsf{Q}^\top$	$\Phi(\mathsf{Q}\mathbf{a}) = \mathsf{Q}\Phi(\mathbf{a})\mathsf{Q}^\top$	$\Phi(\mathsf{QAQ}^\top) = \mathsf{Q}\Phi(\mathsf{A})\mathsf{Q}^\top$

how the outputs of isotropic scalar-, vector-, and tensor-valued functions $\varphi, \boldsymbol{\varphi}$, and Φ, respectively, change in response to these actions of Q:

EXERCISE 168 *Show that the following functions are isotropic:*

1. $\varphi(\mathbf{a}) = \|\mathbf{a}\|$
2. $\varphi(\mathsf{A}) = \det \mathsf{A}$
3. $\Phi(a) = a\,\mathsf{I}$
4. $\Phi(\mathsf{A}) = \mathsf{A}^n$, *for any integer* $n > 0$.

Among several interesting theorems about isotropic functions is the following fact about vector-valued functions of symmetric tensors:

EXERCISE 169 *Suppose that* $\boldsymbol{\varphi}\colon \mathrm{Sym}(\mathbb{E}) \to \mathbb{E}$ *is isotropic. Show that* $\boldsymbol{\varphi}(\mathsf{A}) = \mathbf{0}$ *for every* $\mathsf{A} \in \mathrm{Sym}(\mathbb{E})$.

Another useful theorem about isotropic functions deals with scalar-valued functions of symmetric tensors:

THEOREM 6.4.4 (ISOTROPIC FUNCTIONS). *A function* $\varphi\colon \mathrm{Sym}(\mathbb{E}) \to \mathbb{R}$ *is isotropic if and only if there exists a scalar function* $f\colon \mathbb{R}^3 \to \mathbb{R}$ *such that for any* $\mathsf{A} \in \mathrm{Sym}(\mathbb{E})$, $\varphi(\mathsf{A}) = f(\lambda_1, \lambda_2, \lambda_3)$, *where* λ_1, λ_2, *and* λ_3 *are the eigenvalues of* A.

Before embarking on the proof, notice that we can just as well express the theorem in terms of other invariants associated with A, for example, $\varphi(\mathsf{A}) = f(\mathrm{I}_A, \mathrm{II}_A, \mathrm{III}_A)$ or, by Exercise 54, $\varphi(\mathsf{A}) = f(\mathrm{tr}\,\mathsf{A}, \mathrm{tr}\,\mathsf{A}^2, \mathrm{tr}\,\mathsf{A}^3)$.

PROOF: The proof that the existence of $f(\lambda_1, \lambda_2, \lambda_3)$ is sufficient for $\varphi(\mathsf{A})$ to be isotropic is Exercise 170. To show that it is necessary, assume that φ is isotropic, so that $\varphi(\mathsf{QAQ}^\top) = \varphi(\mathsf{A})$ for every tensor $\mathsf{A} \in \mathrm{Sym}(\mathbb{E})$ and every $\mathsf{Q} \in \mathrm{O}(\mathbb{E})$. By Exercises 50 and 51, $\mathsf{A}, \mathsf{B} \in \mathrm{Sym}(\mathbb{E})$ have the same eigenvalues if and only if there exists a tensor $\mathsf{Q} \in \mathrm{O}(\mathbb{E})$ for which $\mathsf{B} = \mathsf{QAQ}^\top$. Hence, it suffices to show that $\varphi(\mathsf{A}) = \varphi(\mathsf{B})$ whenever A and B have the same eigenvalues λ_1, λ_2, and λ_3. Assume they do. By the spectral decomposition established in Section 3.1, there exist orthonormal bases $\{\mathbf{p}_1, \mathbf{p}_2, \mathbf{p}_3\}$ and $\{\mathbf{q}_1, \mathbf{q}_2, \mathbf{q}_3\}$ such that

$$\mathsf{A} = \sum_{i=1}^{3} \lambda_i \mathbf{p}_i \otimes \mathbf{p}_i, \quad \mathsf{B} = \sum_{i=1}^{3} \lambda_i \mathbf{q}_i \otimes \mathbf{q}_i.$$

Define $\mathsf{Q} \in \mathrm{O}(\mathbb{E})$ by assigning $\mathsf{Q}\mathbf{q}_i = \mathbf{p}_i$ for $i = 1, 2, 3$. Then

$$\mathsf{Q}(\mathbf{q}_i \otimes \mathbf{q}_i)\mathsf{Q}^\top = (\mathsf{Q}\mathbf{q}_i) \otimes (\mathsf{Q}\mathbf{q}_i) = \mathbf{p}_i \otimes \mathbf{p}_i,$$

so $\mathsf{QBQ}^\top = \mathsf{A}$. But φ is isotropic, so $\varphi(\mathsf{A}) = \varphi(\mathsf{QBQ}^\top) = \varphi(\mathsf{B})$. ■

EXERCISE 170 *Prove sufficiency in the theorem above: let φ be a scalar-valued function of symmetric tensors. If there exists a function $f\colon \mathbb{R}^3 \to \mathbb{R}$ such that for every $\mathsf{A} \in \mathrm{Sym}(\mathbb{E})$, $\varphi(\mathsf{A}) = f(\lambda_1, \lambda_2, \lambda_3)$, where λ_1, λ_2, and λ_3 are the eigenvalues A, then φ is isotropic.*

The second remark is that the class of solids admits a wide range of possible behaviors. For example, it follows from the definition that a solid is isotropic at X if there exists a reference configuration $\boldsymbol{\kappa}$ for which $\Gamma_{\kappa}(X) = \mathrm{O}^+(\mathbb{E})$. All pure rotations leave the response T unchanged. For other possibilities, we refer to the classification of symmetry groups used in crystallography. A solid is **triclinic** at X if there exists a reference configuration $\boldsymbol{\kappa}$ for which $\Gamma_{\kappa}(X) = \{\mathsf{I}\}$, that is, if the material exhibits the fewest possible symmetries at X. Figure 6.4(a) illustrates a triclinic crystal. A solid is **orthotropic** at X if there exists a reference configuration $\boldsymbol{\kappa}$ for which $\Gamma_{\kappa}(X)$ contains reflections across the axes spanned by each vector in an orthonormal basis $\{\mathbf{e}_1, \mathbf{e}_2, \mathbf{e}_3\}$, as shown in Figure 6.4(b). For example, it is often reasonable to model wood as an orthotropic solid.

The third remark deals with materials whose symmetry groups for stress satisfy the inclusions

$$\mathrm{O}^+(\mathbb{E}) \subsetneq \Gamma_{\kappa}(X) \subsetneq \mathrm{SL}(\mathbb{E}).$$

We call these materials **fluid crystals**. The following sequence of exercises explores this possibility, beginning with a set of transformations on $\mathbb{E}$ defined as follows with respect to the orthonormal basis $\{\mathbf{e}_1, \mathbf{e}_2, \mathbf{e}_3\}$:

$$\mathrm{FC}_{21;31}(\mathbb{E}) = \left\{ \begin{bmatrix} J_{11} & J_{12} & J_{13} \\ 0 & J_{22} & J_{23} \\ 0 & J_{32} & J_{33} \end{bmatrix} \in \mathrm{L}(\mathbb{E}) \;\middle|\; |J_{11}(J_{22}J_{33} - J_{23}J_{32})| = 1 \right\}.$$

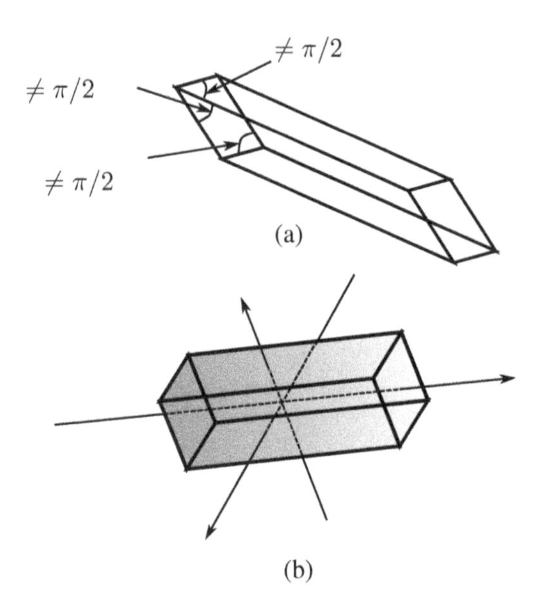

Figure 6.4. (a) A crystal of a triclinic solid having the fewest possible symmetries. (b) Illustration of an orthotropic solid among whose symmetries are reflections across a set of Cartesian coordinate axes.

(The subscript $21;31$ refers to the zero structures, that is, the entries in the matrix representations that must be zero.)

EXERCISE 171 *Show that* $\mathrm{FC}_{21;31}(\mathbb{E})$ *is a proper subgroup of* $\mathrm{SL}(\mathbb{E})$*, that is, it is a group, and* $\mathrm{FC}_{21;31}(\mathbb{E}) \subsetneq \mathrm{SL}(\mathbb{E})$*. Show that* $\mathrm{FC}_{21;31}(\mathbb{E}) \not\subset \mathrm{O}^+(\mathbb{E})$.

EXERCISE 172 *Show that* $\mathbf{e}_1$ *is an eigenvector of every* $\mathsf{J} \in \mathrm{FC}_{21;31}(\mathbb{E})$*. Thus a material for which* $\mathrm{FC}_{21;31}(\mathbb{E})$ *is the symmetry group for stress is one for which* T *remains unchanged under changes in reference configuration that translate particles in the direction* $\mathbf{e}_1$*. The existence of a preferred direction* $\mathbf{e}_1$ *justifies the use of the term "crystal."*

EXERCISE 173 *Find at least two other proper subgroups of* $\mathrm{SL}(\mathbb{E})$ *having the form* $\mathrm{FC}_{ij;kl}(\mathbb{E})$*. Describe the changes in reference configuration that leave the stress function unchanged.*

6.4.4 Implications for Thermoviscous Fluids

At last, we examine implications of the axiom of symmetry for the constitutive relations governing thermoviscous fluids. So far, by applying the axioms of determinism, equipresence, and objectivity, we have found that the generic constitutive function for this type of material has the dependencies $G = G(\mathsf{C}, \mathsf{D}, \theta, \operatorname{grad}\theta)$. Since the material is a fluid at all particles X, the axiom of symmetry requires that

$$G(\mathsf{C}, \mathsf{D}, \theta, \operatorname{grad}\theta) = G(\mathsf{J}^\top \mathsf{C}\mathsf{J}, \mathsf{D}, \theta, \operatorname{grad}\theta),$$

for any $\mathsf{J} \in \mathrm{SL}(\mathbb{E})$. Consider the special choice $\mathsf{J} = \mathsf{U}^{-1}(\det \mathsf{U})^{1/3}$, where U is the right stretch tensor in the polar decomposition $\mathsf{F} = \mathsf{R}\mathsf{U}$.

EXERCISE 174 *Show that, with this choice, $\mathsf{J} \in \mathrm{SL}(\mathbb{E})$. Verify that this definition is equivalent to the equation $\mathsf{J} = (\rho_\kappa/\rho)^{1/3}\sqrt{\mathsf{C}}^{-1}$, where $\sqrt{\mathsf{C}}$ denotes the square root of the symmetric, positive definite tensor C, as discussed in Section 3.1.*

With this choice of J,

$$\begin{aligned} G(\mathsf{C}, \mathsf{D}, \theta, \operatorname{grad}\theta) &= G\left(\left(\frac{\rho_\kappa}{\rho}\right)^{2/3} \sqrt{\mathsf{C}}^{-1}\mathsf{C}\sqrt{\mathsf{C}}^{-1}, \mathsf{D}, \theta, \operatorname{grad}\theta\right) \\ &= G\left(\left(\frac{\rho_\kappa}{\rho}\right)^{2/3}, \mathsf{D}, \theta, \operatorname{grad}\theta\right), \end{aligned}$$

the last line following from the fact that $\mathsf{C} = \mathsf{C}^\top$ and the observation that the action of $(\rho_\kappa/\rho)^{2/3}\mathsf{I}$ reduces to multiplication by a scalar-valued function.

But in examining the possible functional dependencies of G, we regard the density ρ_κ in the reference configuration as being fixed, so we may as well replace $(\rho_\kappa/\rho)^{2/3}$ by $\rho^{-2/3}$—or even by ρ^{-1}—in the argument list of G. Therefore, by using the axiom of symmetry, we have reduced the constitutive function the thermoviscous fluid to the following:

$$G = G(\rho^{-1}, \mathsf{D}, \theta, \operatorname{grad}\theta).$$

Thermodynamicists know ρ^{-1} as the **specific volume**. We reduce this functional relationship even further in the next section.

6.4.5 Summary

The axiom of symmetry states that constitutive functions must be consistent with the known symmetries of the material. These symmetries are elements of the material's symmetry group $\Gamma_\kappa(X)$, which furnishes information about which changes in reference configuration leave the calculated responses of the material at X unchanged.

$\Gamma_\kappa(X)$ is a (possibly improper) subgroup of the group $\mathrm{SL}(\mathbb{E})$ consisting of tensors $\mathsf{J} \in \mathrm{L}(\mathbb{E})$ for which $\det \mathsf{J} = 1$. The nature of the symmetry group for stress provides a system for classifying materials:

- The material is isotropic at X if there exists a reference configuration such that $\Gamma_\kappa(X) \subset \mathrm{O}^+(\mathbb{E})$.
- The material is fluid at X if $\Gamma_\kappa(X) = \mathrm{SL}(\mathbb{E})$.
- The material is solid at X if $\Gamma_\kappa(X) \subset \mathrm{O}^+(\mathbb{E})$. In the nonisotropic case ($\Gamma_\kappa(X) \subsetneq \mathrm{O}^+(\mathbb{E})$), the symmetry group is one of the classic crystallographic groups.

For thermoviscous fluids, the axiom of symmetry yields the following reduced form for each constitutive function:

$$G = G(\rho^{-1}, \mathsf{D}, \theta, \operatorname{grad} \theta).$$

6.5 ADMISSIBILITY

The last axiom of constitutive theory that we discuss in this chapter is the axiom of admissibility.

AXIOM (ADMISSIBILITY). *Constitutive relations must be consistent with the balance laws.*

Viewed in the context of continuum mechanics as a whole, this axiom seems superfluous: inconsistent assertions are inadmissible in any axiomatic system. Indeed, at several junctures in the last few sections we take pains to recognize constraints on constitutive relations imposed by some of the balance equations.

Apparent tautology notwithstanding, the entropy inequality in particular has significant implications for constitutive theory. In this section we examine these implications for thermoviscous fluids in general and for a Newtonian fluid with heat conduction governed by Fourier's law in particular. Then, to end the section, we examine a simple, admissibility-based derivation of Hooke's law, which models stress in a linearly elastic solid.

6.5.1 Implications of the Entropy Inequality

In the form (4.5.5) developed in Section 4.5, the entropy inequality asserts that

$$-\frac{\rho}{\theta}\left(\frac{D\psi}{Dt}+\eta\frac{D\theta}{Dt}\right)+\frac{1}{\theta}\operatorname{tr}(\mathsf{TD})-\frac{1}{\theta^2}\mathbf{q}\cdot\operatorname{grad}\theta\geqslant 0. \tag{6.5.1}$$

Here $\psi=\varepsilon-\theta\eta$ is the specific Helmholtz free energy. Since this form of the entropy inequality explicitly governs ψ and not ε, let us replace the constitutive function $\varepsilon(\rho^{-1},\mathsf{D},\theta,\operatorname{grad}\theta)$ by the specific Helmholtz free energy $\psi(\rho^{-1},\mathsf{D},\theta,\operatorname{grad}\theta)$ throughout. The resulting set $\{\mathsf{T},\psi,\mathbf{q},\eta\}$ of constitutive functions is equivalent to the original set $\{\mathsf{T},\varepsilon,\mathbf{q},\eta\}$, in the sense that we can derive either set from knowledge of the other.

Substituting T, ψ, $\mathbf{q}$, and η into the inequality (6.5.1) and using the chain rule to decompose $D\psi/Dt$ gives

$$-\frac{\rho}{\theta}\left(\frac{\partial\psi}{\partial\rho^{-1}}\frac{D\rho^{-1}}{Dt}+\frac{\partial\psi}{\partial D_{jk}}\frac{DD_{jk}}{Dt}+\frac{\partial\psi}{\partial\theta}\frac{D\theta}{Dt}+\frac{\partial\psi}{\partial\theta_{,j}}\frac{D\theta_{,j}}{Dt}+\eta\frac{D\theta}{Dt}\right)$$
$$+\frac{1}{\theta}T_{jk}D_{jk}-\frac{1}{\theta^2}q_j\theta_{,j}\geqslant 0.$$

Exercise 175 *Show that* $D\rho^{-1}/Dt=\rho^{-1}\operatorname{tr}\mathsf{D}$.

Using the results of Exercise 175, we obtain the equivalent inequality,

$$-\left(\frac{\partial\psi}{\partial D_{jk}}\frac{DD_{jk}}{Dt}+\frac{\partial\psi}{\partial\theta_{,j}}\frac{D\theta_{,j}}{Dt}\right)\frac{\rho}{\theta}-\left(\frac{\partial\psi}{\partial\theta}+\eta\right)\frac{D\theta}{Dt}$$
$$+\left(T_{jk}-\frac{\partial\psi}{\partial\rho^{-1}}\delta_{jk}\right)\frac{D_{jk}}{\theta}-\frac{1}{\theta^2}q_j\theta_{,j}\geqslant 0,$$

since $\operatorname{tr}\mathsf{D}=\delta_{jk}D_{jk}$.

Cast in this form, the entropy inequality is amenable to a type of reasoning that restricts the nature of the functions ψ and η. At its core, the reasoning, developed for continuum mechanics by Coleman and Noll [14], rests a principle easily grasped through the following exercise:

Exercise 176 *If the variable ζ can take arbitrary real values, what constraints does the inequality $a_0+a_1\zeta+a_2\zeta^2\geqslant 0$ impose on the coefficients a_0, a_1, and a_2 of the constant, linear, and quadratic terms?*

In the case at hand, the quantities

$$\frac{DD_{jk}}{Dt},\quad\frac{D\theta_{,j}}{Dt},\quad\frac{D\theta}{Dt},\quad D_{jk},\quad\theta_{,j}$$

can assume arbitrary values *a priori*, depending on the initial–boundary–value problems posed for the balance equations. Therefore, the entropy inequality holds for all possible values of these quantities only if the coefficients of these quantities vanish:

$$\frac{\partial \psi}{\partial D_{jk}} = 0, \qquad \frac{\partial \psi}{\partial \theta_{,j}} = 0, \qquad \eta = -\frac{\partial \psi}{\partial \theta}.$$

Thus the specific Helholtz free energy ψ is independent of both the stretching tensor D and the temperature gradient $\operatorname{grad} \theta$. It follows that

$$\psi = \psi(\rho^{-1}, \theta).$$

Also, the specific entropy η is related to ψ by the equation

$$\eta = -\frac{\partial \psi}{\partial \theta}(\rho^{-1}, \theta),$$

which is a classical thermodynamic relationship.

Having drawn these conclusions about the specific Helmholtz free energy ψ and the specific entropy η, we are left with a reduced form of the entropy inequality,

$$\left(T_{jk} - \frac{\partial \psi}{\partial \rho^{-1}} \delta_{jk} \right) \frac{D_{jk}}{\theta} - \frac{1}{\theta^2} q_j \theta_{,j} \geqslant 0. \tag{6.5.2}$$

This inequality contains information about the remaining constitutive functions T and $\mathbf{q}$, but to extract it requires further dissection. In preparation for this task, we adopt some suggestive notation. Recognizing that the quantity $-\partial \psi / \partial \rho^{-1}$ has the dimensions of force per unit area, let us call

$$p_t(\rho^{-1}, \theta) = -\frac{\partial \psi}{\partial \rho^{-1}}(\rho^{-1}, \theta)$$

the **thermodynamic pressure**. We see shortly how p_t is related to the mechanical pressure p introduced in Section 5.3. Making this substitution in the inequality (6.5.2) and multiplying through by the positive function θ yield

$$\operatorname{tr} \left[\mathsf{T}(\rho^{-1}, \mathsf{D}, \theta, \operatorname{grad} \theta) + p_t(\rho^{-1}, \theta) \mathsf{I} \right] \mathsf{D} - \frac{1}{\theta} \mathbf{q} \cdot \operatorname{grad} \theta \geqslant 0.$$

By analogy with the linear constitutive laws for fluid stresses reviewed in Section 5.3, we decompose the stress T as follows:

$$\mathsf{T}(\rho^{-1}, \mathsf{D}, \theta, \operatorname{grad} \theta) = -p_t(\rho^{-1}, \theta) \mathsf{I} + \mathsf{T}_d(\rho^{-1}, \mathsf{D}, \theta, \operatorname{grad} \theta).$$

The term $-p_t \mathsf{I}$ is clearly an isotropic tensor, mathematically analogous to the stress associated with ideal fluids in Section 5.3. The term T_d is the **extra** or **dissipative**

stress. This decomposition of T reduces the entropy inequality to the following form:

$$\Upsilon = \underbrace{\operatorname{tr}(\mathsf{T}_d\mathsf{D})}_{\text{(I)}} \underbrace{-\frac{1}{\theta}\mathbf{q}\cdot \operatorname{grad}\theta}_{\text{(II)}} \geqslant 0. \tag{6.5.3}$$

Let us call the quantity $\Upsilon(\rho^{-1}, \mathsf{D}, \theta, \operatorname{grad}\theta)$ defined here is the **dissipation rate**. The term labeled (I) represents the rate of dissipation of energy due to viscous heating, associated with shear stress. The term labeled (II) represents the rate of dissipation of energy due to heat flow.

We now analyze the restrictions on the stress T and heat flux $\mathbf{q}$.

6.5.2 Analysis of Equilibrium

The reduced entropy inequality (6.5.3) states that the dissipation rate Υ in a thermoviscous fluid is nonnegative. The dissipation rate clearly vanishes when $\mathsf{D} = \mathsf{0}$ and $\operatorname{grad}\theta = \mathbf{0}$. This observation motivates the following definition.

DEFINITION. *A thermoviscous fluid is in* **thermodynamic equilibrium** *if the stretching tensor* $\mathsf{D} = \mathsf{0}$ *and the temperature gradient* $\operatorname{grad}\theta = \mathbf{0}$.

We now analyze this state and linear departures from it.

As noted, $\Upsilon = 0$ at thermodynamic equilibrium. For this to be its minimum value, as required by the inequality (6.5.3), the constitutive variables T_d and $\mathbf{q}$ must satisfy additional constraints. We explore these using elementary calculus. Fix the values of ρ^{-1} and θ, and let $\mathbf{a} \in \mathbb{E}$ and $\mathsf{A} \in \mathrm{L}(\mathbb{E})$ be arbitrary. Define a real-valued function of a single variable ζ as follows:

$$f(\zeta) = \Upsilon(\rho^{-1}, \zeta\mathsf{A}, \theta, \zeta\mathbf{a}).$$

Provided Υ is sufficiently smooth, the fact that it attains a minimum value at thermodynamic equilibrium implies that $f'(0) = 0$. But

$$f(\zeta) = \zeta \operatorname{tr}\Big(\mathsf{T}_d(\rho^{-1}, \zeta\mathsf{A}, \theta, \zeta\mathbf{a})\mathsf{A}\Big) - \frac{\zeta}{\theta}\mathbf{q}(\rho^{-1}, \zeta\mathsf{A}, \theta, \zeta\mathbf{a})\cdot\mathbf{a},$$

so

$$f'(0) = \operatorname{tr}\Big(\mathsf{T}_d(\rho^{-1}, 0\cdot\mathsf{A}, \theta, 0\cdot\mathbf{a})\mathsf{A}\Big) - \frac{1}{\theta}\mathbf{q}(\rho^{-1}, 0\cdot\mathsf{A}, \theta, 0\cdot\mathbf{a})\cdot\mathbf{a}.$$

Since A and $\mathbf{a}$ are arbitrary, $f'(0) = 0$ only if each term on the right vanishes. We have established the following facts.

THEOREM 6.5.1 (STRESS AND HEAT FLUX AT EQUILIBRIUM). *In a thermoviscous fluid, the dissipative stress* T_d *and the heat flux* $\mathbf{q}$ *satisfy the equations*

$$\mathsf{T}_d(\rho^{-1}, 0, \theta, \mathbf{0}) = 0,$$
$$\mathbf{q}(\rho^{-1}, 0, \theta, \mathbf{0}) = \mathbf{0}. \tag{6.5.4}$$

Let us revisit two of the linear constitutive laws of Section 5.3 in light of these constraints. For the special case of Newtonian fluid obeying Fourier's law of heat conduction, the constitutive relations for T and $\mathbf{q}$ are

$$\mathsf{T}(\rho^{-1}, \mathsf{D}, \theta, \operatorname{grad}\theta) = -p\mathsf{I} + \lambda(\operatorname{tr}\mathsf{D})\mathsf{I} + 2\mu\mathsf{D},$$

$$\mathbf{q}(\rho^{-1}, \mathsf{D}, \theta, \operatorname{grad}\theta) = -k \operatorname{grad}\theta.$$

The dissipation rate in this case is

$$\Upsilon = \operatorname{tr}\left[-(p - p_t)\mathsf{D} + \lambda(\operatorname{tr}\mathsf{D})\mathsf{D} + 2\mu\mathsf{D}^2\right] + \frac{k}{\theta}\|\operatorname{grad}\theta\|^2.$$

To check that the equilibrium constraints (6.5.4) hold for these laws, we must first derive several conditions that the entropy inequality imposes on the parameters p, λ, μ, and k.

THEOREM 6.5.2 (ADMISSIBILITY IN A NEWTONIAN FLUID). *For a Newtonian fluid obeying Fourier's law of heat conduction, the dissipation rate* Υ *remains nonnegative only if all of the following conditions hold:*

1. $k \geqslant 0$;
2. $\mu \geqslant 0$;
3. $3\lambda + 2\mu \geqslant 0$;
4. $p = p_t(\rho^{-1}, \theta)$.

PROOF: We insist that $\Upsilon \geqslant 0$ for all choices of ρ^{-1}, D, θ, and $\operatorname{grad}\theta$. To show (1), choose $\mathsf{D} = 0$ and $\operatorname{grad}\theta \neq \mathbf{0}$. Then, since $\theta > 0$, $\Upsilon = (k/\theta)\,\|\operatorname{grad}\theta\|^2 \geqslant 0$ only if $k \geqslant 0$.

To establish (2), consider any motion for which $\operatorname{tr}\mathsf{D} = 0$ but $\mathsf{D} \neq 0$, and let $\operatorname{grad}\theta = \mathbf{0}$. In this case,

$$\Upsilon = 2\mu\,\mathsf{D} : \mathsf{D} = 2\mu D_{ij} D_{ij},$$

so $\Upsilon \geqslant 0$ only if $\mu \geqslant 0$.

To prove (3) and (4), consider a motion for which $\mathsf{D} = \zeta \mathsf{I}$, where $\zeta \in \mathbb{R}$ stands for an arbitrarily chosen constant, and let $\operatorname{grad} \theta = \mathbf{0}$. One can show that whenever $\zeta \neq 0$,

$$\Upsilon = \left[-(p - p_t)\zeta + (3\lambda + 2\mu)\zeta^2\right] \operatorname{tr} \mathsf{I}.$$

Obviously, $\operatorname{tr} \mathsf{I} = 3$. Also, ζ can be positive or negative, but in either case $\zeta^2 > 0$. Since the magnitude of ζ is otherwise arbitrary, to guarantee that $\Upsilon \geqslant 0$ requires that the first term in square brackets vanish and that the second be nonnegative. In other words, $p = p_t$ and $3\lambda + 2\mu \geqslant 0$. ■

This theorem shows in particular that for a Newtonian fluid,

$$\mathsf{T}_d = \lambda(\operatorname{tr} \mathsf{D})\mathsf{I} + 2\mu \mathsf{D}.$$

Therefore, in the special case when the thermoviscous fluid obeys linear constitutive relations for stress and heat flux, the equilibrium constraints (6.5.4) on T_d and $\mathbf{q}$ hold automatically.

6.5.3 Linear, Isotropic, Thermoelastic Solids

To close this section, we use the axiom of admissibility to derive a form of Hooke's law for an isotropic, linearly elastic solid, discussed in Section 5.4. A **thermoelastic solid** is a material whose constitutive functions have the following dependencies:

$$\begin{aligned}
\psi &= \psi(\mathsf{E}_L, \theta, \operatorname{grad} \theta), \\
\eta &= \eta(\mathsf{E}_L, \theta, \operatorname{grad} \theta), \\
\mathsf{S}_\kappa &= \mathsf{S}_\kappa(\mathsf{E}_L, \theta, \operatorname{grad} \theta), \\
\mathbf{q} &= \mathbf{q}(\mathsf{E}_L, \theta, \operatorname{grad} \theta).
\end{aligned} \tag{6.5.5}$$

Here S_κ denotes the second Piola–Kirchhoff stress, introduced in Equation (4.3.5), and E_L is the Lagrangian strain tensor; see Equation (3.1.13). We examine restrictions arising from the constitutive axioms, then review a set of approximations used to model the material under the following conditions:

1. Departures from a reference temperature θ_0 are small.

2. The reference configuration is a stress-free state, for which

$$\mathsf{S}_\kappa(0, \theta_0, \mathbf{0}) = 0.$$

This condition parallels the equilibrium condition for stress established in Theorem 6.5.1.

3. Departures from the reference configuration are small.
4. Displacement gradients $\partial u_i/\partial X$ are small in magnitude, so $\partial u_i/\partial X \simeq \partial u_i/\partial x_j$ and $\mathsf{E}_L \simeq \mathsf{E}$, as described in Equation (3.2.2).

We treat the second Piola–Kirchhoff stress S_κ as a constitutive variable, in lieu of stress tensor T. In doing so, we anticipate, based on condition (3), the use of the momentum balance in the referential form (4.2.13). Recall from Section 4.3 that $\mathsf{S}_k = \mathsf{F}^{-1}\mathsf{T}_\kappa$, where T_κ is the first Piola–Kirchhoff stress, introduced in Equation (4.2.12).

The dependencies listed above clearly respect the axioms of determinism and equipresence. They also obey the axiom of objectivity: θ and its gradient are both objective, and $\mathsf{E}_L = \frac{1}{2}(\mathsf{C}-\mathsf{I})$, where the right Cauchy–Green tensor C is an allowable independent variable by the reasoning leading to Equation (6.3.6). The axiom of symmetry requires that constitutive relations for this material respect the fact that it is isotropic. We impose a restriction arising from isotropy shortly.

To ensure that the constitutive functions are consistent with the axiom of admissibility, we substitute the relationships (6.5.5) into the referential form of the entropy inequality developed in Exercise 113:

$$-\rho_\kappa\left(\frac{D\psi}{Dt}+\eta\frac{D\theta}{Dt}\right)+\mathsf{S}_\kappa : \frac{D\mathsf{E}_L}{Dt}-\frac{\rho_\kappa}{\rho\theta}\mathbf{q}\cdot \operatorname{grad}\theta \geqslant 0. \tag{6.5.6}$$

Applying the chain rule to $D\psi/Dt$ yields, in index notation,

$$\frac{D\psi}{Dt}=\frac{\partial\psi}{\partial E_{Lij}}\frac{DE_{Lij}}{Dt}+\frac{\partial\psi}{\partial\theta}\frac{D\theta}{Dt}+\frac{\partial\psi}{\partial\theta_{,i}}\frac{D\theta_{,i}}{Dt}.$$

Using this expansion in the inequality (6.5.6) and regrouping yields

$$\left(-\rho_\kappa\frac{\partial\psi}{\partial E_{Lij}}+S_{\kappa ij}\right)\frac{DE_{Lij}}{Dt}-\left(\frac{\partial\psi}{\partial\theta}+\eta\right)\frac{D\theta}{Dt}-\frac{\partial\psi}{\partial\theta_{,i}}\frac{D\theta_{,i}}{Dt}-\frac{\rho_\kappa}{\rho}q_i\theta_{,i}\geqslant 0. \tag{6.5.7}$$

The left side of the inequality (6.5.7) is linear in the functions DE_{Lij}/Dt, $D\theta/Dt$, and $D\theta_{,i}/Dt$, which can assume any real values *a priori*. Therefore, the coefficients of these functions must vanish:

$$S_{\kappa ij}=\rho_\kappa\frac{\partial\psi}{\partial E_{Lij}},\qquad \eta=-\frac{\partial\psi}{\partial\theta},\qquad \frac{\partial\psi}{\partial\theta_{,i}}=0.$$

Three conclusions follow: (1) $\psi=\psi(\mathsf{E}_L,\theta)$; (2) we can compute S_k by differentiating ψ; and (3) the entropy inequality for a thermoelastic solid reduces to

$$-\mathbf{q}\cdot\operatorname{grad}\theta\geqslant 0.$$

This last requirement easily admits Fourier's law of heat conduction, $\mathbf{q} = -k \operatorname{grad} \theta$, provided $k > 0$.

To see what restrictions these conclusions impose on S_κ, first observe that small departures from the reference configuration and small displacement gradients imply that $\mathsf{E}_L \simeq \mathsf{E}$. Also, since $\mathsf{S}_\kappa = \mathsf{F}^{-1}\mathsf{T}_\kappa$, the identity (3.2.1) for F^{-1} gives

$$S_{\kappa ij} = \left(\delta_{ik} - \frac{\partial u_i}{\partial x_k}\right) T_{\kappa kj} \simeq T_{\kappa ij}.$$

Adopting this approximation, we compute $\mathsf{T}_\kappa(\mathsf{E}, \theta)$ by differentiating ψ.

Toward this end, the isotropy of ψ and Theorem 6.4.4 imply the existence of a function $f \colon \mathbb{R}^3 \to \mathbb{R}$ such that

$$\psi(\mathsf{E}, \theta) = f(\operatorname{tr} \mathsf{E}, \operatorname{tr} \mathsf{E}^2, \operatorname{tr} \mathsf{E}^3, \theta).$$

So long as $\theta - \theta_0$ and the entries of E all remain small in magnitude, we can expand ψ in a power series centered at the values $(\mathsf{E}, \theta - \theta_0) = (0, 0)$ and neglect terms of degree greater than 2:

$$\begin{aligned} \psi(\mathsf{E}, \theta) = C_1 &+ C_2 \operatorname{tr} \mathsf{E} + C_3 (\operatorname{tr} \mathsf{E})^2 + C_4 \operatorname{tr} \mathsf{E}^2 \\ &+ C_5(\theta - \theta_0) + C_6(\theta - \theta_0) \operatorname{tr} \mathsf{E} + C_7(\theta - \theta_0)^2. \end{aligned} \tag{6.5.8}$$

Here C_1, C_2, C_3, C_4, C_5, and C_6 are constants.

EXERCISE 177 *Verify the following derivative identities:*

$$\begin{aligned} \frac{\partial}{\partial E_{ij}} \operatorname{tr} \mathsf{E} &= \mathsf{I}, \\ \frac{\partial}{\partial E_{ij}} (\operatorname{tr} \mathsf{E})^2 &= 2 (\operatorname{tr} \mathsf{E})\mathsf{I}, \\ \frac{\partial}{\partial E_{ij}} (\operatorname{tr} \mathsf{E}^2) &= 2\, \mathsf{E}. \end{aligned}$$

Differentiating the expression (6.5.8) with respect to the entries of E and substituting the results of Exercise 177 yield

$$\mathsf{T}_\kappa(\mathsf{E}, \theta) = \rho_\kappa \left[C_2 \mathsf{I} + 2C_3(\operatorname{tr} \mathsf{E})\mathsf{I} + 2C_4\mathsf{E} + C_6(\theta - \theta_0)\mathsf{I}\right].$$

The requirement that $\mathsf{T}_\kappa = 0$ when $\mathsf{E} = 0$ implies that $C_2 = 0$, so we are left with

$$\mathsf{T}_\kappa(\mathsf{E}, \theta) = \lambda(\operatorname{tr} \mathsf{E})\mathsf{I} + 2\mu\mathsf{E} + \beta(\theta - \theta_0)\mathsf{I}. \tag{6.5.9}$$

Equation (6.5.9) serves as an extension of Hooke's law, with Lamé parameters $\lambda = 2\rho_\kappa C_3$ and $\mu = \rho_\kappa C_4$. The parameter $\beta = \rho_\kappa C_6$, called the **coefficient of thermal expansion**, accounts for stress induced when the temperature deviates from its reference value θ_0.

6.5.4 Summary

The axiom of admissibility asserts that constitutive relations must be consistent with the balance laws, including the entropy inequality:

$$-\frac{\rho}{\theta}\left(\frac{D\psi}{Dt}+\eta\frac{D\theta}{Dt}\right)+\frac{1}{\theta}\,\mathrm{tr}\,(\mathsf{TD})-\frac{1}{\theta^2}\mathbf{q}\cdot\mathrm{grad}\,\theta\geqslant 0.$$

To apply this constraint to thermoviscous fluids, we switch from the constitutive functions $\{\mathsf{T},\varepsilon,\mathbf{q},\eta\}$ to the set $\{\mathsf{T},\psi,\mathbf{q},\eta\}$, in which the specific Helmholtz free energy ψ replaces the specific internal energy ε.

Substituting these functions into the entropy inequality, we find that the entropy inequality requires a thermoviscous fluid to obey the following identities:

$$\psi=\psi(\rho^{-1},\theta),\qquad \eta=-\frac{\partial\psi}{\partial\theta}(\rho^{-1},\theta).$$

Also, for materials of this type the stress must have the form

$$\mathsf{T}=-p_t(\rho^{-1},\theta)\mathsf{I}+\mathsf{T}_d(\rho^{-1},\mathsf{D},\theta,\,\mathrm{grad}\,\theta),$$

where $p_t=-\partial\psi/\partial\rho^{-1}$ denotes the thermodynamic pressure and T_d is the dissipative stress. The stress law for Newtonian fluids is a special case.

Examining these results at thermodynamic equilibrium—a state in which $\mathsf{D}=0$ and $\mathrm{grad}\,\theta=\mathbf{0}$—yields further information. In particular, the heat flux and dissipative stress must obey the conditions

$$\mathsf{T}_d(\rho^{-1},0,\theta,\mathbf{0})=0,$$
$$\mathbf{q}(\rho^{-1},0,\theta,\mathbf{0})=\mathbf{0},$$

in agreement with the stress law for Newtonian fluids and Fourier's law for heat flux. If we adopt these two special cases, consistency with the entropy inequality requires that the thermal conductivity k, viscosity coefficients λ and μ, and mechanical pressure p obey the following conditions:

1. $k\geqslant 0$;
2. $\mu\geqslant 0$;
3. $3\lambda+2\mu\geqslant 0$;
4. $p=p_t(\rho^{-1},\theta)$.

An application of the axiom of admissibility also leads to a derivation of a form of Hooke's law,

$$\mathsf{T}_\kappa(\mathsf{E},\theta)=\lambda(\mathrm{tr}\,\mathsf{E})\mathsf{I}+2\mu\mathsf{E}+\beta(\theta-\theta_0)\mathsf{I},$$

for the first Piola–Kirchhoff stress in an isotropic, thermoelastic solid.

CHAPTER 7

MULTICONSTITUENT CONTINUA

The treatment of continua presented so far cannot distinguish among different types of matter in the bodies of interest. In many modeling applications this shortcoming is significant. For example, the theory fails to accommodate such important phenomena as diffusion and multiphase flows, in which particles from different constituents of matter move with respect to each other and can affect each other's motions. This chapter provides a brief introduction to the mechanics of multiconstituent continua.

The literature on this subject is large, varied, and, in some respects, unsettled. There are several different theoretical starting points for the theory, depending on the degree to which one hopes to account for phenomena observable at various scales of observation. These different starting points often lead to macroscopic models that are quite similar. The approach followed here is closest in spirit to **mixture theory**; for introductions to this tradition, see [6] and [27]. Other active and productive traditions exist; see [15] for a review.

Continuum Mechanics: The Birthplace of Mathematical Models, First Edition. M.B. Allen.

7.1 CONSTITUENTS

A **multiconstituent continuum** or **mixture** is a set of bodies $\mathcal{B}_\alpha$, $\alpha = 1, 2, \ldots, N$, called **constituents**. We adopt the following hypothesis.

AXIOM OF OVERLAPPING CONTINUA. *At every point* $\mathbf{x}$ *in the region of interest there is a particle* $X_\alpha \in \mathcal{B}_\alpha$ *from each constituent.*

Examples of materials in which this seemingly paradoxical assumption yields useful models include solutions of different molecular species, such as salt + water, and mixtures of different material phases, such as the solid + fluid mixtures that occur in flows through porous media and the solid + solid mixtures in composite materials. The assumption of overlapping continua is clearly unrealistic from the viewpoint of pure physics: the ions in saltwater are segregated at the molecular scale of observation, about 10^{-10} meters, and the solid matrix and fluid in a typical geologic porous medium are distinguishable at scales observable with a low-power microscope, about 10^{-5} meters. Nevertheless, at sufficiently large scales of observation the idea of overlapping continua often yields useful mathematical models. In this respect the concept is no more objectionable than the modeling assumptions underlying the mechanics of simple continua.

7.1.1 Configurations and Motions

Paralleling the theory of simple continua, for every constituent body $\mathcal{B}_\alpha$ we fix a reference configuration $\boldsymbol{\kappa}_\alpha \colon \mathcal{B}_\alpha \to \mathbb{E}$. The mapping $\boldsymbol{\kappa}_\alpha$ assigns to each particle $X_\alpha \in \mathcal{B}_\alpha$ a unique spatial address $\mathbf{X}_\alpha = \boldsymbol{\kappa}_\alpha(X_\alpha)$, as shown in Figure 7.1. This reference position $\mathbf{X}_\alpha$ serves as a new label for X_α that facilitates analysis using the tools of calculus.

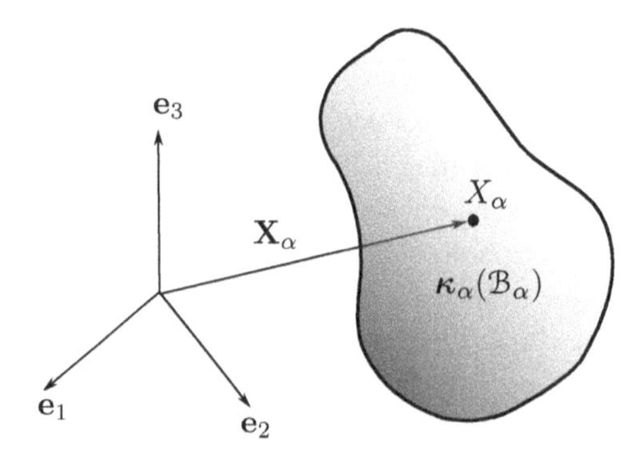

Figure 7.1. A reference configuration $\boldsymbol{\kappa}_\alpha$ associated with a constituent α.

Also associated with each constituent body $\mathcal{B}_\alpha$ is a **motion**, which is a one-parameter family $\chi_\alpha(X_\alpha, t)$ of configurations of $\mathcal{B}_\alpha$. The motion assigns to each particle $X_\alpha \in \mathcal{B}_\alpha$ and each time t a position $\mathbf{x} = \chi_\alpha(X_\alpha, t)$ occupied by the particle X_α at time t. The assumption that the function χ_α is a diffeomorphism (see Section 2.1) guarantees that it maps particles to positions in a continuously differentiable, invertible manner. The inverse motion, also continuously differentiable, gives the particle $X_\alpha = \chi_\alpha^{-1}(\mathbf{x}, t)$ from constituent α that occupies the spatial location $\mathbf{x}$ at time t.

Of greater utility than the motion χ_α, for purposes of calculation, is the **deformation** $\chi_{\alpha,\kappa}$ associated with the reference configuration $\boldsymbol{\kappa}_\alpha$. Formally,

$$\chi_{\alpha,\kappa}(\mathbf{X}_\alpha, t) = \chi_\alpha(\boldsymbol{\kappa}_\alpha^{-1}(\mathbf{X}_\alpha, t), t),$$

which is the analog of the definition given for simple continua in Equation (2.1.1). Conceptually, $\chi_{\alpha,\kappa}(\mathbf{X}_\alpha, t)$ gives the spatial location occupied at time t by the particle whose label in the reference configuration is $\mathbf{X}_\alpha$. Figure 7.2 illustrates. Similarly, $\chi_{\alpha,\kappa}^{-1}(\mathbf{x}, t)$ gives the particle $\mathbf{X}_\alpha$ from constituent α that resides at spatial position $\mathbf{x}$ at time t.

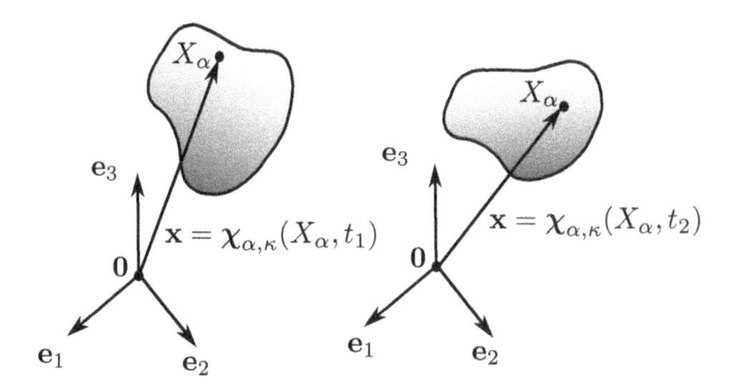

Figure 7.2. A deformation $\chi_{\alpha,\kappa}$ associated with a constituent α, evaluated at distinct times t_1 and t_2.

Using the deformation, we define the **velocity** and **acceleration** of constituent α as

$$\frac{\partial \chi_{\alpha,\kappa}}{\partial t}(\mathbf{X}_\alpha, t), \qquad \frac{\partial^2 \chi_{\alpha,\kappa}}{\partial t^2}(\mathbf{X}_\alpha, t),$$

respectively. The velocity of constituent α at a prescribed spatial position $\mathbf{x}$ is then the **spatial velocity**, given by

$$\mathbf{v}_\alpha(\mathbf{x}, t) = \frac{\partial \chi_{\alpha,\kappa}}{\partial t}(\chi_{\alpha,\kappa}^{-1}(\mathbf{x}, t), t).$$

By analogy with simple continua, we define the **material derivative** of a function f with respect to the motion of constituent α as

$$\frac{D^\alpha f}{Dt}(\mathbf{X}_\alpha, t) = \frac{\partial f}{\partial t}(\mathbf{X}_\alpha, t)$$

when $f = f(\mathbf{X}_\alpha, t)$. As in Theorem 2.2.1, when $f = f(\mathbf{x}, t)$, the chain rule implies that

$$\frac{D^\alpha f}{Dt}(\mathbf{x}, t) = \frac{\partial f}{\partial t}(\mathbf{x}, t) + \mathbf{v}_\alpha(\mathbf{x}, t) \cdot \text{grad}\, f(\mathbf{x}, t).$$

We extend this notation to vector-valued functions $\mathbf{f}(\mathbf{x}, t)$ by writing

$$\frac{D^\alpha \mathbf{f}}{Dt} = \frac{\partial \mathbf{f}}{\partial t} + (\mathbf{v}_\alpha \cdot \text{grad})\mathbf{f}.$$

7.1.2 Volume Fractions and Densities

As for single-constituent continua, we associate with each constituent body $\mathcal{B}_\alpha$ a **mass density** $\rho_\alpha(\mathbf{x}, t)$, giving the mass of constituent α in any region $\mathcal{R}$ as the volume integral

$$\int_{\mathcal{R}} \rho_\alpha(\mathbf{x}, t)\, dv.$$

Engineers commonly refer to ρ_α as the **bulk density** of α; it has dimension [mass of α per unit volume of mixture].

In some multiconstituent continua the microscopic geometry, observed at scales too fine for the axiom of overlapping continua to apply, plays a significant role in the macroscopic mechanics. In such continua there exists a fine scale of observation at which constituents still appear as continua yet at which observable interfaces separate the constituents. Examples include water-saturated sandstone and the suspensions of fine solid particles in air flowing through fluidized-bed combustion chambers. We call such multiconstituent continua **multiphase mixtures** or **immiscible mixtures**, and we refer to the constituents as **phases**.

Multiphase mixtures stand in contrast to **multispecies mixtures**, or **miscible mixtures**, in which there are no continuum-scale interfaces between constituents. Solutions of salt in water are good examples: the segregation between the ions of different chemical species in this system is detectable only at molecular scales of observation, and we model them as having no significant effect on the continuum-scale dynamics. We refer to the constituents in such mixtures as **species**.

To account for the effects of microscopic segregation among phases, we associate with each phase α of a multiphase mixture a **volume fraction** $\phi_\alpha(\mathbf{x}, t)$. For any region $\mathcal{R}$ this function gives the fraction of the volume of $\mathcal{R}$ occupied by phase α as the volume integral

$$\int_{\mathcal{R}} \phi_\alpha(\mathbf{x}, t)\, dv.$$

The dimensions of ϕ_α are [volume of α per unit volume of mixture]. We postulate that

$$\sum_{\alpha=1}^{N} \phi_\alpha = 1$$

in any multiphase mixture. In practice, this assumption amounts to an agreement that we must account for all of the volume occupied by the mixture.

Volume fractions do not arise naturally in the mechanics of multispecies mixtures.

Using mass densities and volume fractions, we define several useful quantities associated with multiconstituent continua. Engineers commonly refer to the quantity

$$\gamma_\alpha = \frac{\rho_\alpha}{\phi_\alpha},$$

having dimension [mass of α per unit volume of α], as the **true density** of species α. The **mixture density** is

$$\overline{\rho} = \sum_{\alpha=1}^{N} \rho_\alpha = \sum_{\alpha=1}^{N} \phi_\alpha \gamma_\alpha.$$

We refer to the mass-weighted average velocity

$$\overline{\mathbf{v}} = \frac{1}{\overline{\rho}} \sum_{\alpha=1}^{N} \rho_\alpha \mathbf{v}_\alpha \tag{7.1.1}$$

as the **barycentric velocity**. By analogy with the material derivatives D^α/Dt associated with individual constituents, it is sometimes useful to refer to a derivative associated with the barycentric velocity of the mixture. For a differentiable function $f(\mathbf{x}, t)$, we define the **barycentric derivative** as

$$\frac{Df}{Dt}(\mathbf{x}, t) = \frac{\partial f}{\partial t}(\mathbf{x}, t) + \overline{\mathbf{v}}(\mathbf{x}, t) \cdot \operatorname{grad} f(\mathbf{x}, t).$$

Finally, the **diffusion velocity** of constituent α is

$$\boldsymbol{\nu}_\alpha = \mathbf{v}_\alpha - \overline{\mathbf{v}}.$$

EXERCISE 178 *Show that*

$$\sum_{\alpha=1}^{N} \rho_\alpha \boldsymbol{\nu}_\alpha = \mathbf{0}.$$

EXERCISE 179 *Show that, for any differentiable, scalar-valued function* $f(\mathbf{x}, t)$,

$$\frac{Df}{Dt} = \frac{D^\alpha f}{Dt} - \boldsymbol{\nu}_\alpha \cdot \operatorname{grad} f. \tag{7.1.2}$$

The definition (7.1.1) of the barycentric velocity is the most important example of a more general construction. If $\Psi_\alpha(\mathbf{x}, t)$ represents a quantity per unit mass of constituent α, then we associate with the overall mixture a corresponding **barycentric average** $\overline{\Psi}$ by defining

$$\overline{\Psi} = \frac{1}{\overline{\rho}} \sum_{\alpha=1}^{N} \rho_\alpha \Psi_\alpha. \tag{7.1.3}$$

In the case of the barycentric velocity, $\Psi_\alpha = \mathbf{v}_\alpha$, which is the momentum per unit mass of constituent α.

EXERCISE 180 *Assuming that* $\overline{\Psi}$ *is a differentiable barycentric average, defined as in Equation (7.1.3), prove the following modified product rule:*

$$\overline{\rho}\frac{D\overline{\Psi}}{Dt} = -\overline{\Psi}\frac{D\overline{\rho}}{Dt} + \sum_{\alpha=1}^{N} \left[\frac{D^\alpha \rho_\alpha}{Dt} \Psi_\alpha + \rho_\alpha \frac{D^\alpha \Psi_\alpha}{Dt} - \boldsymbol{\nu}_\alpha \cdot \operatorname{grad}\left(\rho_\alpha \Psi_\alpha\right) \right]. \tag{7.1.4}$$

7.1.3 Summary

Some applications of continuum mechanics require distinctions among different constituents in a material. One approach to this type of application is through multiconstituent continua or mixtures, in which material from each of a set of constituents $\alpha = 1, 2, \ldots, N$ may be present at any spatial point $\mathbf{x}$. Associated with each constituent α is a body $\mathcal{B}_\alpha$, a motion $\boldsymbol{\chi}_\alpha(X_\alpha, t)$, a reference configuration $\boldsymbol{\kappa}_\alpha \colon \mathcal{B}_\alpha \to \mathbb{E}$, and a deformation $\boldsymbol{\chi}_{\alpha,\kappa} = \boldsymbol{\chi}_\alpha \circ \boldsymbol{\kappa}_\alpha^{-1}$, paralleling the theory of single-constituent continua. The definitions of constituent velocities $\mathbf{v}_\alpha$ and material derivatives D^α/Dt follow.

Two concepts of density arise: the bulk density ρ_α, giving mass of constituent α per unit volume of mixture, and the true density γ_α, giving mass of α per unit volume of constituent α. For multiphase mixtures, the two densities are related by the identity $\gamma_\alpha = \rho_\alpha/\phi_\alpha$, where ϕ_α denotes the volume fraction of constituent α. The mixture density is then

$$\overline{\rho} = \sum_{\alpha=1}^{N} \rho_\alpha = \sum_{\alpha=1}^{N} \phi_\alpha \gamma_\alpha.$$

These concepts give rise to the barycentric velocity

$$\overline{\mathbf{v}} = \frac{1}{\overline{\rho}} \sum_{\alpha=1}^{N} \rho_\alpha \mathbf{v}_\alpha$$

and the diffusion velocity $\boldsymbol{\nu}_\alpha = \mathbf{v}_\alpha - \overline{\mathbf{v}}$ for constituent α.

7.2 MULTICONSTITUENT BALANCE LAWS

Mixtures obey balance laws analogous to those that apply to simple continua. As in the case of simple continua, these laws have primitive global forms expressible as integral equations. When the quantities involved are sufficiently smooth the balance laws reduce to differential equations derivable using localization arguments similar to those used in Chapter 4. For the purposes of this chapter, we examine only the differential equations for the mass, momentum, angular momentum, and energy balances. We also develop an entropy inequality for mixtures. For a more complete review of balance laws for mixtures, see [3] and [6].

The generic form of a balance law for mixtures generalizes Equation (4.6.2):

$$\sum_{\alpha=1}^{N} \left[\rho_\alpha \frac{D^\alpha \Psi_\alpha}{Dt} + \Psi_\alpha \left(\frac{D^\alpha \rho_\alpha}{Dt} + \rho_\alpha \operatorname{div} \mathbf{v}_\alpha \right) - \operatorname{div} \mathbf{\Pi}_\alpha - \rho_\alpha \varpi_\alpha \right] = 0.$$

As in Section 4.6, the symbols Ψ_α, $\mathbf{\Pi}_\alpha$, and ϖ_α represent generic functions having the following interpretations:

- Ψ_α represents the amount, per unit mass of α, of the quantity being balanced;
- $\mathbf{\Pi}_\alpha$ represents the flux of the balanced quantity across mathematical surfaces in the mixture;
- ϖ_α represents the external supply of the balanced quantity.

The function $\mathbf{\Pi}_\alpha$ has tensor order one order higher than Ψ_α and ϖ_α.

In contrast to the single-constituent case, the balance, in the sense of summing to 0, applies to the entire mixture, that is to a sum of expressions over all of the constituents $\alpha = 1, 2, \ldots, N$. It is mathematically trivial but physically meaningful to rewrite the generic balance equation as a set of definitions,

$$\rho_\alpha \frac{D^\alpha \Psi_\alpha}{Dt} + \Psi_\alpha \left(\frac{D^\alpha \rho_\alpha}{Dt} + \rho_\alpha \operatorname{div} \mathbf{v}_\alpha \right) - \operatorname{div} \mathbf{\Pi}_\alpha - \rho_\alpha \varpi_\alpha = \hat{e}_\alpha, \qquad (7.2.1)$$

$$\alpha = 1, 2, \ldots, N,$$

together with the constraint

$$\sum_{\alpha=1}^{N} \hat{e}_\alpha = 0. \tag{7.2.2}$$

In the physical view, the quantity $\hat{e}_\alpha$ defined in this way represents the exchange rate, per unit mass, of the quantity being balanced *into* constituent α from other constituents. The identity (7.2.2) expresses the constraint that the quantity being balanced suffers no net gain or loss through these exchanges.

7.2.1 Multiconstituent Mass Balance

To obtain the **constituent mass balance**, substitute into the generic form (7.2.1) the assignments

$$\Psi_\alpha = 1,$$

$$\mathbf{\Pi}_\alpha = \mathbf{0},$$

$$\varpi_\alpha = 0,$$

as for simple continua. In addition, we identify $\hat{e}_\alpha = \hat{\rho}_\alpha$, which denotes the rate of production of mass of constituent α via reactions involving other constituents. These substitutions yield

$$\frac{D^\alpha \rho_\alpha}{Dt} + \rho_\alpha \operatorname{div} \mathbf{v}_\alpha = \hat{\rho}_\alpha, \quad \alpha = 1, 2, \ldots, N, \tag{7.2.3}$$

subject to the constraint

$$\sum_{\alpha=1}^{N} \hat{\rho}_\alpha = 0.$$

For multispecies mixtures in which mass exchange occurs via chemical reactions, this constraint reflects basic stoichiometry. For multiphase mixtures, it is equivalent to write

$$\frac{D^\alpha}{Dt}(\phi_\alpha \gamma_\alpha) + \phi_\alpha \gamma_\alpha \operatorname{div} \mathbf{v}_\alpha = \hat{\rho}_\alpha, \quad \alpha = 1, 2, \ldots, N,$$

$$\sum_{\alpha=1}^{N} \hat{\rho}_\alpha = 0.$$

For such mixtures, mass exchange among constituents often takes the form of dissolution, adsorption, or phase changes such as freezing, melting, condensation, or evaporation.

EXERCISE 181 *Rewrite the generic constituent balance (7.2.1) in the form*

$$\rho_\alpha \frac{D^\alpha \Psi_\alpha}{Dt} - \operatorname{div} \boldsymbol{\Pi}_\alpha - \rho_\alpha\, \varpi_\alpha = \hat{e}_\alpha - \Psi_\alpha \hat{\rho}_\alpha. \tag{7.2.4}$$

One commonly sees the constituent mass balance in yet a different form. Define the **mass fraction** of constituent α as $c_\alpha = \rho_\alpha/\overline{\rho}$, which has dimension [mass of α per unit mass of mixture]. The identity

$$\sum_{\alpha=1}^{N} c_\alpha = 1$$

follows immediately from the definition of $\overline{\rho}$.

EXERCISE 182 *Show that*

$$\underbrace{\frac{\partial}{\partial t}(\overline{\rho}\, c_\alpha)}_{\text{Accumulation}} + \underbrace{\operatorname{div}(\overline{\rho}\, c_\alpha\, \mathbf{v})}_{\text{Advection}} + \underbrace{\operatorname{div} \mathbf{j}_\alpha}_{\text{Diffusion}} = \underbrace{\hat{\rho}_\alpha}_{\text{Reaction}}, \tag{7.2.5}$$

where $\mathbf{j}_\alpha = \overline{\rho}\, c_\alpha\, \boldsymbol{\nu}_\alpha$.

The **diffusive flux** $\mathbf{j}_\alpha$ identified in Exercise 182 accounts for fluctuations in the velocity of constituent α with respect to the mass-weighted average velocity of the mixture. These fluctuations result in the spreading of mass from constituent α with respect to the mixture. Section 7.4 examines Equation (7.2.5) in more detail.

The following exercise reconciles the mass balance arising in multiconstituent theory with that arising in the theory of simple continua: if we neglect the distinction among constituents, the overall mixture behaves as a simple continuum.

EXERCISE 183 *Show that summing Equation (7.2.3) over α yields the equation*

$$\frac{D\overline{\rho}}{Dt} + \overline{\rho} \operatorname{div} \overline{\mathbf{v}} = 0,$$

which is the mass balance for simple continua.

This reassuring result leads to the observations in the next three exercises.

EXERCISE 184 *Using the constituent mass balance and assuming that $\overline{\Psi}$ denotes a differentiable barycentric average, rewrite Equation (7.1.4) as follows:*

$$\overline{\rho} \frac{D\overline{\Psi}}{Dt} = \sum_{\alpha=1}^{N} \left[\rho_\alpha \frac{D^\alpha \Psi_\alpha}{Dt} - \operatorname{div}(\rho_\alpha \Psi_\alpha \boldsymbol{\nu}_\alpha) + \hat{\rho}_\alpha \Psi_\alpha \right]. \tag{7.2.6}$$

EXERCISE 185 *When $\Psi_\alpha = \mathbf{v}_\alpha$, the divergence terms in the summand of Exercise 184 take the form*

$$\operatorname{div}(\rho_\alpha \mathbf{v}_\alpha \otimes \boldsymbol{\nu}_\alpha) = \operatorname{div}(\rho_\alpha \mathbf{v} \otimes \boldsymbol{\nu}_\alpha) + \operatorname{div}(\rho_\alpha \boldsymbol{\nu}_\alpha \otimes \boldsymbol{\nu}_\alpha).$$

(See Exercise 33.) Show that the first term on the right vanishes upon summation over α.

EXERCISE 186 *If Ψ_α is a differentiable quantity per unit mass, show that*

$$\sum_\alpha \rho_\alpha \frac{D^\alpha \Psi_\alpha}{Dt} = \sum_\alpha \frac{D^\alpha}{Dt}(\rho_\alpha \Psi_\alpha) + \sum_\alpha \rho_\alpha \Psi_\alpha \mathsf{I} : \mathsf{D}_\alpha - \sum_\alpha \Psi_\alpha \hat{\rho}_\alpha.$$

Here $\mathsf{D}_\alpha = \frac{1}{2}[\operatorname{grad}\mathbf{v}_\alpha + (\operatorname{grad}\mathbf{v}_\alpha)^\top]$ is the stretching tensor for constituent α, and $\mathsf{I} : \mathsf{D}_\alpha$ denotes the double-dot product defined in Equation (1.2.12).

7.2.2 Multiconstituent Momentum Balance

For the multiconstituent momentum balance,

$$\Psi_\alpha = \mathbf{v}_\alpha,$$

$$\boldsymbol{\Pi}_\alpha = \mathsf{T}_\alpha,$$

$$\varpi_\alpha = \mathbf{b}_\alpha,$$

where T_α denotes the stress tensor for constituent α, and $\mathbf{b}_\alpha$ represents the body force per unit mass acting on constituent α. The exchange term in this case is $\hat{e}_\alpha = \hat{\mathbf{m}}_\alpha$, which represents the rate of exchange of momentum into constituent α via mechanisms other than direct exchange of momentum-bearing mass among constituents. Equation (7.2.4) becomes

$$\rho_\alpha \frac{D^\alpha \mathbf{v}_\alpha}{Dt} - \operatorname{div}\mathsf{T}_\alpha - \rho_\alpha \mathbf{b}_\alpha = \hat{\mathbf{m}}_\alpha - \hat{\rho}_\alpha \mathbf{v}_\alpha, \quad \alpha = 1, 2, \ldots, N, \tag{7.2.7}$$

subject to the constraint

$$\sum_{\alpha=1}^{N} \hat{\mathbf{m}}_\alpha = \mathbf{0}.$$

The term $-\hat{\rho}_\alpha \mathbf{v}_\alpha$ on the right side of Equation (7.2.7) accounts for the momentum imported into constituent α from other constituents through the exchange of mass.

As with the mass balance, we may reconcile the constituent momentum balance, together with the constraint on momentum exchanges, with the theory of simple continua.

EXERCISE 187 *Sum the constituent momentum balance (7.2.7) and use Equation (7.2.6) to get*

$$\overline{\rho}\frac{D\overline{\mathbf{v}}}{Dt} - \operatorname{div}\overline{\mathsf{T}} - \overline{\rho}\,\overline{\mathbf{b}} = \mathbf{0},$$

where

$$\overline{\mathbf{b}} = \frac{1}{\overline{\rho}}\sum_{\alpha=1}^{N} \rho_\alpha \mathbf{b}_\alpha$$

represents the barycentric average body force per unit mass and

$$\overline{\mathsf{T}} = \sum_{\alpha=1}^{N}(\mathsf{T}_\alpha - \rho_\alpha \boldsymbol{\nu}_\alpha \otimes \boldsymbol{\nu}_\alpha)$$

represents the **total stress**. *Note that* $\overline{\mathsf{T}}$ *is not a barycentric average.*

Two remarks about the total stress are in order. First, it includes a straightforward contribution,

$$\mathsf{T}_I = \sum_{\alpha=1}^{N} \mathsf{T}_\alpha,$$

arising from the stresses of the individual constituents. We call this contribution the **inner stress**. Second, there is an additional contribution,

$$\mathsf{T}_R = -\sum_{\alpha=1}^{N} \rho_\alpha \boldsymbol{\nu}_\alpha \otimes \boldsymbol{\nu}_\alpha,$$

called the **Reynolds stress**. This term accounts for the transport of momentum via the relative motions of constituents. Observe that T_R is symmetric, and hence so is $\overline{\mathsf{T}}$ in the case that each constituent stress tensor T_α is symmetric.

Equation (7.2.7) serves as a multiconstituent continuum analog of Newton's second law of motion. As in the case of simple continua, we refer to the quantities

$$\rho_\alpha \frac{D^\alpha \mathbf{v}_\alpha}{Dt} = \rho_\alpha \left[\frac{\partial \mathbf{v}_\alpha}{\partial t} + (\mathbf{v}_\alpha \cdot \operatorname{grad})\mathbf{v}_\alpha\right]$$

as **inertial terms**.

7.2.3 Multiconstituent Angular Momentum Balance

For the constituent angular momentum balance, we consider spatial moments of the quantities appearing in the momentum balance:

$$\Psi_\alpha = \mathbf{x} \times \mathbf{v}_\alpha,$$

$$\mathbf{\Pi}_\alpha = \mathbf{x} \times \mathsf{T}_\alpha,$$

$$\varpi_\alpha = \mathbf{x} \times \mathbf{b}_\alpha.$$

Denote the exchange term as $\hat{\boldsymbol{\mu}}_\alpha$, a quantity that represents the rate of angular momentum transfer into constituent α via mechanisms other than direct mass exchange. Substituting these quantities into Equations (7.2.4) and (7.2.2) yields

$$\rho_\alpha \frac{D^\alpha}{Dt}(\mathbf{x} \times \mathbf{v}_\alpha) - \operatorname{div}(\mathbf{x} \times \mathsf{T}_\alpha) - \rho_\alpha (\mathbf{x} \times \mathbf{b}_\alpha) = \hat{\boldsymbol{\mu}}_\alpha - \hat{\rho}_\alpha(\mathbf{x} \times \mathbf{v}_\alpha), \quad (7.2.8)$$

for $\alpha = 1, 2, \ldots, N$, together with the constraint

$$\sum_{\alpha=1}^{N} \hat{\boldsymbol{\mu}}_\alpha = \mathbf{0}.$$

An argument similar to that given in Section 4.3 reduces Equation (7.2.8) to a simpler form. The identities

$$\frac{D^\alpha}{Dt}(\mathbf{x} \times \mathbf{v}_\alpha) = \mathbf{x} \times \frac{D^\alpha \mathbf{v}_\alpha}{Dt}$$

and

$$\operatorname{div}(\mathbf{x} \times \mathsf{T}_\alpha) = (\operatorname{grad} \mathbf{x}) \times \mathsf{T}_\alpha + \mathbf{x} \times \operatorname{div} \mathsf{T}_\alpha$$

reduce the constituent angular momentum balance to the form

$$\mathbf{x} \times \left[\rho_\alpha \frac{D^\alpha \mathbf{v}_\alpha}{Dt} - \operatorname{div} \mathsf{T}_\alpha - \rho_\alpha \mathbf{b}_\alpha\right] - (\operatorname{grad} \mathbf{x}) \times \mathsf{T}_\alpha = \hat{\boldsymbol{\mu}}_\alpha - \mathbf{x} \times \hat{\rho}_\alpha \mathbf{v}_\alpha.$$

The quantity in square brackets equals $\hat{\mathbf{m}}_\alpha - \hat{\rho}_\alpha \mathbf{v}_\alpha$ by the constituent momentum balance. Therefore,

$$-(\operatorname{grad} \mathbf{x}) \times \mathsf{T}_\alpha = \hat{\boldsymbol{\mu}}_\alpha - \mathbf{x} \times \hat{\mathbf{m}}_\alpha.$$

In Cartesian coordinates, evaluating $\operatorname{grad} \mathbf{x}$ as in Section 4.3 yields

$$\begin{bmatrix} T_{32} - T_{23} \\ T_{13} - T_{31} \\ T_{21} - T_{12} \end{bmatrix}_\alpha = \hat{\boldsymbol{\mu}}_\alpha - \mathbf{x} \times \hat{\mathbf{m}}_\alpha. \quad (7.2.9)$$

This equation shows that the stress tensors T_α for individual constituents need not be symmetric. Still the constraints on $\hat{\mathbf{m}}_\alpha$ and $\hat{\boldsymbol{\mu}}_\alpha$ imply that the inner part T_I of the total stress, and hence the total stress T, of the mixture must be symmetric.

In many models of multiphase mixtures, the stress tensor for each constituent remains symmetric, as if the constituent were isolated in a simple continuum. Nevertheless, the angular momentum balance for mixtures allows for more general behavior. For the remainder of this chapter, we assume that each $\mathsf{T}_\alpha = \mathsf{T}_\alpha^\top$ for every constituent.

EXERCISE 188 *Show that, under this assumption,* $\mathsf{T}_\alpha : \mathsf{L}_\alpha = \mathsf{T}_\alpha : \mathsf{D}_\alpha$, *where* L_α *is the velocity gradient of constituent* α.

7.2.4 Multiconstituent Energy Balance

For the constituent energy balance we identify

$$\begin{aligned}
\Psi_\alpha &= \varepsilon_\alpha + \tfrac{1}{2}\mathbf{v}_\alpha \cdot \mathbf{v}_\alpha, \\
\boldsymbol{\Pi}_\alpha &= \mathsf{T}_\alpha \mathbf{v}_\alpha - \mathbf{q}_\alpha, \\
\varpi_\alpha &= r_\alpha + \mathbf{v}_\alpha \cdot \mathbf{b}_\alpha, \quad \alpha = 1, 2, \ldots, N,
\end{aligned}$$

where the physical interpretations of these quantities parallel those discussed in Section 4.4. Equation (7.2.4) becomes

$$\begin{aligned}
\rho_\alpha \frac{D^\alpha}{Dt}\left(\varepsilon_\alpha + \tfrac{1}{2}\mathbf{v}_\alpha \cdot \mathbf{v}_\alpha\right) &- \operatorname{div}\left(\mathsf{T}_\alpha \mathbf{v}_\alpha - \mathbf{q}_\alpha\right) - \rho_\alpha\left(r_\alpha + \mathbf{v}_\alpha \cdot \mathbf{b}_\alpha\right) \\
&= \hat{\varepsilon}_\alpha - \hat{\rho}_\alpha\left(\varepsilon_\alpha + \tfrac{1}{2}\mathbf{v}_\alpha \cdot \mathbf{v}_\alpha\right). \qquad (7.2.10)
\end{aligned}$$

Here $\hat{e}_\alpha = \hat{\varepsilon}_\alpha$ represents the exchange of energy into constituent α from other constituents, beyond that attributable to direct mass exchange. These exchange rates obey the constraint

$$\sum_{\alpha=1}^{N} \hat{\varepsilon}_\alpha = 0. \qquad (7.2.11)$$

EXERCISE 189 *Mimicking the development in Section 4.4, reduce Equation (7.2.10) to the following thermal energy balance for constituent* α*:*

$$\begin{aligned}
\rho_\alpha \frac{D^\alpha \varepsilon_\alpha}{Dt} &- \operatorname{tr}\left(\mathsf{T}_\alpha \mathsf{D}_\alpha\right) + \operatorname{div}\mathbf{q}_\alpha - \rho_\alpha r_\alpha \\
&= \hat{\varepsilon}_\alpha - \left(\varepsilon_\alpha - \tfrac{1}{2}\mathbf{v}_\alpha \cdot \mathbf{v}_\alpha\right)\hat{\rho}_\alpha - \mathbf{v}_\alpha \cdot \hat{\mathbf{m}}_\alpha. \qquad (7.2.12)
\end{aligned}$$

Interpret the terms.

Retrieving the energy balance for simple continua from the mixture energy balance (7.2.11) is more tedious than enlightening. It requires the following definitions:

$$\overline{\varepsilon} = \frac{1}{\overline{\rho}} \sum_{\alpha} \rho_\alpha (\varepsilon_\alpha + \tfrac{1}{2} \mathbf{v}_\alpha \cdot \mathbf{v}_\alpha),$$

$$\overline{\mathbf{q}} = \sum_{\alpha} [\mathbf{q}_\alpha + \rho_\alpha (\varepsilon_\alpha + \tfrac{1}{2} \mathbf{v}_\alpha \cdot \mathbf{v}_\alpha) \boldsymbol{\nu}_\alpha - \mathsf{T}_\alpha \boldsymbol{\nu}_\alpha],$$

$$\overline{r} = \frac{1}{\overline{\rho}} \sum_{\alpha} \rho_\alpha (r_\alpha + \mathbf{b}_\alpha \cdot \boldsymbol{\nu}_\alpha),$$

$$\overline{\mathsf{L}} = \operatorname{grad} \overline{\mathbf{v}},$$

none of which represents a simple barycentric average. Summing the energy balance (7.2.12) and using these definitions yield

$$\overline{\rho} \frac{D\overline{\varepsilon}}{Dt} - \overline{\mathsf{T}} : \overline{\mathsf{L}} + \operatorname{div} \overline{\mathbf{q}} - \overline{\rho}\, \overline{r} = 0.$$

7.2.5 Multiconstituent Entropy Inequality

Applications of mixture mechanics that involve constitutive theory often require the use of an entropy inequality, as illustrated in the last two sections of this chapter. To establish such an inequality, we start with the generic balance law (7.2.4), summed over all constituents and written as an inequality. We identify the generic quantities in equation (7.2.4) as follows:

$\Psi_\alpha = \eta_\alpha$ represents the specific entropy of constituent α;

$\mathbf{\Pi}_\alpha = \mathbf{q}_\alpha / \theta_\alpha$ represents the entropy flux;

$\varpi_\alpha = r_\alpha / \theta_\alpha$ represents the external supply of entropy.

These assignments yield

$$\sum_{\alpha=1}^{N} \left[\rho_\alpha \frac{D^\alpha \eta_\alpha}{Dt} + \operatorname{div} \left(\frac{\mathbf{q}_\alpha}{\theta_\alpha} \right) - \frac{\rho_\alpha r_\alpha}{\theta_\alpha} + \eta_\alpha \hat{\rho}_\alpha \right] \geqslant 0. \tag{7.2.13}$$

The inequality (7.2.13) allows the temperature $\theta_\alpha(\mathbf{x}, t)$ to differ from the temperatures at $(\mathbf{x}, t)$ of other constituents.

We have no need to track entropy production in individual constituents: it suffices to demand that the overall entropy production of the mixture be nonnegative. Hence, we examine only the sum over all constituents of the terms in square brackets, and the only exchange terms that appear are the terms $\eta_\alpha \hat{\rho}_\alpha$ associated with mass transfer among constituents.

As a form of shorthand, we identify the sum in Equation (7.2.13) as the **total dissipation rate** Υ. With this notation, we abbreviate the mixture entropy inequality as $\Upsilon \geqslant 0$.

7.2.6 Isothermal, Nonreacting Multiphase Mixtures

Sections 7.3 and 7.4 examine models of mixtures in which the following assumptions apply:

1. $\hat{\rho}_\alpha = 0$ for each constituent α, meaning there is no mass exchange attributable to chemical reactions or interphase mass transfer, such as melting, condensation, or evaporation;
2. all constituents have the same temperature, $\theta_\alpha = \theta$, which is constant in space and time;
3. external heat supplies, such as radiative heat transfer, are absent: $r_\alpha = 0$ for each constituent α.

Some may find the assumption that constituents share a common temperature physically compelling, since by hypothesis they are overlapping continua. By this reasoning one might expect the heat transfer required for interconstituent equilibration to be instantaneous. However, this modeling assumption may not always be justified. In some applications, such as the rapid movement of water vapor through snow under a large imposed temperature gradient, useful models require that each constituent have its own temperature, which can vary in time and space.

Under the assumptions listed above, the mass balance for each constituent $\alpha = 1, 2, \ldots, N$ in a multiphase mixture reduces to the following form:

$$\frac{D^\alpha}{Dt}(\phi_\alpha \gamma_\alpha) + \phi_\alpha \gamma_\alpha \mathrm{div}\, \mathbf{v}_\alpha = 0. \tag{7.2.14}$$

For the momentum balance, we obtain

$$\phi_\alpha \gamma_\alpha \frac{D^\alpha \mathbf{v}_\alpha}{Dt} - \mathrm{div}\, \mathsf{T}_\alpha - \phi_\alpha \gamma_\alpha \mathbf{b}_\alpha = \hat{\mathbf{m}}_\alpha, \tag{7.2.15}$$

subject to the constraint

$$\sum_{\alpha=1}^{N} \hat{\mathbf{m}}_\alpha = \mathbf{0}. \tag{7.2.16}$$

The angular momentum balance for constituent α reduces to the assumption, made earlier, that the stress tensor for each constituent is symmetric:

$$\mathsf{T}_\alpha = \mathsf{T}_\alpha^\top. \tag{7.2.17}$$

For a multiphase mixture having a common temperature θ, there is no need for a separate equation for each constituent temperature, so it typically suffices to consider the total energy balance,

$$\sum_{\alpha=1}^{N} \left(\phi_\alpha \gamma_\alpha \frac{D^\alpha \varepsilon_\alpha}{Dt} - \mathsf{T}_\alpha : \mathsf{D}_\alpha + \operatorname{div} \mathbf{q}_\alpha \right) = \sum_{\alpha=1}^{N} \mathbf{v}_\alpha \cdot \hat{\mathbf{m}}_\alpha. \tag{7.2.18}$$

Here, fortified by Exercise 188, we have used the symmetry of each stress tensor T_α to write $\mathsf{T}_\alpha : \mathsf{L}_\alpha = \mathsf{T}_\alpha : \mathsf{D}_\alpha$. Finally, for the entropy inequality, the assumption that the temperature is a constant common to all constituents yields

$$\theta \Upsilon = \sum_{\alpha=1}^{N} \left(\theta \phi_\alpha \gamma_\alpha \frac{D^\alpha \eta_\alpha}{Dt} + \operatorname{div} \mathbf{q}_\alpha \right) \geqslant 0. \tag{7.2.19}$$

Paralleling the development in Section 4.5, in isothermal settings we rewrite the entropy inequality in terms of the specific Helmholtz free energies $\psi_\alpha = \varepsilon_\alpha - \theta \eta_\alpha$. Subtracting the energy balance (7.2.18) from the inequality (7.2.19) yields

$$\theta \Upsilon = -\sum_{\alpha=1}^{N} \left(\phi_\alpha \gamma_\alpha \frac{D^\alpha \psi_\alpha}{Dt} + \mathsf{T}_\alpha : \mathsf{D}_\alpha - \mathbf{v}_\alpha \cdot \hat{\mathbf{m}}_\alpha \right) \geqslant 0. \tag{7.2.20}$$

Here we have used the identity

$$\frac{D^\alpha \varepsilon_\alpha}{Dt} - \theta \frac{D^\alpha \eta_\alpha}{Dt} = \frac{D^\alpha \psi_\alpha}{Dt}.$$

With this version of the entropy inequality in hand, it proves useful in turn to rewrite the mixture energy balance (7.2.18) in terms of the specific Helmholtz free energies. Substituting for ε_α, we get

$$\sum_{\alpha=1}^{N} \left[\phi_\alpha \gamma_\alpha \frac{D^\alpha}{Dt} (\psi_\alpha + \theta \eta_\alpha) - \mathsf{T}_\alpha : \mathsf{D}_\alpha + \operatorname{div} \mathbf{q}_\alpha \right] = \sum_{\alpha=1}^{N} \mathbf{v}_\alpha \cdot \hat{\mathbf{m}}_\alpha.$$

By the identity (7.1.4) with each $\hat{\rho}_\alpha = 0$,

$$\sum_{\alpha=1}^{N} \phi_\alpha \gamma_\alpha \frac{D^\alpha \eta_\alpha}{Dt} = \overline{\rho} \frac{D\overline{\eta}}{Dt} + \operatorname{div} \sum_{\alpha=1}^{N} \phi_\alpha \gamma_\alpha \eta_\alpha \boldsymbol{\nu}_\alpha,$$

where

$$\overline{\eta} = \frac{1}{\overline{\rho}} \sum_{\alpha=1}^{N} \phi_\alpha \gamma_\alpha \eta_\alpha.$$

Substitution yields

$$\sum_{\alpha=1}^{N} \phi_\alpha \gamma_\alpha \frac{D^\alpha \psi_\alpha}{Dt} + \overline{\rho}\theta \frac{D\overline{\eta}}{Dt} - \sum_{\alpha=1}^{N} \mathsf{T}_\alpha : \mathsf{D}_\alpha + \operatorname{div} \overline{\overline{\mathbf{q}}} = - \sum_{\alpha=1}^{N} \mathbf{v}_\alpha \cdot \hat{\mathbf{m}}_\alpha, \quad (7.2.21)$$

where

$$\overline{\overline{\mathbf{q}}} = \sum_{\alpha=1}^{N} (\mathbf{q}_\alpha + \phi_\alpha \gamma_\alpha \theta \eta_\alpha \boldsymbol{\nu}_\alpha) \quad (7.2.22)$$

represents a total heat flux for the mixture. This version of the mixture energy balance serves as a starting point for the derivation in the next section.

7.2.7 Summary

The general balance law for a constituent α in a mixture takes the form

$$\rho_\alpha \frac{D^\alpha \Psi_\alpha}{Dt} + \Psi_\alpha \left(\frac{D^\alpha \rho_\alpha}{Dt} + \rho_\alpha \operatorname{div} \mathbf{v}_\alpha \right) - \operatorname{div} \mathbf{\Pi}_\alpha - \rho_\alpha \varpi_\alpha = \hat{e}_\alpha,$$

where

- Ψ_α represents the amount, per unit mass of α, of the quantity being balanced;
- $\mathbf{\Pi}_\alpha$ represents the flux of the balanced quantity across mathematical surfaces in the mixture;
- ϖ_α represents the external supply of the balanced quantity;
- $\hat{e}_\alpha$ represents the exchange of the quantity Ψ into α.

These equations, valid for $\alpha = 1, 2, \ldots, N$, obey the constraint

$$\sum_{\alpha=1}^{N} \hat{e}_\alpha = 0.$$

Table 7.1, which generalizes Table 4.1, lists the assignments of specific variables to the generic quantities Ψ_α, $\mathbf{\Pi}_\alpha$, ϖ_α, and $\hat{e}_\alpha$.

Table 7.1. Assignments of Physical Values to Generic Quantities in Mixture Balance Laws

Balance Law	Ψ_α	$\mathsf{\Pi}_\alpha$	ϖ_α	$\hat{e}_\alpha$
Mass	1	$\mathbf{0}$	0	$\hat{\rho}_\alpha$
Momentum	$\mathbf{v}_\alpha$	T_α	$\mathbf{b}_\alpha$	$\hat{\mathbf{m}}_\alpha$
Angular Momentum	$\mathbf{x}\times\mathbf{v}_\alpha$	$\mathbf{x}\times\mathsf{T}_\alpha$	$\mathbf{x}\times\mathbf{b}_\alpha$	$\hat{\boldsymbol{\mu}}_\alpha$
Energy	$\varepsilon_\alpha+\frac{1}{2}\mathbf{v}_\alpha\cdot\mathbf{v}_\alpha$	$\mathsf{T}_\alpha\mathbf{v}_\alpha-\mathbf{q}_\alpha$	$r_\alpha+\mathbf{v}_\alpha\cdot\mathbf{b}_\alpha$	$\hat{\varepsilon}_\alpha$

Rewriting the constituent mass balance in terms of the mass fractions $c_\alpha = \gamma_\alpha/\overline{\rho}$ and the diffusive flux $\mathbf{j}_\alpha = \overline{\rho}c_\alpha\boldsymbol{\nu}_\alpha$ yields the advection–diffusion–reaction equation,

$$\frac{\partial}{\partial t}(\overline{\rho}c_\alpha) + \mathrm{div}\,(\rho c_\alpha\mathbf{v}) + \mathrm{div}\,\mathbf{j}_\alpha = \hat{\rho}_\alpha.$$

Simplifying the constituent balance using the constituent mass balance yields

$$\rho_\alpha\frac{D^\alpha\mathbf{v}_\alpha}{Dt} - \mathrm{div}\,\mathsf{T}_\alpha - \rho_\alpha\mathbf{b}_\alpha = \hat{\mathbf{m}}_\alpha - \hat{\rho}_\alpha\mathbf{v}_\alpha.$$

Although simplification of the constituent angular momentum balance does not lead to the conclusion that the stress tensors T_α are symmetric, in many applications we model them as symmetric. Simplification of the constituent energy balance leads to the following equation:

$$\rho_\alpha\frac{D^\alpha\varepsilon_\alpha}{Dt} - \mathrm{tr}\,(\mathsf{T}_\alpha\mathsf{D}_\alpha) + \mathrm{div}\,\mathbf{q}_\alpha - \rho_\alpha r_\alpha$$
$$= \hat{\varepsilon}_\alpha - \left(\varepsilon_\alpha - \tfrac{1}{2}\mathbf{v}_\alpha\cdot\mathbf{v}_\alpha\right)\hat{\rho}_\alpha - \mathbf{v}_\alpha\cdot\hat{\mathbf{m}}_\alpha.$$

For the entropy inequality we adopt the form

$$\sum_{\alpha=1}^{N}\left[\rho_\alpha\frac{D^\alpha\eta_\alpha}{Dt} + \mathrm{div}\left(\frac{\mathbf{q}_\alpha}{\theta_\alpha}\right) - \frac{\rho_\alpha r_\alpha}{\theta_\alpha} + \eta_\alpha\hat{\rho}_\alpha\right] \geqslant 0,$$

having neither need nor, some argue, justification for an entropy inequality for each constituent.

7.3 FLUID FLOW IN A POROUS SOLID

The flow of a single fluid in a porous solid provides fertile ground for the principles reviewed in the previous two sections and for concepts from the constitutive theory outlined in Chapter 6. Flows of this type arise in groundwater hydrology, where the fluid of interest is usually water. The solid is a porous rock, such as sandstone, in an underground formation called an aquifer. Flows in porous media also arise in many other settings, such as natural oil and gas reservoirs, chromatographic columns, and biomedical applications in which the media of interest may be human tissues.

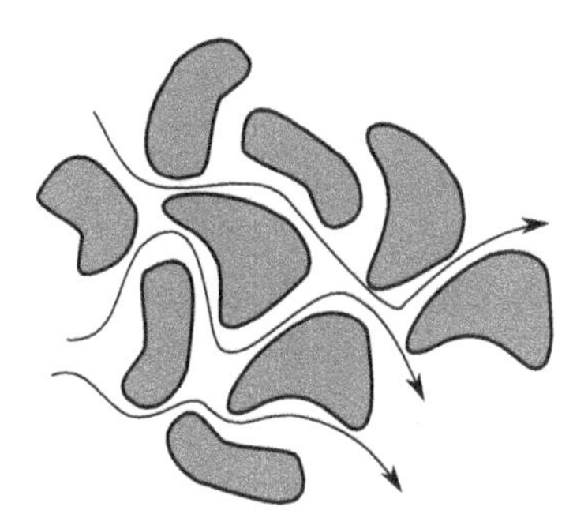

Figure 7.3. Schematic illustration of a fluid flowing through the interstices of a porous rock matrix.

In geologic porous media, the solid phase S often consists of grains cemented into a matrix. Within this matrix resides a system of connected interstices called pores. In this section we treat the simple case in which a single-phase fluid F flows in the pores, illustrated schematically in Figure 7.3. Although the fluid and solid may act as overlapping continua at macroscopic scales of observation—say 10^{-2} m or larger—they appear as distinct continua in their own right at microscopic scales of observation, roughly 10^{-5} m. In addition, the interface between these continua plays a significant, macroscopically observable role in the mechanics of the flow. Therefore, we treat the system as a multiphase mixture in which the volume fractions ϕ_F and ϕ_S are important variables.

With this foundation, we examine the mass, momentum, and energy balance laws for a mixture of this type, under certain restrictive assumptions. We apply constitutive theory to establish functional relationships needed to define a closed system of governing equations, then review additional physical assumptions that yield the most commonly used version of the fluid momentum balance, **Darcy's law** [16].

7.3.1 Modeling Assumptions for Porous Media

For simplicity, consider an isothermal, two-phase mixture consisting of a rigid solid matrix S having incompressible grains and containing a single, isotropic fluid F, with no external heat supplies in either constituent. The assumption of a rigid solid matrix serves the present purpose, namely, to introduce concepts. But in practice it is too restrictive for many applications to underground flow, in which compressibility of the rock matrix influences the fluid motion. One can relax this assumption and still derive the most common mathematical models of subsurface flow, with more realistic representations of the solid matrix physics and a corresponding increase in complexity.

The rigidity of the solid has the following consequences:

1. The deformation gradient for the solid takes the form $\mathsf{F}_S(\mathbf{X}_S,t) = \mathsf{Q}(t)$, for some one-parameter family $\mathsf{Q}(t)$ of orthogonal tensors. Provided the deformation is continuously differentiable, the corresponding velocity gradient is $\mathsf{L}_S(\mathbf{X}_S,t) = \mathsf{Q}'(t)\mathsf{Q}^\top(t)$, which by Exercise 67 is skew. It follows that the solid-phase stretching tensor $\mathsf{D}_S = 0$.

2. There exists a frame of reference in which $\mathbf{v}_S = \mathbf{0}$. Adopting this frame makes $\mathbf{v}_F = \mathbf{v}_F - \mathbf{v}_S$, which is a relative velocity. See Exercise 190.

3. The volume fraction and true density of the solid are constant:

$$\frac{D^S\phi_S}{Dt} = \frac{\partial\phi_S}{\partial t} = 0; \qquad \frac{D^S\gamma_S}{Dt} = \frac{\partial\gamma_S}{\partial t} = 0.$$

   Therefore, all terms in the mass balance for the solid S vanish, and $\phi_F = 1-\phi_S$ is constant in time.

4. The mass balance for the fluid F reduces to the following identity:

$$\frac{D^F\gamma_F}{Dt} = -\frac{\gamma_F}{\phi_F}(\mathbf{v}_F \cdot \operatorname{grad}\phi_F + \phi_F\mathsf{I} : \mathsf{D}_F). \tag{7.3.1}$$

   See Exercise 191.

5. We have no use for the momentum balance for the solid phase S, regarded as an equation for the velocity $\mathbf{v}_S$.

EXERCISE 190 *Show that the relative velocity* $\mathbf{v}_F - \mathbf{v}_S$ *is objective.*

EXERCISE 191 *Prove Equation (7.3.1).*

In addition to these observations, since the solid phase is rigid and isothermal, we may treat its specific Helmholtz free energy as constant, so $D^S\psi_S/Dt = 0$. Also, because the mixture itself is isothermal with no external heat supplies, we assume for simplicity that the total mixture heat flux $\overline{\overline{\mathbf{q}}}$, defined in Equation (7.2.22), vanishes.

7.3.2 Balance Laws for the Fluid and Solid Phases

The modeling assumptions listed above yield a reduced set of balance laws. For the fluid phase F we have the mass balance established in Equation (7.3.1),

$$\frac{D^F \gamma_F}{Dt} = -\frac{\gamma_F}{\phi_F}(\mathbf{v}_F \cdot \operatorname{grad} \phi_F + \phi_F \mathsf{I} : \mathsf{D}_F);$$

the momentum balance,

$$\phi_F \gamma_F \frac{D^F \mathbf{v}_F}{Dt} - \operatorname{div} \mathsf{T}_F - \phi_F \gamma_F \mathbf{b}_F = \hat{\mathbf{m}}_F; \tag{7.3.2}$$

and the angular momentum balance,

$$\mathsf{T}_F = \mathsf{T}_F^\top. \tag{7.3.3}$$

We also have the energy balance for the mixture, which we obtain from Equation (7.2.21) by applying the modeling assumptions adopted above:

$$\phi_F \gamma_F \frac{D^F \psi^F}{Dt} + \overline{\rho}\theta \frac{D\overline{\eta}}{Dt} - \mathsf{T}_F : \mathsf{D}_F = -\mathbf{v}_F \cdot \hat{\mathbf{m}}_F. \tag{7.3.4}$$

Equations (7.3.1) through (7.3.4) furnish eight scalar equations, which we regard as governing the eight scalar functions needed to define γ_F, $\mathbf{v}_F$, three off-diagonal components of T_F, and $\overline{\eta}$. (We can calculate D_F knowing $\mathbf{v}_F$.) Solutions to all of these equations must be consistent with the entropy inequality (7.2.20), which under our modeling assumptions reduces to the following form:

$$\theta \Upsilon = -\phi_F \gamma_F \frac{D^F \psi_F}{Dt} + \mathsf{T}_F : \mathsf{D}_F - \mathbf{v}_F \cdot \hat{\mathbf{m}}_F \geqslant 0. \tag{7.3.5}$$

The system (7.3.1) through (7.3.4) involves 22 scalar functions, namely, those needed to determine ϕ_F, γ_F, $\mathbf{v}_F$, ψ_F, T_F, $\mathbf{b}_F$, $\hat{\mathbf{m}}_F$, and $\overline{\eta}$. For the remainder of this section, we drop the subscript F and write these variables using the more streamlined notation

$$\phi, \gamma, \mathbf{v}, \psi, \mathsf{T}, \mathbf{b}, \hat{\mathbf{m}}, \overline{\eta}.$$

We treat the function ϕ, called the **porosity** of the porous medium, as a prescribed function of position, and we assume that the body force per unit mass $\mathbf{b} = \mathbf{g}$, a temporally constant, spatially uniform vector field giving the gravitational acceleration near the Earth's surface.

Constitutive Relations

Given 22 variables, eight balance laws, and the four prescribed scalar functions defining ϕ and $\mathbf{g}$, we must adopt $22 - 8 - 4 = 10$ constitutive relations to match the number of equations with the number of variables. These relations take the form of constitutive functions for six components of T, three components of $\hat{\mathbf{m}}$, and the scalar function ψ. For the model developed in this section, we examine constitutive relations depending *a priori* on fluid density γ and velocity $\mathbf{v}$ and their first-order spatial derivatives:

$$\begin{aligned} \mathsf{T} &= \mathsf{T}(\gamma, \operatorname{grad}\gamma, \mathbf{v}, \mathsf{D}) \\ \hat{\mathbf{m}} &= \hat{\mathbf{m}}(\gamma, \operatorname{grad}\gamma, \mathbf{v}, \mathsf{D}) \\ \psi &= \psi(\gamma, \operatorname{grad}\gamma, \mathbf{v}, \mathsf{D}). \end{aligned} \tag{7.3.6}$$

This set of functional dependencies satisfies four of the five axioms discussed in Chapter 6: determinism, equipresence, objectivity (by Exercise 190 and the fact that D is objective), and symmetry (since the fluid is isotropic, with no preferred directions).

Left to impose is the axiom of admissibility. For this purpose, we substitute the constitutive relations (7.3.6) into the entropy inequality (7.3.5) and apply the chain rule. In index notation with the summation convention, this step yields

$$\begin{aligned} -\phi\gamma\frac{\partial\psi}{\partial\gamma}\frac{D^F\gamma}{Dt} - \phi\gamma\frac{\partial\psi}{\partial\gamma_{,i}}\frac{D^F\gamma_{,i}}{Dt} - \phi\gamma\frac{\partial\psi}{\partial v_i}\frac{D^F v_i}{Dt} - \phi\gamma\frac{\partial\psi}{\partial D_{ij}}\frac{D^F D_{ij}}{Dt} \\ + T_{ij}D_{ij} - v_i\hat{m}_i \geqslant 0. \end{aligned} \tag{7.3.7}$$

By the fluid mass balance,

$$-\phi\gamma\frac{\partial\psi}{\partial\gamma}\frac{D^F\gamma}{Dt} = \gamma^2\frac{\partial\psi}{\partial\gamma}\phi_{,i}v_i + \phi\gamma^2\frac{\partial\psi}{\partial\gamma}\delta_{ij}D_{ij},$$

and substituting this identity into the inequality (7.3.7) yields

$$\begin{aligned} -\phi\gamma\frac{\partial\psi}{\partial\gamma_{,i}}\frac{D^F\gamma_{,i}}{Dt} - \phi\gamma\frac{\partial\psi}{\partial v_i}\frac{D^F v_i}{Dt} - \phi\gamma\frac{\partial\psi}{\partial D_{ij}}\frac{D^F D_{ij}}{Dt} \\ + \left(\phi\gamma^2\frac{\partial\psi}{\partial\gamma}\delta_{ij}D_{ij} + T_{ij}\right)D_{ij} + \left(\gamma^2\frac{\partial\psi}{\partial\gamma}\phi_{,i} - \hat{m}_i\right)v_i \geqslant 0. \end{aligned} \tag{7.3.8}$$

The left side of this inequality is linear in the functions $D^F\gamma_{,i}/Dt$, $D^F v_i/Dt$, and $D^F D_{ij}/Dt$, which can assume any sign. According to the Coleman–Noll procedure, for the inequality to hold for all possible values of these derivatives, their coefficients must vanish. Hence,

$$\frac{\partial\psi}{\partial\gamma_{,i}} = \frac{\partial\psi}{\partial v_i} = \frac{\partial\psi}{\partial D_{ij}} = 0.$$

It follows that $\psi = \psi(\gamma)$.

As in Section 6.5, let us identify the derivative $\gamma^2\partial\psi/\partial\gamma = -\partial\psi/\partial\gamma^{-1}$ as a pressure. We denote by

$$p = \gamma^2\frac{\partial\psi}{\partial\gamma} \tag{7.3.9}$$

the fluid **pore pressure**, interpreting the quantity $\phi p = \phi\gamma^2\partial\psi/\partial\gamma$ as the fluid pressure acting on the fraction of the area of a surface occupied by fluid in the pores. Implicit in this interpretation—but not essential to the mathematics—is an assumption that the area fraction of fluid on any mathematical surface in the porous medium equals the porosity.

With these observations, we deduce from the condition (7.3.8) a reduced entropy inequality,

$$\begin{aligned}\theta\Upsilon &= (\phi\, p\, \delta_{ij} + T_{ij})D_{ij} + (p\,\phi_{,i} - \hat{m}_i)v_i \\ &= (\phi\, p\, \mathsf{I} + \mathsf{T}) : \mathsf{D} + (p \operatorname{grad}\phi - \hat{\mathbf{m}})\cdot\mathbf{v} \geqslant 0.\end{aligned} \tag{7.3.10}$$

The left side may be nonlinear in $\mathbf{v}$ and D.

7.3.3 Equilibrium Constraints

Paralleling the logic of Section 6.5, we glean further information by examining the entropy inequality at equilibrium, which in this context refers to any state in which $\mathbf{v} = \mathbf{0}$ and $\mathsf{D} = 0$. In such states, $\theta\Upsilon = 0$ by Equation (7.3.10). Since the entropy inequality requires $\theta\Upsilon \geqslant 0$, these states must yield a minimum value for $\theta\Upsilon$. To deduce the consequences of this observation when $\theta\Upsilon$ is continuously differentiable with respect to $\mathbf{v}$ and D, fix the values of γ and $\operatorname{grad}\gamma$, and define the following function of a single variable ζ:

$$\begin{aligned}f(\zeta) = &\left[\phi\, p(\gamma, \operatorname{grad}\gamma, \zeta\mathbf{a}, \zeta\mathsf{A})\,\mathsf{I} + \mathsf{T}(\gamma, \operatorname{grad}\gamma, \zeta\mathbf{a}, \zeta\mathsf{A})\right] : \zeta\mathsf{A} \\ &+ \left[p(\gamma, \operatorname{grad}\gamma, \zeta\mathbf{a}, \zeta\mathsf{A}) \operatorname{grad}\phi - \hat{\mathbf{m}}(\gamma, \operatorname{grad}\gamma, \zeta\mathbf{a}, \zeta\mathsf{A})\right]\cdot\zeta\mathbf{a}.\end{aligned}$$

Here $\mathbf{a} \in \mathbb{E}$ and $\mathsf{A} \in L(\mathbb{E})$ are arbitrary but nonzero. The value $\zeta = 0$ yields $f(0) = 0$ and corresponds to the fact that $\theta\Upsilon = 0$ at equilibrium. To guarantee that

$\zeta = 0$ yields a minimum value for f, the derivative must also vanish there:

$$f'(0) = \left[\phi\, p(\gamma, \operatorname{grad}\gamma, \mathbf{0}, 0)\, \mathsf{I} + \mathsf{T}(\gamma, \operatorname{grad}\gamma, \mathbf{0}, 0)\right] : \mathsf{A} + \left[p(\gamma, \operatorname{grad}\gamma, \mathbf{0}, 0) \operatorname{grad}\phi - \hat{\mathbf{m}}(\gamma, \operatorname{grad}\gamma, \mathbf{0}, 0)\right] \cdot \mathbf{a} = 0.$$

It follows that the expressions in square brackets vanish:

$$\mathsf{T}(\gamma, \operatorname{grad}\gamma, \mathbf{0}, 0) = -\phi\, p(\gamma, \operatorname{grad}\gamma, \mathbf{0}, 0)\, \mathsf{I}, \tag{7.3.11}$$

and

$$\hat{\mathbf{m}}(\gamma, \operatorname{grad}\gamma, \mathbf{0}, 0) = p(\gamma, \operatorname{grad}\gamma, \mathbf{0}, 0) \operatorname{grad}\phi. \tag{7.3.12}$$

Equation (7.3.11) closely resembles the inviscid term derived in Section 6.5 for stress in a Newtonian fluid. The additional factor ϕ has some physical motivation in this setting: since in a porous medium T represents force on the fluid per unit area of the mixture, multiplying the pore pressure by the porosity yields an appropriate model for the pressure. The conclusion (7.3.12) for the momentum exchange term $\hat{\mathbf{m}}$ deserves further comment below.

7.3.4 Linear Extensions From Equilibrium

The most common field equation for fluid flow in porous media arises through linear extensions from equilibrium of the constitutive relations for the fluid stress T and the momentum exchange $\hat{\mathbf{m}}$. For the stress, the linear extension parallels the Newtonian model derived in Section 6.5:

$$\mathsf{T} = -\phi\, p\, \mathsf{I} + \lambda(\operatorname{tr}\mathsf{D})\, \mathsf{I} + 2\,\mu\, \mathsf{D}. \tag{7.3.13}$$

Here λ and μ are the fluid's coefficients of viscosity.

For the momentum exchange, we adopt the following linear extension:

$$\hat{\mathbf{m}} = p \operatorname{grad}\phi - \Lambda^{-1}\,\mathbf{v}. \tag{7.3.14}$$

Here Λ^{-1} denotes a tensor-valued function of spatial position called the **resistivity**. We assume that Λ^{-1} is symmetric, to guarantee that it has real eigenvalues and eigenvectors. In a moment we justify the use of notation suggesting that this tensor function is invertible at every point in space. The model (7.3.14) generalizes the Stokes drag exerted by a flowing fluid on a stationary solid sphere, discussed in Section 5.3. The Stokes drag model for momentum exchange is less closely tied to the observer's frame of reference than the notation suggests. Here and in more general settings, the Stokes drag is proportional to the relative velocity of the fluid with respect to the solid, which by Exercise 190 is objective.

These linear models must remain consistent with the entropy inequality. Substituting Equations (7.3.13) and (7.3.14) into the inequality (7.3.10) yields

$$\theta \Upsilon = \lambda (\mathrm{tr}\,\mathsf{D})^2 + 2\mu \mathsf{D} : \mathsf{D} + \mathbf{v} \cdot \Lambda^{-1} \mathbf{v} \geqslant 0.$$

EXERCISE 192 *Show that, for this inequality to hold,* $\mu \geqslant 0$ *and* $3\lambda + 2\mu \geqslant 0$*, as we found in Theorem 6.5.2.*

EXERCISE 193 *Show that the resistivity* Λ^{-1} *is* **positive semidefinite**, *that is,* $\mathbf{v} \cdot \Lambda^{-1}\mathbf{v} \geqslant 0$ *for all vectors* $\mathbf{v} \in \mathbb{E}$.

We assume, in fact, that $\mathbf{v} \cdot \Lambda^{-1}\mathbf{v} = 0$ only if $\mathbf{v} = \mathbf{0}$, so that $\Lambda^{-1}\colon \mathbb{E} \to \mathrm{SPD}(\mathbb{E})$. Postulating that Λ^{-1} is symmetric and positive definite at every spatial point ensures that the solid matrix exerts nonzero resistance to fluid flow in every direction, at all points in the porous medium. It also implies that Λ^{-1} is invertible.

Substituting the linear constitutive equations (7.3.13) and (7.3.14) into the fluid momentum balance (7.3.2) yields

$$\phi\gamma \frac{D^F \mathbf{v}}{Dt} + \mathrm{grad}\,(\phi\, p) - \mathrm{div}\,[\lambda\,\mathrm{tr}\,(\mathsf{D})\mathsf{I} + 2\mu\mathsf{D}] - \phi\gamma\mathbf{g} \tag{7.3.15}$$

$$= p\,\mathrm{grad}\,\phi - \Lambda^{-1}\mathbf{v}. \tag{7.3.16}$$

We now restrict attention to slow flows, in which two additional assumptions hold. First, the inertial term is negligible:

$$\phi\gamma \frac{D^F \mathbf{v}}{Dt} \simeq \mathbf{0}.$$

Second, the drag on the fluid exerted by the solid matrix dominates the momentum transfer within the fluid, so that the contribution of viscous effects is also negligible:

$$\mathrm{div}\,[\lambda\,\mathrm{tr}\,(\mathsf{D})\mathsf{I} + 2\mu\mathsf{D}] \simeq \mathbf{0}.$$

Imposing these assumptions on the fluid momentum balance gives

$$\mathrm{grad}\,(\phi\, p) - \phi\gamma\mathbf{g} = p\,\mathrm{grad}\,\phi - \Lambda^{-1}\mathbf{v}.$$

We now apply the product rule to $\mathrm{grad}\,(\phi\, p)$ and rearrange the result to obtain **Darcy's law**,

$$\mathbf{v} = -\phi\,\Lambda(\mathrm{grad}\,p - \gamma\mathbf{g}). \tag{7.3.17}$$

It is common to decompose $\mathsf{\Lambda} = \mathsf{K}/(\phi^2\mu)$, where μ is the fluid viscosity and K is a rock property called the **permeability**. This hypothesis yields the following form for Darcy's law:

$$\phi\mathbf{v} = -\frac{\mathsf{K}}{\mu}(\operatorname{grad} p - \gamma\mathbf{g}), \tag{7.3.18}$$

where $\phi\mathbf{v}$ is called the **filtration velocity**. This quantity represents the volume per unit time crossing a unit of surface area in the porous medium oriented in the direction normal to $\mathbf{v}$. Todd [48] refers to the field $\mathbf{v}$ as the **average interstitial velocity**.

EXERCISE 194 *Determine the dimension of* K *in Equation (7.3.18). Referring to Equation (5.3.17), heuristically justify the appearance of the fluid viscosity, based on the constitutive models adopted here.*

EXERCISE 195 *Show that adding the porosity* ϕ *to the list* $(\gamma, \operatorname{grad}\gamma, \mathbf{v}, \mathsf{D})$ *of independent variables results in the following reduced entropy inequality:*

$$\theta\Upsilon = \left(\phi\gamma^2\frac{\partial\psi}{\partial\gamma}\delta_{ij} + T_{ij}\right)D_{ij} + \left[\left(\gamma^2\frac{\partial\psi}{\partial\gamma} - \phi\gamma\frac{\partial\psi}{\partial\phi}\right)\phi_{,i} - m_i\right]v_i \geqslant 0. \tag{7.3.19}$$

7.3.5 Commentary

Several aspects of this derivation deserve further discussion. One is the nature of the equilibrium momentum exchange term (7.3.12). From a physical standpoint this term seems inscrutable. It appears to describe a rate of momentum transfer from solid to fluid to compensate for gradients in porosity. Raats [43] refers to this effect as a "buoyancy term," even though the most common uses of the term buoyancy involve a body force—gravity—instead of interphase momentum exchange. While it may be tempting to argue heuristically for momentum exchange proportional to $\operatorname{grad}\phi$, the axiom of admissibility yields exactly the term needed to offset a contribution arising from the flux term $\operatorname{grad}(\phi p) = \phi \operatorname{grad} p + p \operatorname{grad}\phi$ of the fluid momentum balance. Perhaps the best explanation for this term is that consistency with the entropy inequality requires that fluid movements not arise purely in response to porosity gradients.

A remark is also in order about the exercise above, which adds the volume fraction ϕ as an independent variable. In flows involving more than one fluid phase in a porous medium, the curvature of fluid interfaces gives rise to differences in fluid pressures, known as capillary pressures. In applications of mixture theory to this setting, the capillary pressures involve factors of the form $\partial\psi_\alpha/\partial\phi_\alpha$. Hence, the use of volume fractions as independent variables is essential to the physics of multifluid flows in porous media. See [10] for more details.

Finally, this application of the theory of multiconstituent continua is at once reassuring and disappointing. It reassuringly illustrates that a common field equation, derived in the nineteenth century from one-dimensional experiments in sand columns, possesses a rational explanation based on principles of mechanics. It also reveals the disappointing features of the hypothesis of overlapping continua. The starkest limitation is the opaqueness of the factor $\mathsf{\Lambda}$ appearing in Darcy's law.

Exercise 194 suggests the decomposition $\phi^2\mathsf{\Lambda} = \mathsf{K}/\mu$. But the common interpretation of the permeability K as a material property only of the solid remains largely phenomenological. A true understanding of the resistance imposed by the solid matrix on fluid flow requires detailed study of pore geometry and fluid-solid interactions at the microscopic scale. Yet in the formulation of this section the sole variable available to quantify these effects is the porosity ϕ, which captures only the crudest attribute of pore-scale geometry. A more sophisticated theory of flow in porous media almost certainly requires more details of the pore geometry and physics. See [22] for one approach.

7.3.6 Potential Formulation of Darcy's Law

In a 1956 paper, Hubbert [26] observed that one can often cast Darcy's law in a form that involves a scalar potential, thereby enabling tools from potential theory reviewed in Section 5.2. To explore this possibility in a simple setting, we restrict attention to the case when the tensor K is isotropic, so $\mathsf{K} = K\mathsf{I}$ for a scalar field $K(\mathbf{x})$, and the flow is steady, so time derivatives vanish. In this case Equation (7.3.18) becomes

$$\phi\mathbf{v} = -\frac{K}{\mu}(\operatorname{grad} p - \gamma\mathbf{g}).$$

Now adopt a Cartesian coordinate system in which the gravitational force $\mathbf{g} = -g\,\mathbf{e}_3 = g\operatorname{grad} z$, where $z(\mathbf{x}) = -\mathbf{x}\cdot\mathbf{e}_3$ denotes the depth of the point $\mathbf{x}$ below some prescribed datum. Following Hubbert, we write Darcy's law in the form

$$\phi\mathbf{v} = \frac{\gamma K}{\mu}\mathbf{E},$$

where

$$\mathbf{E} = g\operatorname{grad} z - \frac{\operatorname{grad} p}{\gamma}. \tag{7.3.20}$$

The vector field $\mathbf{E}(\mathbf{x})$ represents the force per unit mass acting on the fluid as a result of pressure gradients and gravity.

As discussed in Section 3.3, for $\mathbf{E}$ to be the gradient of a scalar potential, $\operatorname{curl}\mathbf{E}$ must vanish. This can happen in several ways:

EXERCISE 196 *Using the vector-field identity* $\operatorname{curl}(c\,\mathbf{a}) = (\operatorname{grad} c) \times \mathbf{a} + c \operatorname{curl}\mathbf{a}$, *show that any one of the following conditions ensures that* $\operatorname{curl}\mathbf{E}$ *vanishes identically:*

1. *The pressure p is uniform in space.*
2. *The true fluid density γ is uniform in space.*
3. *The vector field* $\operatorname{grad}\gamma^{-1}$ *is everywhere parallel to the vector field* $\operatorname{grad} p$.

If we exclude the uninteresting case of uniform pressure, this exercise reveals that unless the fluid density γ is uniform, $\mathbf{E}$ is a gradient field only if the level sets of γ coincide with those of p. This condition holds when $\gamma = \gamma(p)$, a relationship that excludes nonisothermal flows, in which γ varies with temperature, and flows in which γ varies with the fluid's chemical composition. A fluid for which $\gamma = \gamma(p)$ is **barotropic**.

In the barotropic case, integrating $\mathbf{E}$ yields a continuously differentiable scalar field $\Phi(\mathbf{x})$ for which $\mathbf{E} = -\operatorname{grad}\Phi$. To see how, pick a spatial point $\mathbf{x}_{\text{ref}} \in \mathbb{E}$ whose vertical coordinate is the prescribed datum, that is, for which $z(\mathbf{x}_{\text{ref}}) = -\mathbf{x}_{\text{ref}} \cdot \mathbf{e}_3 = 0$. To define $\Phi(\mathbf{x})$ for an arbitrary point $\mathbf{x}$ in the domain, consider any continuously differentiable arc $\boldsymbol{\zeta}\colon [0,1] \to \mathbb{E}$ such that $\boldsymbol{\zeta}(0) = \mathbf{x}_{\text{ref}}$ and $\boldsymbol{\zeta}(1) = \mathbf{x}$, as illustrated in Figure 7.4. By the fundamental theorem of calculus and Equation (7.3.20),

$$\begin{aligned}
\Phi(\mathbf{x}) &= \Phi(\mathbf{x}_{\text{ref}}) + \int_{\boldsymbol{\zeta}} \operatorname{grad}\Phi \cdot d\mathbf{x} \\
&= \Phi(\mathbf{x}_{\text{ref}}) - \int_{\boldsymbol{\zeta}} \mathbf{E} \cdot d\mathbf{x} \\
&= \Phi(\mathbf{x}_{\text{ref}}) \underbrace{- g \int_0^1 \operatorname{grad} z(\boldsymbol{\zeta}(s)) \cdot \boldsymbol{\zeta}'(s)\, ds}_{\text{(I)}} + \underbrace{\int_0^1 \frac{\operatorname{grad} p(\boldsymbol{\zeta}(s))}{\gamma(p(\boldsymbol{\zeta}(s)))} \cdot \boldsymbol{\zeta}'(s)\, ds}_{\text{(II)}}.
\end{aligned}$$

The integrals on the right reduce as follows:

$$\text{(I)} = -g \int_0^1 \frac{d}{ds} z(\boldsymbol{\zeta}(s))\, ds = -g[z(\mathbf{x}) - z(\mathbf{x}_{\text{ref}})] = -g\, z(\mathbf{x}),$$

$$\text{(II)} = \int_{p(\mathbf{x}_{\text{ref}})}^{p(\mathbf{x})} \frac{d\varpi}{\gamma(\varpi)}.$$

In constructing Φ we impose the arbitrary but convenient condition that $\Phi(\mathbf{x}_{\text{ref}}) = 0$, in which case

$$\Phi(\mathbf{x}) = -g\,z + \int_{p(\mathbf{x}_{\text{ref}})}^{p(\mathbf{x})} \frac{d\varpi}{\gamma(\varpi)}. \tag{7.3.21}$$

This scalar potential has dimension energy/mass, or [L^2T^{-2}].

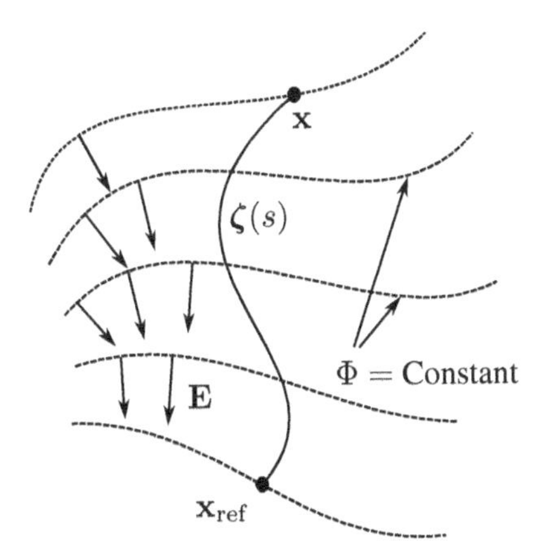

Figure 7.4. Construction used to define the Hubbert potential for Darcy's law.

Thus, for a barotropic fluid, Darcy's law takes the form

$$\phi\mathbf{v} = -\frac{\gamma K}{\mu}\,\text{grad}\,\Phi,$$

where $\phi\mathbf{v}$ is the filtration velocity and Φ is the **Hubbert potential** defined by Equation (7.3.21). In the special case when the fluid is incompressible, γ is constant, and $\Phi = -gz + p/\gamma$.

This potential formulation leads to a classic problem in groundwater hydrology, whose solution provided one of the earliest practical methods for measuring the parameter K. Consider a rigid aquifer of radius R in which the material properties ϕ, γ, and Λ are constant and spatially uniform and in which water flows in a region surrounding a vertical well of radius r_w. The mass balance and Darcy's law for this case are

$$\text{div}\,(\phi\mathbf{v}) = 0, \qquad \phi\mathbf{v} = -\frac{\gamma K}{\mu}\,\text{grad}\,\Phi.$$

These equations, together with the spatial uniformity of γ, K, and μ, imply that the Hubbert potential is harmonic: $-\Delta\Phi = 0$.

If the aquifer is horizontal with constant thickness b, we may adopt a Cartesian coordinate system in which the wellbore has centerline $(0, 0, x_3)$, as shown in Figure 7.5. If, in addition, no fluid flows across the horizontal bounding planes $x_3 = 0$ and

$x_3 = b$, then we may exploit Exercise 129 and pose the equation for the potential in two space dimensions:

$$-\Delta_2 \tilde{\Phi} = 0, \quad \|\mathbf{x}\| < R, \tag{7.3.22}$$

where $\Delta_2 = \partial^2/\partial x_1^2 + \partial^2/\partial x_2^2$ denotes the two-dimensional Laplace operator and

$$\tilde{\Phi} = \frac{1}{b}\int_0^b \Phi(\mathbf{x})\, dx_3$$

is the vertically averaged Hubbert potential.

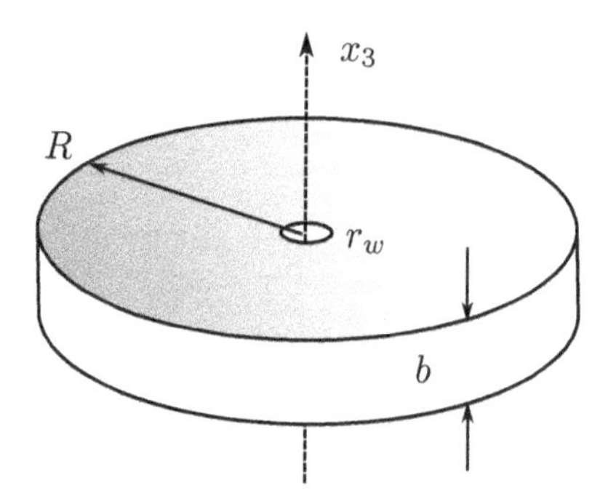

Figure 7.5. Horizontal aquifer having a vertical well centered at $(x_1, x_2) = (0, 0)$.

To close the problem, we impose boundary conditions at the outer edge $\|\mathbf{x}\| = R$ and at the wellbore $\|\mathbf{x}\| = r_w$. On the outer edge, the potential remains constant at a known value Φ_R. At the wellbore, a pump withdraws water at the total rate Q, with dimension $[\mathrm{L}^3\mathrm{T}^{-1}]$. The rate of water withdrawal per unit area across the wellbore is then $Q/(2\pi r_w b)$. Thus the boundary conditions are as follows:

$$\begin{aligned} \tilde{\Phi}(\mathbf{x}) &= \Phi_R, & \|\mathbf{x}\| &= R, \\ \phi \mathbf{v} \cdot \mathbf{n} &= \frac{Q}{2\pi r_w b}, & \|\mathbf{x}\| &= r_w. \end{aligned} \tag{7.3.23}$$

EXERCISE 197 *Show that the outward unit normal vector field at the wellbore is* $\mathbf{n} = -(x_1, x_2, 0)/r_w$, *and hence the boundary condition at the wellbore reduces to*

$$-\frac{\gamma K}{\mu}\left(x_1 \frac{\partial \tilde{\Phi}}{\partial x_1} + x_2 \frac{\partial \tilde{\Phi}}{\partial x_2}\right) = \frac{Q}{2\pi b}.$$

EXERCISE 198 *Verify that a function of the form*

$$\tilde{\Phi} = C_1 \log\left(\frac{1}{\|\mathbf{x}\|}\right) + C_2$$

is a solution to $-\Delta_2\tilde{\Phi} = 0$ *on the domain* $r_w < \|\mathbf{x}\| < R$ *for arbitrary constants* C_1 *and* C_2.

EXERCISE 199 *Use the boundary conditions to determine* C_1 *and* C_2*, obtaining the following solution:*

$$\tilde{\Phi}(x_1, x_2) = \frac{Q\mu}{2\pi\gamma b K} \log\left(\frac{R}{\|\mathbf{x}\|}\right) + \Phi_R. \tag{7.3.24}$$

Thiem [47] derived this solution in 1906. Equation (7.3.24) furnishes a method for determining K by pumping water at the rate Q and observing the value of $\tilde{\Phi}$ at an observation well located at some distance $\|\mathbf{x}\| < R$ from the origin. The method remains useful, although the advent of high-performance computers, high-accuracy numerical approximations, and greater understanding of parameter identification have spurred the development of more sophisticated methods.

7.3.7 Summary

Mixture theory furnishes one avenue to the derivation of equations governing flow of an isothermal, single-phase fluid F through a porous solid S. If the solid matrix is rigid with incompressible grains and porosity ϕ, the main focus is on the momentum balance for the fluid,

$$\phi\gamma\frac{D^F\mathbf{v}}{Dt} - \operatorname{div}\mathsf{T} - \phi\gamma\mathbf{g} = \hat{\mathfrak{m}},$$

where $\mathbf{g}$ denotes the gravitational acceleration. We adopt constitutive relations that, *a priori*, give $\mathsf{T}, \hat{\mathfrak{m}}$ as functions of $(\gamma, \operatorname{grad}\gamma, \mathbf{v}, \mathsf{D})$. Applying the entropy inequality and analyzing the behavior under the conditions $v_i = D_{ij} = 0$ yield the following results for the fluid stress and momentum exchange at equilibrium:

$$\mathsf{T}(\gamma, \operatorname{grad}\gamma, \mathbf{0}, 0) = -\phi\, p(\gamma, \operatorname{grad}\gamma, \mathbf{0}, 0)\,\mathsf{I},$$

$$\hat{\mathfrak{m}}(\gamma, \operatorname{grad}\gamma, \mathbf{0}, 0) = p(\gamma, \operatorname{grad}\gamma, \mathbf{0}, 0)\operatorname{grad}\phi.$$

Assuming simple linear extensions from equilibrium of the form

$$\begin{aligned} \mathsf{T} &= -\phi\, p\,\mathsf{I} \\ \hat{\mathfrak{m}} &= p\operatorname{grad}\phi - \Lambda^{-1}\mathbf{v}, \end{aligned}$$

restricting attention to slow flows in which the inertial terms are negligible, and neglecting momentum transfer within the fluid reduce the fluid momentum balance to Darcy's law,

$$\mathbf{v} = -\phi\,\Lambda(\operatorname{grad} p - \gamma\mathbf{g}).$$

Under certain circumstances, for example, when the fluid density depends only on the fluid pressure, it is possible to recast Darcy's law in terms of a scalar potential field, yielding simplified flow models that are amenable to classical mathematical analysis.

7.4 DIFFUSION IN A BINARY FLUID MIXTURE

The diffusion of a dilute solute in a binary fluid mixture furnishes a second prominent application of multiconstituent continuum mechanics. For example, consider what happens when you open a bottle of perfume at one end of a room. Although some of the perfume immediately evaporates, its concentration in the air never reaches levels high enough to affect the density appreciably. Its presence in the air remains undetectable at the other end of the room for a short period of time. But the perfume vapor spreads, until its odor is noticeable—eventually even overwhelming—throughout the room.

In this setting, the mixture is miscible. Since the segregation between chemical species occurs at the molecular level, there is no need in the macroscopic mechanics to account for volume fractions of the constituents. Impossible to model using the mechanics of simple continua, miscible diffusion occurs in gases, liquids, and solids. It plays a central role in environmental engineering, groundwater contaminant hydrology, spatial models of epidemics, snow mechanics, and many other fields.

The most common model for diffusion is **Fick's law**, which asserts that the diffusive flux $\mathbf{j}_\alpha$ of species α, identified in Exercise 182, is proportional to the density gradient of the diffusing species and points in the direction of the steepest density decrease:

$$\mathbf{j}_\alpha = -D \,\mathrm{grad}\,(\overline{\rho}\, c_\alpha). \tag{7.4.1}$$

Here $c_\alpha = \rho_\alpha/\overline{\rho}$ is the mass fraction of species α, and D_α denotes the positive **diffusion coefficient**, with dimension $[\mathrm{L}^2\mathrm{T}^{-1}]$.

As geometrically appealing as this linear, gradient-based relationship may be, it rests on several assumptions that may appear to be quite restrictive. The next subsection enumerates some of these modeling assumptions. Later subsections review a set of balance laws and constitutive relationships that yield Equation (7.4.1), following a line of reasoning presented in [3]. Many other treatments of diffusion exist; see [8] and [34] for examples

7.4.1 Modeling Assumptions for Binary Diffusion

Consider a miscible, two-fluid mixture consisting of a dilute, isotropic solute A and an isotropic host fluid B. We adopt the following assumptions:

1. The mixture is isothermal, with a common temperature θ for which $\partial\theta/\partial t = 0$ and $\operatorname{grad}\theta = \mathbf{0}$. Heat transfer effects are negligible: $\bar{\bar{\mathbf{q}}} = \mathbf{0}$; see Equation (7.2.22).

2. Effects of body forces are negligible: $\mathbf{b}_A = \mathbf{b}_B = \mathbf{0}$.

3. The mixture is sufficiently dilute that the solute has negligible effect on the total mixture density. As consequences,

$$\frac{\partial}{\partial t}(\rho_A + \rho_B) = 0, \qquad \operatorname{grad}(\rho_A + \rho_B) = \mathbf{0}.$$

4. There are no chemical reactions: $\hat{\rho}_A = -\hat{\rho}_B = 0$.

Further assumptions arise as the derivation unfolds.

7.4.2 Balance Laws for the Two Species

The modeling assumptions listed above yield reduced forms of the balance laws developed in Section 7.2. In the absence of chemical reactions, the mass balances take the form

$$\frac{D^\alpha \rho_\alpha}{Dt} + \rho_\alpha \operatorname{div}\mathbf{v}_\alpha = 0, \qquad \alpha = A, B,$$

which we rewrite in terms of the stretching tensors using the identity $\operatorname{div}\mathbf{v}_\alpha = \mathsf{I} : \mathsf{D}_\alpha$:

$$\begin{aligned}
\frac{D^A \rho_A}{Dt} &= -\rho_A \mathsf{I} : \mathsf{D}_A, \\
\frac{D^B \rho_B}{Dt} &= -\rho_B \mathsf{I} : \mathsf{D}_B.
\end{aligned} \tag{7.4.2}$$

In the absence of body forces, the momentum balance equations are as follows:

$$\begin{aligned}
\rho_A \frac{D^A \mathbf{v}_A}{Dt} - \operatorname{div}\mathsf{T}_A &= \hat{\mathbf{m}}, \\
\rho_B \frac{D^B \mathbf{v}_B}{Dt} - \operatorname{div}\mathsf{T}_B &= -\hat{\mathbf{m}},
\end{aligned} \tag{7.4.3}$$

where $\hat{\mathbf{m}} = \hat{\mathbf{m}}_A = -\hat{\mathbf{m}}_B$ denotes the rate of momentum exchange per unit mass into species A from species B. For the angular momentum balance, we assume that both species have symmetric stress tensors:

$$\mathsf{T}_A = \mathsf{T}_A^\top, \qquad \mathsf{T}_B = \mathsf{T}_B^\top. \tag{7.4.4}$$

Since the mixture is isothermal with a common temperature θ for the two species, it suffices to use the total energy balance, which reduces to a special case of Equation (7.2.21) with ρ_α taking the place of $\phi_\alpha \gamma_\alpha$:

$$\sum_\alpha \rho_\alpha \frac{D^\alpha \psi^\alpha}{Dt} + \overline{\rho}\theta \frac{D\overline{\eta}}{Dt} - \sum_\alpha \mathsf{T}_\alpha : \mathsf{D}_\alpha = (\mathbf{v}_B - \mathbf{v}_A) \cdot \hat{\mathbf{m}}. \tag{7.4.5}$$

Equations (7.4.2) through (7.4.5) furnish 15 scalar equations for the 32 scalar functions needed to specify

$$\rho_A, \rho_B, \mathbf{v}_A, \mathbf{v}_B, \mathsf{T}_A, \mathsf{T}_B, \hat{\mathbf{m}}, \psi_A, \psi_B, \overline{\eta}.$$

We treat the temperature θ as a prescribed parameter, and as in Section 7.3 we treat the stretching tensors D_A and D_B as calculable once $\mathbf{v}_A$ and $\mathbf{v}_B$ are known. Solutions to these equations must be consistent with the entropy inequality (7.2.20), which for the binary mixture of interest here takes the following form:

$$-\rho_A \frac{D^A \psi_A}{Dt} - \rho_B \frac{D^B \psi_B}{Dt} + \mathsf{T}_A : \mathsf{D}_A + \mathsf{T}_B : \mathsf{D}_B + (\mathbf{v}_B - \mathbf{v}_A) \cdot \hat{\mathbf{m}} \geqslant 0. \tag{7.4.6}$$

7.4.3 Constitutive Relationships for Diffusion

To close this system of balance laws, we adopt $17 = 32 - 15$ constitutive relations, giving the remaining scalar components of

$$\mathsf{T}_A, \mathsf{T}_B, \hat{\mathbf{m}}, \psi_A, \psi_B$$

as functions of other quantities. (By the angular momentum balance equations (7.4.3), only the diagonal entries and, say, the subdiagonal entries of T_A and T_B require such relations.) To keep the exposition as simple as possible, we examine constitutive functions having the following forms:

$$\begin{aligned}
\mathsf{T}_\alpha &= \mathsf{T}_\alpha(\rho_A, \rho_B), \qquad \alpha = A, B; \\
\hat{\mathbf{m}} &= \hat{\mathbf{m}}(\rho_A, \rho_B); \\
\psi_\alpha &= \psi_\alpha(\rho_A, \rho_B), \qquad \alpha = A, B.
\end{aligned} \tag{7.4.7}$$

These functions greatly simplify the behavior observed in most natural settings, where the constitutive variables may depend on the velocities and perhaps the gradients of density and velocity. But they allow us to focus on the specific mechanics that give rise to density-driven diffusion. They also automatically satisfy the axioms of determinism, equipresence, objectivity, and symmetry (since the fluids are isotropic).

We now enforce the axiom of admissibility by substituting the constitutive relationships (7.4.7) into the entropy inequality (7.4.6). Applying the chain rule to the material derivative of the specific Helmholtz free energy ψ_A yields

$$\frac{D^A\psi_A}{Dt} = \frac{\partial\psi_A}{\partial\rho_A}\frac{D^A\rho_A}{Dt} + \frac{\partial\psi_A}{\partial\rho_B}\frac{D^A\rho_B}{Dt}. \tag{7.4.8}$$

By the mass balances and the definitions of D^α/Dt,

$$\begin{aligned}\frac{D^A\rho_A}{Dt} &= -\rho_A \mathsf{I} : \mathsf{D}_A,\\ \frac{D^A\rho_B}{Dt} &= -\rho_B \mathsf{I} : \mathsf{D}_B - (\mathbf{v}_B - \mathbf{v}_A)\cdot \mathrm{grad}\rho_B,\end{aligned}$$

and therefore

$$\frac{D^A\psi_A}{Dt} = -\rho_A\frac{\partial\psi_A}{\partial\rho_A}\mathsf{I} : \mathsf{D}_A - \rho_B\frac{\partial\psi_A}{\partial\rho_B}\mathsf{I} : \mathsf{D}_B - \frac{\partial\psi_A}{\partial\rho_B}(\mathbf{v}_B - \mathbf{v}_A)\cdot \mathrm{grad}\,\rho_B. \tag{7.4.9}$$

Similarly,

$$\frac{D^B\psi_B}{Dt} = -\rho_A\frac{\partial\psi_B}{\partial\rho_A}\mathsf{I} : \mathsf{D}_A - \rho_B\frac{\partial\psi_B}{\partial\rho_B}\mathsf{I} : \mathsf{D}_B + \frac{\partial\psi_B}{\partial\rho_A}(\mathbf{v}_B - \mathbf{v}_A)\cdot \mathrm{grad}\,\rho_A. \tag{7.4.10}$$

Substituting for $D^A\psi_A/Dt$ and $D^B\psi_B/Dt$ in the entropy inequality using Equations (7.4.9) and (7.4.10) and grouping terms with common factors produces an expanded entropy inequality,

$$\begin{aligned}&\left(p_A\mathsf{I} + \rho_A\rho_B\frac{\partial\psi_B}{\partial\rho_A}\mathsf{I} + \mathsf{T}_A\right) : \mathsf{D}_A + \rho_A\frac{\partial\psi_A}{\partial\rho_B}(\mathbf{v}_B - \mathbf{v}_A)\cdot \mathrm{grad}\,\rho_B\\ &+\left(p_B\mathsf{I} + \rho_A\rho_B\frac{\partial\psi_A}{\partial\rho_B}\mathsf{I} + \mathsf{T}_B\right) : \mathsf{D}_B - \rho_B\frac{\partial\psi_B}{\partial\rho_A}(\mathbf{v}_B - \mathbf{v}_A)\cdot \mathrm{grad}\,\rho_A\\ &+(\mathbf{v}_B - \mathbf{v}_A)\cdot\hat{\mathbf{m}} \geqslant 0.\end{aligned} \tag{7.4.11}$$

In writing this inequality, we identify the species pressures

$$p_\alpha = \rho_\alpha^2\frac{\partial\psi_\alpha}{\partial\rho_\alpha}$$

by analogy with Equation (7.3.9).

We now apply the Coleman–Noll procedure. The left side of the inequality (7.4.11) is linear in the functions D_A, D_B, $\operatorname{grad}\rho_A$, $\operatorname{grad}\rho_B$, and $\mathbf{v}_B - \mathbf{v}_A$. For the inequality to remain valid for all possible values of these functions, their coefficients must vanish. It follows that

$$\frac{\partial \psi_A}{\partial \rho_B} = \frac{\partial \psi_B}{\partial \rho_A} = 0,$$

that is, the specific Helmholtz free energies obey functional relationships of the form $\psi_\alpha = \psi_\alpha(\rho_\alpha)$. In addition,

$$\mathsf{T}_A = -p_A(\rho_A)\mathsf{I},$$

$$\mathsf{T}_B = -p_B(\rho_B)\mathsf{I}.$$

These dependencies show that viscous effects do not arise when the stresses depend only on the fluid densities, with no allowance for dependencies on the fluid velocities or their derivatives.

At sufficiently low densities, a common constitutive relations for the fluid pressures is the **ideal gas law**, $p_\alpha(\rho_\alpha) = C\rho_\alpha$, where C denotes a positive, temperature-dependent parameter. In this case, each of the pressures $p_\alpha(\rho_\alpha)$ is continuously differentiable, with $p'_\alpha(\rho_\alpha) > 0$.

These restricted constitutive relations yield an unusually simple residual entropy inequality:

$$(\mathbf{v}_B - \mathbf{v}_A) \cdot \hat{\mathbf{m}} \geqslant 0. \tag{7.4.12}$$

In this case, analysis of equilibrium states, in which $\mathbf{v}_B - \mathbf{v}_A = \mathbf{0}$ and $\mathsf{D}_A = \mathsf{D}_B = 0$, is uninformative. The simplest linear model of interspecies momentum transfer that guarantees the inequality (7.4.12) is a scalar analog of the Stokes drag model adopted in the previous section:

$$\hat{\mathbf{m}} = -\Lambda^{-1}(\mathbf{v}_B - \mathbf{v}_A),$$

where Λ^{-1} denotes a nonnegative parameter that we call the **resistivity**. We restrict attention to the case when the momentum transfer is nontrivial, so $\Lambda^{-1} > 0$.

Now consider the momentum balance laws for species A and B using these results. If the mixture moves sufficiently slowly, inertial effects are negligible:

$$\frac{D^A \mathbf{v}_A}{Dt} \simeq \mathbf{0} \simeq \frac{D^B \mathbf{v}_B}{Dt},$$

and the surviving terms yield

$$\operatorname{grad} p_A = \Lambda^{-1}(\mathbf{v}_B - \mathbf{v}_A),$$

$$\operatorname{grad} p_B = -\Lambda^{-1}(\mathbf{v}_B - \mathbf{v}_A).$$

By the chain rule, $\text{grad}\, p_\alpha = p'_\alpha(\rho_\alpha)\,\text{grad}\,\rho_\alpha$, and the fact that the mixture is so dilute that $\rho_A + \rho_B$ remains constant implies that $\text{grad}\,\rho_B = -\text{grad}\,\rho_A$. Hence,

$$\begin{aligned} p'_A(\rho_A)\,\text{grad}\,\rho_A &= \Lambda^{-1}(\boldsymbol{\nu}_B - \boldsymbol{\nu}_A), \\ -p'_B(\rho_B)\,\text{grad}\,\rho_A &= -\Lambda^{-1}(\boldsymbol{\nu}_B - \boldsymbol{\nu}_A). \end{aligned} \tag{7.4.13}$$

EXERCISE 200 *Use the identity $\rho_A\boldsymbol{\nu}_A + \rho_B\boldsymbol{\nu}_B = \mathbf{0}$, established in Exercise 178, to rewrite the system (7.4.13) as follows:*

$$\begin{aligned} \rho_B\, p'_A\,\text{grad}\,\rho_A &= -\Lambda^{-1}\overline{\rho}\,\boldsymbol{\nu}_A, \\ \rho_A\, p'_B\,\text{grad}\,\rho_A &= -\Lambda^{-1}\overline{\rho}\,\boldsymbol{\nu}_B. \end{aligned} \tag{7.4.14}$$

Subtracting the second of Equations (7.4.14) from the first and simplifying gives

$$(\rho_B p'_A + \rho_A p'_B)\,\text{grad}\,\rho_A = -\frac{\Lambda^{-1}\overline{\rho}^2}{\rho_B}\boldsymbol{\nu}_A. \tag{7.4.15}$$

EXERCISE 201 *Using the identity $\rho_\alpha = \overline{\rho}c_\alpha$, rewrite Equation (7.4.15) as follows:*

$$-\Lambda\frac{\rho_A\rho_B}{\overline{\rho}^2}(\rho_B p'_A + \rho_A p'_B)\,\text{grad}\,(\overline{\rho}c_A) = \overline{\rho}c_A\boldsymbol{\nu}_A. \tag{7.4.16}$$

Finally, identifying the **diffusion coefficient**

$$D = \Lambda\frac{\rho_A\rho_B}{\overline{\rho}^2}(\rho_B p'_A + \rho_A p'_B) > 0$$

and the diffusive flux of the solute

$$\mathbf{j}_A = \overline{\rho}\,c_A\,\boldsymbol{\nu}_A,$$

we see that the momentum balance equations reduce to Fick's law,

$$\mathbf{j}_A = -D\,\text{grad}\,(\overline{\rho}\,c_A).$$

7.4.4 Modeling Solute Transport

According to Fick's law, mass from the solute A diffuses from regions where the density of A is large to regions where it is small, at a rate proportional to the density gradient. If the mixture is a nonreacting solution sufficiently dilute that the mixture density is a constant independent of $(\mathbf{x}, t)$ and solute mass fraction c_A, then the mass balance for species α takes the form

$$\frac{\partial c_A}{\partial t} + \operatorname{div}(c_A \mathbf{v}) - \operatorname{div}(D \operatorname{grad} c_A) = 0, \tag{7.4.17}$$

known as the **advection-diffusion equation**.

The mathematical and numerical analysis of Equation (7.4.17) hinges on the relative contributions of advection and diffusion to the motion. To quantify this relationship, we cast the equation in dimensionless form. Consider a one-dimensional setting in which v and D are positive constants:

$$\frac{\partial c}{\partial t} + v\frac{\partial c}{\partial x} - D\frac{\partial^2 c}{\partial x^2} = 0. \tag{7.4.18}$$

In problems having a characteristic distance L, such as a channel length, we define dimensionless spatial and temporal variables as follows:

$$\xi = \frac{x}{L}, \qquad \tau = \frac{vt}{L}.$$

By the chain rule,

$$\frac{\partial}{\partial x} = \frac{1}{L}\frac{\partial}{\partial \xi}, \qquad \frac{\partial}{\partial t} = \frac{v}{L}\frac{\partial}{\partial \tau}.$$

Hence, in terms of the new variables, Equation (7.4.18) becomes

$$\frac{\partial c}{\partial \tau} + \frac{\partial c}{\partial \xi} - \mathrm{Pe}^{-1}\frac{\partial^2 c}{\partial \xi^2} = 0.$$

The dimensionless quantity $\mathrm{Pe} = vL/D$ is the **Péclet number**.

When Pe is comparable to or smaller than 1, the diffusion term exerts a significant effect, and Equation (7.4.18) has many qualitative features in common with the heat equation discussed in Section 5.1. In particular, in the large-diffusion case the equation tends to exert a damping or smoothing effect on initial solute distributions that exhibit sharp fronts or large gradients.

However, when Pe is much greater than 1, advection dominates the transport, and any initial mass fraction profile $c_0(x) = c(x, t_0)$ yields subsequent profiles $c(x, t)$ in which smoothing effects are much less pronounced. To get a feel for this case, consider as a limiting case the diffusion-free **advection equation**,

$$\frac{\partial c}{\partial t} + v\frac{\partial c}{\partial x} = 0. \tag{7.4.19}$$

In the (t, x) plane, along any sufficiently smooth arc $(t(s), x(s))$, the chain rule gives

$$\frac{dt}{ds}\frac{\partial c}{\partial t} + \frac{dx}{ds}\frac{\partial c}{\partial x} = \frac{dc}{ds}.$$

For the special choice of arcs for which

$$\frac{dt}{ds} = 1, \qquad \frac{dx}{ds} = v, \tag{7.4.20}$$

consistency with the original differential equation (7.4.19) requires

$$\frac{dc}{ds} = 0. \tag{7.4.21}$$

In short, c remains constant along arcs where $dx/dt = v$. For Equation (7.4.19), these arcs are lines along which $x - vt$ is constant, as sketched in Figure 7.6.

The ordinary differential equations (7.4.20) define **characteristic curves** for Equation (7.4.19). We call the differential equation (7.4.21) the **characteristic equation**; it describes how $c(x,t)$ changes along characteristic curves. This reasoning furnishes a solution technique called the **method of characteristics**, and it has implications for solutions to Equation (7.4.18) when Pe is large. Given an initial concentration profile $c(x,0) = c_0(x)$, the fact that c remains constant along arcs for which $x - vt$ is constant implies that $c(x,t) = c_0(x - vt)$. That is, the solution profile undergoes pure translation with speed v, preserving any sharp fronts or gradients in the initial profile. Resolving such features remains a significant challenge in numerical models of advection-dominated solute transport, especially in two and three space dimensions.

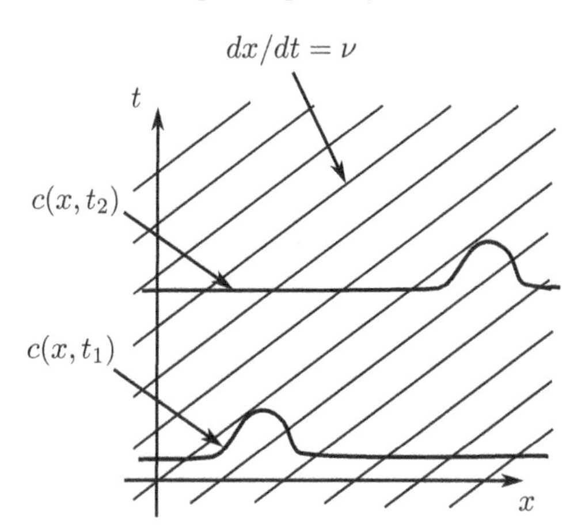

Figure 7.6. Characteristic curves for the advection equation showing how the equation propagates an initial concentration distribution by pure translation with velocity v.

EXERCISE 202 *Solve the initial-value problem*

$$\frac{\partial c}{\partial t} + v\frac{\partial c}{\partial x} = -q\,c, \qquad c(x,0) = c_0(x),$$

where the positive constant q represents the rate of a decay reaction.

7.4.5 Summary

Binary diffusion is the simplest case of a phenomenon that pervades engineering and environmental science. Under a restrictive set of assumptions about the diluteness of the mixture and the effects of the solute on the total density, one can apply constitutive theory to derive Fick's law, which is the most common model of diffusion. When we incorporate this model into the mass balance for the solute, we obtain the advection-diffusion equation. This partial differential equation serves as a prototype for balancing the competing effects of sharp-front preservation, associated with pure advection, and smoothing of the concentration profile, associated with diffusion.

APPENDIX A

GUIDE TO NOTATION

While much of the notation used in this text is standard, some notational conventions are peculiar to continuum mechanics, and some are peculiar to this text. This appendix provides a general guide to the notational conventions used throughout the book.

A.1 GENERAL CONVENTIONS

Scalars (elements of $\mathbb{R}$ or $\mathbb{C}$ or scalar-valued functions): lower- and upper-case Greek and Roman letters, such as a, b, f, α, β, M, N, Φ, and Ψ.

Vectors (elements of $\mathbb{E}$ or vector-valued functions): lower- and upper-case boldface Greek and Roman letters, such as $\mathbf{u}$, $\mathbf{v}$, $\boldsymbol{\mu}$, $\boldsymbol{\nu}$, $\mathbf{F}$, $\mathbf{X}$, $\boldsymbol{\Phi}$, and $\boldsymbol{\Psi}$.

Continuum Mechanics: The Birthplace of Mathematical Models, First Edition. M.B. Allen.

Tensors (elements of $\mathrm{L}(\mathbb{E})$ or tensor-valued functions): upper-case Roman or Greek sans-serif letters, such as S, T, and Λ.

Sets of points, either in the Euclidean point space $\mathbb{X}$ or in the associated space $\mathbb{E}$ of 3-vectors: upper-case script letters, such as $\mathcal{R}$ and $\mathcal{V}$.

Points in the Euclidean point space $\mathbb{X}$: upper-case Roman letters, such as P, Q, and X.

Spaces of tensors: upper-case upright Roman or Greek letters, such as $\mathrm{O}(\mathbb{E})$, or abbreviations using Roman letters, such as $\mathrm{Sym}(\mathbb{E})$ and $\mathrm{SPD}(\mathbb{E})$.

Special-purpose mappings: lower-case boldface italic letters, such as $\boldsymbol{d}$ and $\boldsymbol{f}$.

A.2 LETTERS RESERVED FOR DEDICATED USES

Many types of variables and functions appear so commonly that it is useful to reserve symbols exclusively to denote them. In other cases, a symbol may have different meanings in different parts of the exposition. In the latter cases, the meanings can be determined from the context. The following is a reasonably complete list of dedicated symbols used in this book.

$\mathbf{b}$	body force
B	left Cauchy–Green tensor
$\mathcal{B}$	body
C	right Cauchy–Green tensor
$\mathbb{C}$	complex number system
D	stretching tensor
$\mathbf{e}$	unit vector
E	infinitesimal strain tensor
E_E	Eulerian strain tensor
E_L	Lagrangian strain tensor
$\mathbb{E}$	three-dimensional Euclidean vector space
F	deformation gradient
H	displacement gradient
I	identity tensor
$\mathbf{j}$	diffusive flux
L	velocity gradient
$\mathbf{n}$	unit normal vector field

$\mathcal{P}$	part of a body
$\mathbf{q}$	heat flux
$\bar{\bar{\mathbf{q}}}$	total heat flux for a mixture
r	external heat supply
R	orthogonal part of a polar decomposition
$\mathbb{R}$	real number system
$\mathbb{R}^n$	vector space of ordered n-tuples of real numbers
S_κ	Second Piola–Kirchhoff stress
t	time
T	stress
T_κ	first Piola–Kirchhoff stress
$\mathbb{T}$	time interval in space-time
U	symmetric, positive definite part of a polar decomposition
V	symmetric, positive definite part of a polar decomposition
$\mathbb{V}, \mathbb{W}$	general Euclidean 3-spaces
$\mathbf{w}$	vorticity
W	spin tensor
X	material point
$\mathbf{x}$	spatial position
$\mathbf{X}$	referential position
$\mathbb{X}$	Euclidean point space
$\boldsymbol{\gamma}, \boldsymbol{\Gamma}$	parametrized curves
δ_{ij}	Kronecker symbol
Δ	Laplace operator
ε_{ijk}	Levi-Civita symbol
ε	specific internal energy
θ	temperature
η	specific entropy
$\boldsymbol{\chi}$	configuration or, as a function of t, the motion
$\boldsymbol{\chi}_\kappa$	deformation
$\boldsymbol{\kappa}, \boldsymbol{\lambda}$	reference configurations
ρ	mass density
ψ	specific Helmholtz free energy

A.3 SPECIAL SYMBOLS

The following special symbols—some nearly standard, others not—appear throughout the text.

$\mathsf{A}^\top$	transpose of the tensor A
$a \in A$	a is an element of the set A
$A \subset B$	$x \in A$ implies $x \in B$
$\mathcal{A} \subset \mathcal{B}$	$\mathcal{A}$ is a part of the body $\mathcal{B}$
$A \subsetneq B$	$A \subset B$ and $A \neq B$
$[a, b]$	interval containing real numbers $a \leqslant x \leqslant b$
$A \times B$	the set of all ordered pairs (x, y), where $x \in A$ and $y \in B$
$\mathsf{A} : \mathsf{B}$	double-dot product of two tensors, $\operatorname{tr}(\mathsf{A}^\top \mathsf{B})$
$\mathbf{a} \otimes \mathbf{b}$	dyadic product of two vectors
$[\mathbf{a}, \mathbf{b}, \mathbf{c}]$	scalar triple product of three vectors
curl	curl operator with respect to spatial coordinates
Curl	curl operator with respect to referential coordinates
D/Dt	material derivative operator
det	determinant
div	divergence operator with respect to spatial coordinates
Div	divergence operator with respect to referential coordinates
dv, dV	elements of volume integration
$d\boldsymbol{\sigma}$, $d\boldsymbol{\Sigma}$	elements of surface integration
$f\colon A \to B$	for every $x \in A$, $f(x) \in B$
$f\colon a \mapsto b$	$f(a) = b$, that is, f maps the element a to the element b.
$\mathrm{GL}(\mathbb{E})$	invertible tensors
grad	gradient operator with respect to spatial coordinates
Grad	gradient operator with respect to referential coordinates
$\mathrm{I}_A, \mathrm{II}_A, \mathrm{III}_A$	principal invariants of the tensor A
$\mathrm{L}(\mathbb{E})$	space of second-order tensors (linear transformations) on $\mathbb{E}$
$[\mathrm{L}^{\alpha_1}\mathrm{M}^{\alpha_2}\mathrm{T}^{\alpha_3}]$	physical dimensions, as powers of length, mass, and time
$\mathrm{O}(\mathbb{E})$	orthogonal tensors
$\mathrm{O}^+(\mathbb{E})$	orthogonal tensors with positive determinant
$\mathrm{Proj}_{\mathbf{x}}(\mathbf{y})$	orthogonal projection of $\mathbf{y}$ onto $\mathbf{x}$
$\mathrm{SL}(\mathbb{E})$	tensors having determinant 1
$\mathrm{SPD}(\mathbb{E})$	symmetric, positive definite tensors
$\mathrm{Sym}(\mathbb{E})$	symmetric tensors
tr	trace operator

APPENDIX B

VECTOR INTEGRAL THEOREMS

This appendix reviews several integral theorems of multidimensional calculus. The fundamental theorem of calculus lies at the heart of the first four, which relate integrals of derivatives to function values on boundaries. The following theorem about integrals along arcs in $\mathbb{E}$ furnishes the simplest example.

THEOREM B.0.1 (FUNDAMENTAL THEOREM OF CALCULUS FOR ARCS). *Suppose $\gamma\colon [a,b] \to \mathbb{E}$ is a continuously differentiable arc and $f\colon \mathbb{E} \to \mathbb{R}$ is a continuously differentiable scalar field defined on an open set containing the image $\gamma([a,b])$ of $[a,b]$ under γ. Then*

$$\int_{\gamma} \operatorname{grad} f \cdot d\mathbf{x} = f(\gamma(b)) - f(\gamma(a)).$$

This theorem follows directly from the definition of the arc integral and the fundamental theorem of calculus:

$$\int_\gamma \operatorname{grad} f \cdot d\mathbf{x} = \int_a^b \operatorname{grad} f(\gamma(s)) \cdot \gamma'(s)\, ds$$
$$= \int_a^b (f \circ \gamma)'(s)\, ds = f(\gamma(b)) - f(\gamma(a)).$$

One can generalize this argument to arcs γ satisfying fewer restrictions, for example, arcs that are piecewise continuously differentiable, and to functions satisfying less stringent continuity requirements.

Two extensions of this idea to higher dimensions—Stokes's theorem and the divergence theorem—involve integrals over surfaces and volumes in $\mathbb{E}$. These theorems require much more work than the theorem just proved. In particular, they require greater attention to the nature of the sets over which the integrals are taken. Nevertheless, they rest ultimately on the fundamental theorem of calculus.

In the same spirit as the statement of the theorem above, the exposition below does not attempt to state the theorems in their most general form. Instead, we identify useful sufficient conditions for the theorems to hold. For an extensive—and classic—discussion of the theorems, along with proofs, readers may refer to [31, Chapter IV].

B.1 STOKES'S THEOREM

For purposes of this text, a **parametrized smooth surface** is the image $\mathcal{S} \subset \mathbb{E}$ of a **parameter domain** $\Omega \subset \mathbb{R}^2$ under a twice continuously differentiable function $\varphi \colon \Omega \to \mathbb{E}$. This **parametrization** must satisfy the following conditions:

1. Ω is a nonempty, connected, open, bounded subset of the plane $\mathbb{R}^2$, together with its boundary $\partial\Omega$.

2. For any $\boldsymbol{\xi} = (\xi_1, \xi_2) \in \Omega$, the vector $(\partial\varphi/\partial\xi_1) \times (\partial\varphi/\partial\xi_2) \neq \mathbf{0}$, and the unit-length vector-valued function

$$\mathbf{n}(\boldsymbol{\xi}) = \frac{\dfrac{\partial\varphi}{\partial\xi_1}(\boldsymbol{\xi}) \times \dfrac{\partial\varphi}{\partial\xi_2}(\boldsymbol{\xi})}{\left\| \dfrac{\partial\varphi}{\partial\xi_1}(\boldsymbol{\xi}) \times \dfrac{\partial\varphi}{\partial\xi_2}(\boldsymbol{\xi}) \right\|}$$

defines a continuous unit normal vector field on $\mathcal{S}$ that points to one side of $\mathcal{S}$.

3. The boundary $\partial\mathcal{S}$ of the set $\mathcal{S}$ can be parametrized by a piecewise continuously differentiable function $\gamma \colon [a, b] \to \mathbb{E}$ that is **positively oriented**, that is, as the

parameter $s \in [a, b]$ increases, $\gamma(s)$ traces $\partial\mathcal{S}$ counterclockwise when viewed from the side of $\mathcal{S}$ toward which $\mathbf{n}$ points.

Figure B.1 illustrates. Section 2.4 reviews the definition of an integral over a surface of this type.

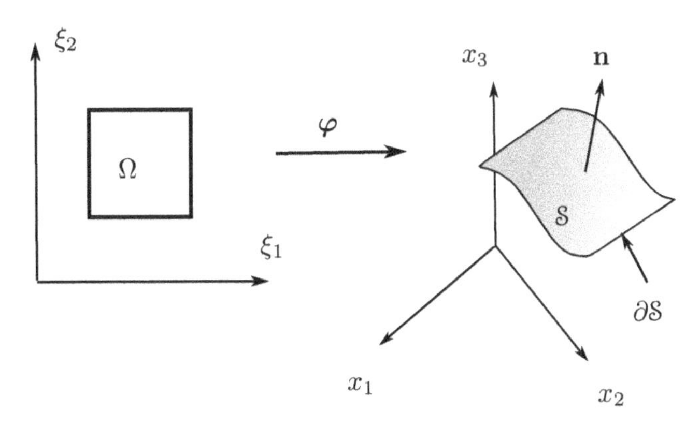

Figure B.1. A smooth surface $\mathcal{S} \in \mathbb{E}$ with unit normal vector field $\mathbf{n}$ and positively oriented boundary $\partial\mathcal{S}$.

Stokes's theorem relates the surface integral of a certain type of derivative over $\mathcal{S}$ to an integral over the arc that parametrizes $\partial\mathcal{S}$:

THEOREM B.1.1 (STOKES'S THEOREM). *Let $\mathcal{S}$ be a smooth surface in $\mathbb{E}$, and suppose $\mathbf{f}$ is a continuously differentiable vector field defined on an open set in $\mathbb{E}$ that contains $\mathcal{S}$. Then*

$$\int_{\partial\mathcal{S}} \operatorname{curl}\mathbf{f} \cdot \mathbf{n}\, d\sigma = \int_{\partial\mathcal{S}} \mathbf{f} \cdot d\mathbf{x}.$$

EXERCISE 203 *Some surfaces have no boundary. Consider, for example, the surface $\mathcal{S}$ that bounds a sphere in $\mathbb{E}$. Assuming that $\mathbf{f}$ is sufficiently smooth, give an argument leading to the conclusion that in this case*

$$\int_{\partial\mathcal{S}} \operatorname{curl}\mathbf{f} \cdot \mathbf{n}\, d\sigma = 0.$$

B.2 THE DIVERGENCE THEOREM

Let us call a region $\mathcal{R} \subset \mathbb{E}$ **normal** if it has the following properties:

1. $\mathcal{R}$ is nonempty, open, and bounded.

2. $\mathcal{R}$ is connected: for any two points $\mathbf{x}, \mathbf{y} \in \mathcal{R}$ there is a continuous arc $\boldsymbol{\gamma}\colon [0,1] \to \mathcal{R}$ for which $\boldsymbol{\gamma}(0) = \mathbf{x}$ and $\boldsymbol{\gamma}(1) = \mathbf{y}$.

3. The boundary $\partial\mathcal{R}$ is a piecewise continuously differentiable, orientable surface with unit normal vector field $\mathbf{n}$ pointing outward, as drawn in Figure B.2.

4. The set $\overline{\mathcal{R}} = \mathcal{R} \cup \partial\mathcal{R}$ is a union $\mathcal{R}_1 \cup \cdots \cup \mathcal{R}_n$ of finitely many subsets of $\mathbb{E}$, each bounded by graphs of continuously differentiable functions

$$\xi_1(x_1, x_2), \eta_1(x_1, x_2), \xi_2(x_2, x_3), \eta_2(x_2, x_3), \xi_3(x_3, x_1), \eta_3(x_3, x_1),$$

   in some Cartesian coordinate system on $\mathbb{E}$, and for any two such sets $\mathcal{R}_i \cap \mathcal{R}_j$ consists at most of shared bounding surfaces. Figure B.3 illustrates such a component region.

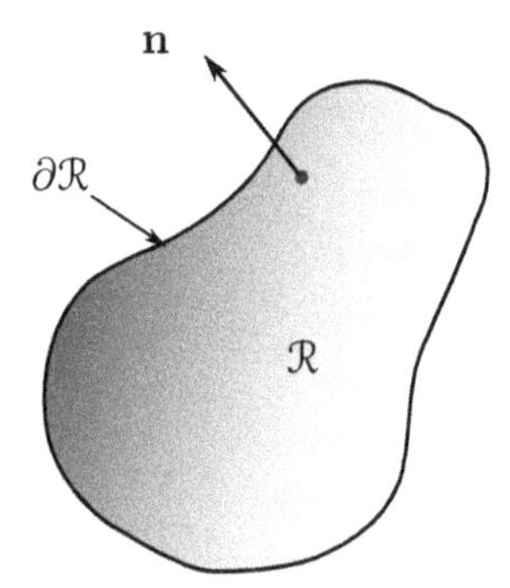

Figure B.2. A normal region $\mathcal{R} \subset \mathbb{E}$ with bounding surface $\partial\mathcal{R}$ and outward unit normal vector field $\mathbf{n}$.

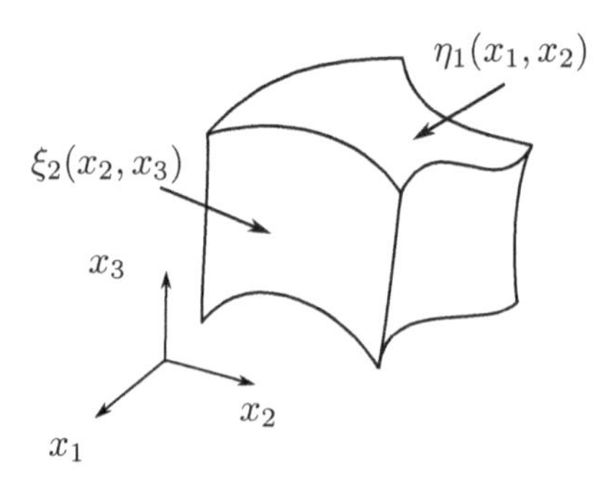

Figure B.3. A component region $\mathcal{R}_i$ of a normal region in $\mathbb{E}$ showing the graphs of continuously differentiable functions of two variables that form bounding surfaces.

The divergence theorem relates the volume integral of the divergence of a vector field $\mathbf{f}$ over a normal region $\mathcal{R}$ to the flux of $\mathbf{f}$ across the boundary $\partial\mathcal{R}$.

THEOREM B.2.1 (DIVERGENCE THEOREM). *Suppose $\mathcal{R} \subset \mathbb{E}$ is a normal region and $\mathbf{f} \colon \overline{\mathcal{R}} \to \mathbb{E}$ is a continuously differentiable vector field. Then*

$$\int_{\partial\mathcal{R}} \operatorname{div} \mathbf{f}\, dv = \int_{\partial\mathcal{R}} \mathbf{f} \cdot \mathbf{n}\, d\sigma.$$

EXERCISE 204 *Let $\mathcal{R}$ be a normal region in $\mathbb{E}$, and suppose that $f, g \colon \overline{\mathcal{R}} \to \mathbb{R}$ are twice continuously differentiable scalar functions. Use the divergence theorem to prove* **Green's identities**,

$$\int_{\mathcal{R}} \Delta f\, dv = \int_{\partial\mathcal{R}} \operatorname{grad} f \cdot \mathbf{n}\, d\sigma,$$

$$\int_{\mathcal{R}} (f\,\Delta g + \operatorname{grad} f \cdot \operatorname{grad} g)\, dv = \int_{\partial\mathcal{R}} f \operatorname{grad} g \cdot \mathbf{n}\, d\sigma \tag{B.2.1}$$

$$\int_{\mathcal{R}} (f\,\Delta g - g\,\Delta f)\, dv = \int_{\partial\mathcal{R}} (f \operatorname{grad} g - g \operatorname{grad} f) \cdot \mathbf{n}\, d\sigma.$$

Here $\Delta = \operatorname{div}(\operatorname{grad})$ is the Laplace operator.

The divergence theorem extends to second-order tensor fields, for which Equation (1.2.24) defines the divergence:

THEOREM B.2.2 (DIVERGENCE THEOREM FOR TENSOR FIELDS). *Suppose $\mathcal{R} \subset \mathbb{E}$ is a normal region and $\mathsf{A} \colon \overline{\mathcal{R}} \to \mathrm{L}(\mathbb{E})$ is a continuously differentiable tensor field. Then*

$$\int_{\partial\mathcal{R}} \mathsf{A}\mathbf{n}\, d\sigma = \int_{\mathcal{R}} \operatorname{div} \mathsf{A}\, dv.$$

PROOF: It suffices to show that the vectors on the left and right sides have the same inner product with an arbitrary vector. For any vector $\mathbf{b} \in \mathbb{E}$,

$$\begin{aligned}
\int_{\partial\mathcal{R}} \mathsf{A}\mathbf{n}\, d\sigma \cdot \mathbf{b} &= \int_{\partial\mathcal{R}} (\mathsf{A}\mathbf{n}) \cdot \mathbf{b}\, d\sigma \\
&= \int_{\partial\mathcal{R}} \mathsf{A}^{\top}\mathbf{b} \cdot \mathbf{n}\, d\sigma \\
&= \int_{\mathcal{R}} \operatorname{div}(\mathsf{A}^{\top}\mathbf{b})\, dv \\
&= \int_{\partial\mathcal{R}} (\operatorname{div} \mathsf{A}) \cdot \mathbf{b}\, dv = \int_{\partial\mathcal{R}} \operatorname{div} \mathsf{A}\, dv \cdot \mathbf{b}
\end{aligned}$$

by the definition of the transpose and the relationship (1.2.24) defining the divergence of a tensor field. ∎

B.3 THE CHANGE-OF-VARIABLES THEOREM

Several theorems concerning integrals under motion rest on the idea of changing variables between a body's current configuration and its reference configuration. The following theorem shows how integrals transform under such changes, provided the transformation is sufficiently well behaved.

THEOREM B.3.1 (CHANGE-OF-VARIABLES THEOREM). *Let $\mathcal{R}$ be a normal region in $\mathbb{E}$, and suppose the transformation $\mathbf{F}\colon \mathbb{E} \to \mathbb{E}$ has the following properties:*

1. *$\mathbf{F}$ is continuously differentiable on an open set containing $\overline{\mathcal{R}} = \mathcal{R} \cup \partial\mathcal{R}$;*
2. *$\mathbf{F}$ is one-to-one on $\overline{\mathcal{R}}$;*
3. *$\det \mathbf{F}'(\mathbf{x}) \neq 0$ for every $\mathbf{x} \in \overline{\mathcal{R}}$.*

Then for any bounded, continuous function $f\colon \mathbf{F}(\mathcal{R}) \to \mathbb{R}$,

$$\int_{\mathbf{F}(\mathcal{R})} f(\mathbf{y})\, dv = \int_{\mathcal{R}} f(\mathbf{F}(\mathbf{x}))\, |\det \mathbf{F}'(\mathbf{x})|\, dv.$$

Figure B.4 shows the regions involved. See [54, p. 467–482] for proof.

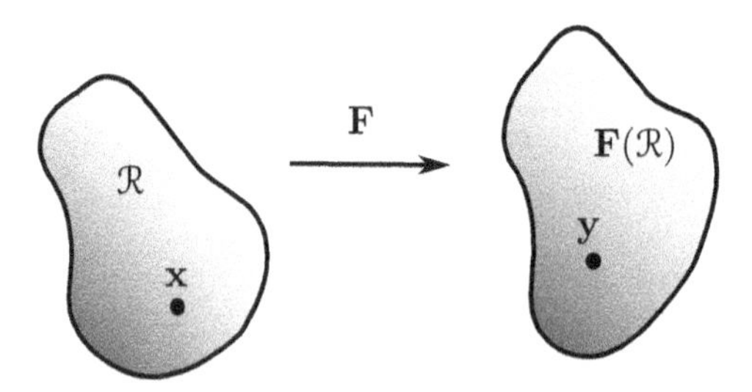

Figure B.4. A normal region $\mathcal{R} \subset \mathbb{E}$ and its image $\mathbf{F}(\mathcal{R})$ under a change of variables $\mathbf{F}\colon \mathbb{E} \to \mathbb{E}$.

APPENDIX C

HINTS AND SOLUTIONS TO EXERCISES

This appendix lists hints and sketches of solutions to the exercises that appear throughout the text.

CHAPTER 1

EXERCISE 1. Dot the identity $\mathbf{x} = \sum_{j=1}^{3} x_j \mathbf{e}_j$ with $\mathbf{e}_i$.

EXERCISE 2. Expand $\|\mathbf{x} + \mathbf{y}\|^2$ and $\|\mathbf{x} - \mathbf{y}\|^2$.

EXERCISE 3. $\mathbf{0} = \boldsymbol{d}(P, P) = \boldsymbol{d}(P, Q) + \boldsymbol{d}(Q, P)$.

EXERCISE 4. The inner product is linear.

EXERCISE 5. By orthonormality, $\hat{x}_i = \hat{\mathbf{e}}_i \cdot \sum_{j=1}^{3} \hat{x}_j \hat{\mathbf{e}}_j = \hat{\mathbf{e}}_i \cdot \sum_{j=1}^{3} x_j \mathbf{e}_j$.

Continuum Mechanics: The Birthplace of Mathematical Models, First Edition. M.B. Allen.

EXERCISE 6. Use the hint in Exercise 5. The quantity $\hat{\mathbf{e}}_i \cdot \mathbf{e}_j$ is the cosine of the angle between $\hat{\mathbf{e}}_i$ and $\mathbf{e}_j$.

EXERCISE 7. $x_j = \sum_{i=1}^{3} x_i \mathbf{e}_i \cdot \mathbf{e}_j = \sum_{l=1}^{3} \hat{x}_l \hat{\mathbf{e}}_l \cdot \mathbf{e}_j = \sum_{l=1}^{3} Q_{lj} \hat{x}_l = \sum_{l=1}^{3} Q_{jl}^{\top} \hat{x}_l$.

EXERCISE 8. It suffices to check linearity.

EXERCISE 9. For $\mathbf{a} \otimes \mathbf{b}$ it suffices to check linearity. $\mathrm{Proj}_{\mathbf{x}} = \mathbf{x} \otimes \mathbf{x}/\|\mathbf{x}\|^2$.

EXERCISE 10. Use the definition of A_{ij}.

EXERCISE 11. First show that, for any $\mathbf{x} \in \mathbb{E}$, $\mathsf{B}\mathbf{x} = (B_{kl}\mathbf{e}_k \otimes \mathbf{e}_l)x_m\mathbf{e}_m$. Then show that $\mathsf{AB}\mathbf{x} = (A_{jk}\mathbf{e}_j \otimes \mathbf{e}_k)B_{kl}x_k\mathbf{e}_k = A_{jk}B_{kl}x_l\mathbf{e}_j = (A_{jk}B_{kl}\mathbf{e}_j \otimes \mathbf{e}_l)(x_m\mathbf{e}_m)$.

EXERCISE 12. $\hat{y}_l = \mathbf{y} \cdot \hat{\mathbf{e}}_l = y_i \mathbf{e}_i \cdot \hat{\mathbf{e}}_l = A_{ij}x_jQ_{li} = Q_{li}A_{ij}Q_{jm}^{\top}\hat{x}_m = \hat{A}_{lm}\hat{x}_m$.

EXERCISE 13. Examine $\mathsf{A}(\mathbf{x} - \mathbf{y})$.

EXERCISE 14. Check the cases.

EXERCISE 15. If $\{\mathbf{v}_1, \mathbf{v}_2, \mathbf{v}_3\}$ is linearly dependent, then it is possible to write one of the vectors $\mathbf{v}_i$ as a linear combination of the other two.

EXERCISE 16. Part 1 follows from the observation that the entries in the triple product defining $\det \mathbf{a} \otimes \mathbf{b}$ are all multiples of $\mathbf{a}$. To show invertibility it suffices to show that A maps linearly independent triples onto linearly independent triples. To show $\det \mathsf{A} \neq 0$, use A^{-1} to show that $\det \mathsf{A} \det \mathsf{A}^{-1} = 1$. To show $\det(\mathsf{AB}) = \det \mathsf{A} \det \mathsf{B}$, apply AB to a triple product. With respect to any orthonormal basis, the triple product equals the determinant of the 3×3 matrix having the vectors' representations in $\mathbb{R}^3$ as columns or rows. The identity $\det(c\mathsf{A}) = c^3 \det \mathsf{A}$ follows from 3-linearity.

EXERCISE 17. Use properties established in Exercise 16.

EXERCISE 18. Use the definition of the matrix of I and the fact that $\mathbf{e}_i \cdot \mathbf{e}_j = \delta_{ij}$.

EXERCISE 19. Use the values of $\varepsilon_{ijk}\varepsilon_{lmk}$ established in Exercise 14.

EXERCISE 20. $[\mathbf{a} \times (\mathbf{b} \times \mathbf{c})]_l = \varepsilon_{lmi}a_m\varepsilon_{ijk}b_jc_k$. Use the $\varepsilon\delta$ identity.

EXERCISE 21. Examine the alternating 3-linear form $[\mathbf{a}, \mathbf{b}, \mathbf{c}]$ using the vector triple product identity (1.2.10).

EXERCISE 22. If Q is orthogonal, so is $\mathsf{Q}^{\top}$.

EXERCISE 23. If Q is proper orthogonal, so is $\mathsf{Q}^{\top}$.

EXERCISE 24. The fact that $\mathsf{Q}^{\top} = \mathsf{Q}^{-1}$ plays a central role.

EXERCISE 25. Observe that $[\mathsf{Q}\mathbf{a}, \mathsf{Q}\mathbf{b}, \mathsf{Q}\mathbf{c}] = \det \mathsf{Q}\,[\mathbf{a}, \mathbf{b}, \mathbf{c}]$ and $\mathsf{Q}(\mathbf{a} \times \mathbf{b}) \cdot \mathsf{Q}\mathbf{c} = \mathsf{Q}^{\top}\mathsf{Q}(\mathbf{a} \times \mathbf{b}) \cdot \mathbf{c}$.

EXERCISE 26. Because Q is an isometry, it suffices to show that $\mathsf{Q}\mathbf{p}_1$, $\mathsf{Q}\mathbf{p}_2$, and $\mathsf{Q}\mathbf{p}_3$ are mutually orthogonal. But $\mathsf{Q}\mathbf{p}_i \cdot \mathsf{Q}\mathbf{p}_j = \mathsf{Q}^\top\mathsf{Q}\mathbf{p}_i \cdot \mathbf{p}_j = \mathbf{p}_i \cdot \mathbf{p}_j$.

EXERCISE 27. Given any $\mathbf{y} = y_i\mathbf{p}_i \in \mathbb{E}$, $\mathsf{Q}^\top\mathsf{Q}\mathbf{y} \cdot \mathbf{p}_j = \mathsf{Q}\mathbf{y} \cdot \mathsf{Q}\mathbf{p}_j = y_i\mathsf{Q}\mathbf{p}_i \cdot \mathsf{Q}\mathbf{p}_j = y_i\delta_{ij} = y_j = \mathsf{I}\mathbf{y} \cdot \mathbf{p}_j$. Therefore, $\mathsf{Q}^\top\mathsf{Q}$ has the same action as I on every basis vector $\mathbf{p}_j$. Similarly, $\mathsf{Q}\mathsf{Q}^\top$ has the same action as I. It follows that $\mathsf{Q}^\top\mathsf{Q} = \mathsf{I} = \mathsf{Q}\mathsf{Q}^\top$.

EXERCISE 28. With the summation convention, $\operatorname{tr} A_{ij}\mathbf{e}_i \otimes \mathbf{e}_j = A_{ij}\mathbf{e}_i \cdot \mathbf{e}_j = A_{ii}$.

EXERCISE 29. From Exercise 11, $\operatorname{tr}(\mathsf{AB}) = A_{jk}B_{kl}\operatorname{tr}(\mathbf{e}_j \otimes \mathbf{e}_l) = A_{jk}B_{kl}\delta_{jl} = A_{jk}B_{kj} = B_{jk}A_{kj} = \operatorname{tr}(\mathsf{BA})$.

EXERCISE 30. (a) Using the orthonormal basis, $\operatorname{tr}(\mathsf{A}^\top\mathsf{B}) = A_{ij}B_{ik}\mathbf{e}_j \otimes \mathbf{e}_k = A_{ij}B_{ik}\delta_{jk}$. (b) Symmetry follows from the observation that $\operatorname{tr}(\mathsf{C}^\top) = \operatorname{tr}(\mathsf{C})$; additivity and homogeneity follow from the linearity of the trace; positive definiteness follows from part (a).

EXERCISE 31. $p_{\mathsf{A}}(\lambda) = \det(\lambda\mathsf{I} - \mathsf{A})$.

EXERCISE 32. With respect to $\{\mathbf{e}_1, \mathbf{e}_2, \mathbf{e}_3\}$, the matrix entries of $\lambda\mathsf{I} - \mathsf{A}$ have the form $\lambda\delta_{ij} - a_{ij}$.

EXERCISE 33. By definition, $[\operatorname{div}(\mathbf{a} \otimes \mathbf{b})] \cdot \mathbf{e}_i = \operatorname{div}[(\mathbf{b} \otimes \mathbf{a})\mathbf{e}_i] = \operatorname{div}[\mathbf{b}(\mathbf{a} \cdot \mathbf{e}_i) = \operatorname{div}(a_i\mathbf{b})$.

EXERCISE 34. Expand $\mathbf{a}$ and A in terms of $\{\mathbf{e}_1, \mathbf{e}_2, \mathbf{e}_3\}$ and use the definitions.

CHAPTER 2

EXERCISE 35. Show that both $(\boldsymbol{\kappa} \circ \boldsymbol{\chi}^{-1}) \circ (\boldsymbol{\chi} \circ \boldsymbol{\kappa}^{-1})$ and $(\boldsymbol{\chi} \circ \boldsymbol{\kappa}^{-1}) \circ (\boldsymbol{\kappa} \circ \boldsymbol{\chi}^{-1})$ equal the identity function.

EXERCISE 36. There are off-diagonal terms of the form $\mathbf{e}_2 \otimes \mathbf{e}_j$ and $\mathbf{e}_3 \otimes \mathbf{e}_j$, respectively. Each choice of j yields a family of glide planes.

EXERCISE 37. When does multiplication of the matrix representations commute?

EXERCISE 38. The composition $\boldsymbol{\kappa}_2 \circ \boldsymbol{\kappa}_1^{-1}$ maps one reference configuration $\boldsymbol{\kappa}_1$ onto another $\boldsymbol{\kappa}_2$.

EXERCISE 39. The results are $\mathbf{c}'(t)$, $\mathsf{Q}'(t)\mathbf{X}$, $\alpha'(t)(\mathbf{e}_1 \otimes \mathbf{e}_2)\,\mathbf{X}$.

EXERCISE 40. For every $\mathbf{b} \in \mathbb{E}$, $(\operatorname{Grad}\mathbf{x})^\top\mathbf{b} = \operatorname{Grad}(x_ib_i) = (\partial x_i/\partial X_j)\hat{\mathbf{e}}_jb_i = [(\partial x_i/\partial X_j)\hat{\mathbf{e}}_j \otimes \mathbf{e}_i]b_k\mathbf{e}_k = [(\partial x_i/\partial X_j)\hat{\mathbf{e}}_j \otimes \mathbf{e}_i]\mathbf{b} = [(\partial x_i/\partial X_j)\mathbf{e}_i \otimes \hat{\mathbf{e}}_j]^\top\mathbf{b}$.

EXERCISE 41. The differentiations are straightforward.

EXERCISE 42. Expand the matrix representation for $\det(\mathsf{I} + \mathsf{H})$ and neglect terms that are quadratic and cubic in partial derivatives $\partial u_i/\partial X_j$.

EXERCISE 43. Recall the definition of div $\mathbf{v}$.

EXERCISE 44. The streamlines at $t = 0$ are straight lines with slope X_2/X_1.

EXERCISE 45. The pathlines at $t_0 = 1$ are $\mathbf{x} = (2X_1, X_2 + 1, X_3)^\top$, so $x_1 = X_1(X_2 + 1) + X_1 - X_1X_2 = X_1x_2 + (X_1 - X_1X_2)$. Substituting for $(X_1, X_2, X_3)^\top$ in the equation $\overline{\mathbf{x}} = (X_1(1 + t_0), X_2 + t_0, X_3)^\top = (1, 1, 0)^\top$ and solving yield $x_1 = 2/(3 - x_2)$, $x_3 = 0$ for the streaklines.

EXERCISE 46. The deformation is a diffeomorphism.

EXERCISE 47. Set $\Psi = 1$. Changes in the volume of $\mathcal{R}(t)$ arise from net movement of the boundary.

EXERCISE 48. Impose volume invariance on the change-of-variables formula.

CHAPTER 3

EXERCISE 49. For any eigenvector $\mathbf{p}$, $\mathbf{p} \cdot \mathsf{A}\mathbf{p} = \lambda \|\mathbf{a}\|^2$.

EXERCISE 50. Let λ_1, λ_2, and λ_3 be the eigenvalues of A with associated orthonormal eigenvectors $\mathbf{p}_1$, $\mathbf{p}_2$, and $\mathbf{p}_3$, respectively. Then $\lambda_i \mathbf{p}_i = \mathsf{A}\mathbf{p}_i = \mathsf{QBQ}^\top \mathbf{p}_i$, so $\lambda_i \mathsf{Q}^\top \mathbf{p}_i = \mathsf{BQ}^\top \mathbf{p}_i$. Also, $\|\mathsf{Q}^\top \mathbf{p}_i\| = \|\mathbf{p}_i\| = 1$, so $\mathsf{Q}\mathbf{p}_i \neq \mathbf{0}$. Therefore, $\mathsf{Q}^\top \mathbf{p}_i$ is an eigenvector of B with associated eigenvalue λ_i.

EXERCISE 51. Let $\{\mathbf{p}_1, \mathbf{p}_2, \mathbf{p}_3\}$ be an orthonormal basis of eigenvalues of A corresponding to the eigenvalues $\lambda_1, \lambda_2, \lambda_3$, respectively, and let $\{\mathbf{q}_1, \mathbf{q}_2, \mathbf{q}_3\}$ be a corresponding orthonormal basis of eigenvalues of B. Define $\mathsf{Q}\mathbf{q}_i = \mathbf{p}_i$, and show that $\mathsf{QB}\sum_{i=1}^3 y_i \mathbf{q}_i = \mathsf{AQ}\sum_{i=1}^3 y_i \mathbf{q}_i$ for all vectors $\sum_{i=1}^3 y_i \mathbf{q}_i$.

EXERCISE 52. Any $\mathbf{v} \in \mathbb{E}$ has an expansion $\mathbf{v} = \sum_{i=1}^3 v_i \mathbf{p}_i$.

EXERCISE 53. Use the matrix representation for A with respect to $\{\mathbf{p}_1, \mathbf{p}_2, \mathbf{p}_3\}$.

EXERCISE 54. Examine matrix representations of A, A^2, and A^3 with respect to the orthonormal basis of eigenvectors.

EXERCISE 55. The actions of the two expressions on $\sum_{i=1}^3 v_i \mathbf{p}_i$ are identical.

EXERCISE 56. Apply $\sqrt{\mathsf{A}}$ to $\sum_{i=1}^3 v_i \mathbf{p}_i$.

EXERCISE 57. Check closure.

EXERCISE 58. Symmetry is straightforward. Apply the definition of $\mathsf{A}^\top$ to $\mathbf{a} \cdot \mathsf{A}^\top \mathsf{A}\mathbf{a}$.

EXERCISE 59. If $\mathsf{U}\mathbf{p} = \lambda \mathbf{p}$, then $\mathsf{RU}\mathbf{p} = \lambda \mathsf{R}\mathbf{p}$.

EXERCISE 60. $\det \mathsf{F} = \det \mathsf{R} \det \mathsf{U}$.

EXERCISE 61. Compute $\det \mathsf{F}$.

EXERCISE 62. Recall $\mathsf{H} = \mathsf{F} - \mathsf{I}$.

EXERCISE 63. B and C are symmetric and have positive eigenvalues.

EXERCISE 64. Show that $R_\kappa^2 = \|\mathbf{X} - \mathbf{X}_0\|^2 = (\mathbf{x} - \mathbf{x}_0) \cdot \mathsf{B}^{-1}(\mathbf{x} - \mathbf{x}_0)$ and $R_t^2 = (\mathbf{X} - \mathbf{X}_0) \cdot \mathsf{C}(\mathbf{X} - \mathbf{X}_0)$.

EXERCISE 65. Formally multiply by F the proposed expression for F^{-1}.

EXERCISE 66. Differentiate $\mathsf{F} = \mathsf{RU} = \mathsf{VR}$.

EXERCISE 67. Examine $(\mathsf{QQ}^\top)'$.

EXERCISE 68. It suffices to show that the identity holds in Cartesian coordinates.

EXERCISE 69. Differentiate the identity $\mathsf{E}_L = \frac{1}{2}(\mathsf{F}^\top\mathsf{F} - \mathsf{I})$ and use the identity (2.2.6).

EXERCISE 70. Start with the results of Exercise 66.

EXERCISE 71. The differentiation is straightforward.

EXERCISE 72. The differentiation is straightforward.

EXERCISE 73. All diagonal entries $\mathbf{p}_i \cdot \mathsf{S}\mathbf{p}_i$ vanish, as do all entries involving $\mathbf{p}_1$.

EXERCISE 74. By definition $\mathbf{s}$ is an eigenvector of S. Check that $\mathbf{s}$ gives the action of S using representations with respect to the basis $\{\mathbf{e}_1, \mathbf{e}_2, \mathbf{e}_3\}$.

EXERCISE 75. Show that $(\mathbf{a} \times \mathbf{b}) \times \mathbf{c} = (\mathbf{b} \otimes \mathbf{a} - \mathbf{a} \otimes \mathbf{b})\mathbf{c} = \mathbf{b}(\mathbf{a} \cdot \mathbf{c}) - \mathbf{c}(\mathbf{a} \cdot \mathbf{b})$ by expanding the vectors with respect to the standard orthonormal basis.

EXERCISE 76. Examine the entries in $\mathbf{w} = (W_{32}, W_{31}, W_{12})^\top$.

EXERCISE 77. $\mathbf{w} = \frac{1}{2}(0, 0, 2 - \alpha)/\|\mathbf{x}\|^\alpha$.

EXERCISE 78. Use the equality of mixed partial derivatives.

EXERCISE 79. The gradient of a continuously differentiable scalar field is everywhere orthogonal to its level sets. For an isochoric motion, $\operatorname{div} \mathbf{v} = 0$.

EXERCISE 80. The differential operators grad and $\partial/\partial t$ commute.

EXERCISE 81. Show that $\mathbf{w} \cdot \mathbf{n} = 0$ at time t_0 implies that the circulation around the boundary of the surface vanishes at t_0.

EXERCISE 82. Use the fact that $\mathrm{O}(\mathbb{E})$ is a group.

EXERCISE 83. Recall $\mathsf{Q}^\top = \mathsf{Q}^{-1}$.

EXERCISE 84. Substitute for $\mathbf{x}$ in Equation (3.5.2).

EXERCISE 85. Differentiate Equation (3.5.4).

EXERCISE 86. $\mathbf{X} = \mathsf{Q}^\top(t_0)[\hat{\mathbf{X}} - \mathbf{c}(t_0)] + \mathbf{x}_0$.

EXERCISE 87. $\det \mathsf{F}(\mathbf{X}, t) = \det \mathsf{Q}(t) \det \mathsf{F}(\mathbf{X}, t) \det \mathsf{Q}^\top(t_0)$, and because of continuity $\det \mathsf{Q}(t)$ does not change sign.

EXERCISE 88. By the chain rule, $\partial \hat{A}_{ij}/\partial \hat{x}_j = (\partial \hat{A}_{ij}/\partial x_k)(\partial x_k/\partial \hat{x}_j) = (\partial \hat{A}_{ij}/\partial x_k)Q^\top_{kj}$ $= (\partial \hat{A}_{ij}/\partial x_k)Q_{jk} = Q_{im}(\partial A_{mn}/\partial x_k)Q^\top_{nj}Q_{jk} = Q_{im}(\partial A_{mn}/\partial x_k)\delta_{nk}$. The last expression is $Q_{im}(\partial A_{mk}/\partial x_k)$.

EXERCISE 89. $\text{grad}_{\hat{\mathbf{x}}} \hat{f} = \text{grad}_{\mathbf{x}} f$ by objectivity of gradients, and $\|\text{grad}_{\hat{\mathbf{x}}} \hat{f}\| = \|\mathsf{Q}\, \text{grad}_{\mathbf{x}} f\|$ $= \|\text{grad}_{\mathbf{x}} f\|$ since Q is an isometry.

EXERCISE 90. Compute the skew part of $\hat{\mathsf{L}}$.

EXERCISE 91. $\hat{\mathsf{B}} = \hat{\mathsf{F}}\hat{\mathsf{F}}^\top$.

EXERCISE 92. Establish the identities for $\hat{\mathsf{R}}$, $\hat{\mathsf{U}}$, and $\hat{\mathsf{V}}$.

CHAPTER 4

EXERCISE 93. Since the integrand is continuous, if it is nonzero at any point, then it is nonzero in a neighborhood $\mathcal{P}$ of the point.

EXERCISE 94. The spatial form of the mass balance shows that the field is solenoidal. Use the referential form to show that $\det \mathsf{F} = 1$.

EXERCISE 95. Work in Cartesian coordinates.

EXERCISE 96. $\text{mass}\,(\mathcal{P})$ is constant.

EXERCISE 97. Apply the Reynolds transport theorem. Some resulting terms vanish by the spatial mass balance.

EXERCISE 98. As $h \to 0$ in the momentum balance, the volume integrals tend to 0 faster than the surface integral, which tends to a sum of integrals over the planar segments.

EXERCISE 99. The base of face j is related to that of face $\mathbf{n}$ by the cosine of the angle between $\mathbf{e}_j$ and $\mathbf{n}$.

EXERCISE 100. Use the facts that $\mathbf{t}$ is objective, $\mathbf{t} = \mathsf{T}\mathbf{n}$ for all unit vectors $\mathbf{n}$, $\hat{\mathbf{t}} = \hat{\mathsf{T}}\hat{\mathbf{n}}$ for all unit vectors $\hat{\mathbf{n}}$, and $\mathbf{n}$ is objective since it is the unit normal vector to some surface. For all unit vectors $\hat{\mathbf{n}}$, $\hat{\mathsf{T}}\hat{\mathbf{n}} = \hat{\mathbf{t}} = \mathsf{Q}\mathbf{t} = \mathsf{Q}\mathsf{T}\mathsf{Q}^\top\hat{\mathbf{n}}$, so $\hat{\mathsf{T}} = \mathsf{Q}\mathsf{T}\mathsf{Q}^\top$.

EXERCISE 101. The inertial term (mass $\times$ acceleration) appears on the left.

EXERCISE 102. $D^2\mathbf{x}/Dt^2 = \mathsf{Q}^\top D^2\hat{\mathbf{x}}/Dt^2$; $\rho = \hat{\rho}$; $\text{div}_{\mathbf{x}}\, \mathsf{T} = \mathsf{Q}^\top \text{div}_{\hat{\mathbf{x}}}\, \hat{\mathsf{T}}$; $\mathbf{b} = \mathsf{Q}^\top\hat{\mathbf{b}}$. Substitute these expressions into Equation (4.2.8) and multiply on the left by Q.

EXERCISE 103. Apply the tools used for the momentum balance to express all terms as volume integrals, then localize.

EXERCISE 104. Use Cartesian coordinates to prove the identity for $D(\mathbf{x} \times \mathbf{v})/Dt$.

EXERCISE 105. Use the definition of T_κ to show that both sides equal $(\det \mathsf{F})\mathsf{T}$.

EXERCISE 106. By definition, $\hat{\mathsf{S}}_\kappa = (\det \hat{\mathsf{F}})\hat{\mathsf{F}}^{-1}\hat{\mathsf{T}}\hat{\mathsf{F}}^{-\top}$. Since $\det \mathsf{F} > 0$, $\det \hat{\mathsf{F}} = \det \mathsf{F}$, and $\hat{\mathsf{F}} = \mathsf{Q}\mathsf{F}\mathsf{Q}_0^\top$. Use this latter relationship to refer $\hat{\mathsf{F}}^{-1}$ and $\hat{\mathsf{F}}^{-\top}$ to the unhatted frame. The result of the exercise shows that S_κ is objective.

EXERCISE 107. Each term in Equation (4.4.3) has dimension $[\text{energy}/\text{time}] = [\mathrm{ML}^2\mathrm{T}^{-3}]$, or power.

EXERCISE 108. It suffices to work in Cartesian coordinates.

EXERCISE 109. Change variables to an integral over $\boldsymbol{\kappa}(\mathcal{P})$, using techniques developed in Section 2.4; substitute for D using Exercise 69; and observe that

$$(\det \mathsf{F}) T_{ij} \frac{DF_{ik}}{Dt} F_{kj}^{-1} = (\det \mathsf{F}) T_{ij} F_{jk}^{-\top} \frac{DF_{ik}}{Dt}.$$

EXERCISE 110. Reason as for Exercise 109 and observe that

$$(\det \mathsf{F}) T_{ij} F_{il}^{-\top} \frac{DE_{Llm}}{Dt} F_{mj}^{-1} = (\det \mathsf{F}) F_{li}^{-1} T_{ij} F_{jm}^{-\top} \frac{DE_{Llm}}{Dt}.$$

EXERCISE 111. The first inequality follows from the quotient rule.

EXERCISE 112. The substitution is straightforward.

EXERCISE 113. Integrate (4.5.5) over $\boldsymbol{\chi}(\mathcal{P}, t)$, then change variables to the reference configuration $\boldsymbol{\kappa}(\mathcal{P})$. Use Exercise 110 for the integral of $\mathrm{tr}(\mathsf{TD}) = \mathsf{T} : \mathsf{D}$, then apply the localization principle.

EXERCISE 114. Follow the derivation given, for example, in Section 4.2.

EXERCISE 115. Take spatial moments of the quantities used for momentum.

EXERCISE 116. Expand the expressions using the definition of $[\![\cdot]\!]$ and rearrange terms.

EXERCISE 117. Split $\mathcal{R}$ into two regions, separated by $\mathcal{S}$, then apply the divergence theorem to each.

EXERCISE 118. Replace $\rho\Psi$ by 1 in the derivation of the jump condition and observe that $[\![\mathbf{v}_S \cdot \mathbf{n}]\!] = 0$.

EXERCISE 119. At a vortex sheet, $[\![\mathbf{v}_S \cdot \mathbf{n}]\!] = 0$ and $[\![\mathbf{v} \cdot \mathbf{n}]\!] = 0$.

EXERCISE 120. Using the identity $[\![\rho V]\!] = 0$ twice yields

$$\begin{aligned}[\![\rho V v_n]\!] &= \rho^+ V^+ v_n^+ - \rho^+ V^+ v_n^- + \rho^+ V^+ v_n^- - \rho^- V^- v_n^- \\ &= \rho^+ V^+ [\![v_n]\!] = -\rho^+ V^+ [\![V]\!] \\ &= -\rho^+ (V^+)^2 + \rho^- (V^-)^2 - \rho^- (V^-)^2 + \rho^+ V^+ V^- = -[\![\rho V^2]\!].\end{aligned}$$

CHAPTER 5

EXERCISE 121. Examine D under rigid motion.

EXERCISE 122. Substitute and clear the nonzero factor $\exp(i\beta x)$ from the terms that result.

EXERCISE 123. Use the chain rule to compute $\partial/\partial\tau$.

EXERCISE 124. Straightfoward substitution works.

EXERCISE 125. The previous exercise shows that $|u - u_p|$ can be as small as we wish at $\tau = 0$ but still grow without bound for $\tau > 0$.

EXERCISE 126. In a steady state, $\partial\theta/\partial\tau = 0$.

EXERCISE 127. This is an exercise in the chain rule.

EXERCISE 128. Consider low-degree polynomials in the Cartesian coordinates x_1, x_2, and x_3.

EXERCISE 129. Integrate $-\Delta u$ from 0 to b and use the boundary conditions to evaluate the term arising from $\partial^2 u/\partial x_3^2$.

EXERCISE 130. Use the chain rule.

EXERCISE 131. The difference function $w = u_1 - u_2$ satisfies $\Delta w = 0$ on $\mathcal{R}$ and $w = 0$ on $\mathcal{R}$, so by the Green identity $\operatorname{grad} w = \mathbf{0}$ on $\mathcal{R}$.

EXERCISE 132. If u is a solution, so is $u + C$ for any constant C.

EXERCISE 133. Use the divergence theorem.

EXERCISE 134. Consider $-u$.

EXERCISE 135. Since $u(\mathbf{x})_M > 0$, $\mathbf{x}_M$ must lie in $\mathcal{R}$, not on the boundary. Hence, $-\Delta u(\mathbf{x}_M) \geqslant 0$, a contradiction.

EXERCISE 136. Since $-\Delta(u - M) = -\Delta u \leqslant 0$ in $\mathcal{R}$ and $u - M \leqslant 0$ on $\partial\mathcal{R}$, $u - M \leqslant 0$ in $\mathcal{R}$. A similar argument shows that $u - m \geqslant 0$ in $\mathcal{R}$.

EXERCISE 137. A harmonic function is both subharmonic and superharmonic.

EXERCISE 138. The gradient is orthogonal to the level sets.

EXERCISE 139. Use the mean value theorem for integrals.

EXERCISE 140. A rotation of coordinates corresponds to a matrix multiplication $\mathsf{Q}^\top\mathsf{T}\mathsf{Q}$, where Q is the matrix of direction cosines.

EXERCISE 141. $-\frac{1}{2}\operatorname{grad}\|\boldsymbol{\omega}\times\mathbf{x}\|^2 = -\frac{1}{2}\operatorname{grad}[\|\boldsymbol{\omega}\|^2\|\mathbf{x}\|^2 - (\boldsymbol{\omega}\cdot\mathbf{x})^2] = -\|\boldsymbol{\omega}\|^2\mathbf{x} + (\boldsymbol{\omega}\cdot\mathbf{x})\operatorname{grad}(\boldsymbol{\omega}\cdot\mathbf{x}) = -\|\boldsymbol{\omega}\|^2\mathbf{x} + (\boldsymbol{\omega}\cdot\mathbf{x})\boldsymbol{\omega} = -\mathbf{x}(\boldsymbol{\omega}\cdot\boldsymbol{\omega}) + \boldsymbol{\omega}(\boldsymbol{\omega}\cdot\mathbf{x}) = -\boldsymbol{\omega}\times(\mathbf{x}\times\boldsymbol{\omega}) = \boldsymbol{\omega}\times(\boldsymbol{\omega}\times\mathbf{x})$.

EXERCISE 142. Use the chain rule to establish the derivative identities, then substitute for $\partial/\partial t$ and $\operatorname{grad}_{\mathbf{x}}$ in the momentum balance.

EXERCISE 143. Using $\|\boldsymbol{\omega}\| \simeq 1.16\times 10^{-5}\mathrm{s}^{-1}$ and converting the velocity units to m/s yield $\mathrm{Ro}\simeq 2.4$ for the jet stream and $\mathrm{Ro}\simeq 5.8\times 10^3$ for the bathtub drain.

EXERCISE 144. $|\operatorname{div}\mathbf{v}| = |\mathsf{T} : \operatorname{grad}\mathbf{v}|$.

EXERCISE 145. Subtract ρ_0 times the divergence of the momentum balance from the time derivative of the mass balance.

EXERCISE 146. Use the chain rule.

EXERCISE 147. The inner product $\mathbf{x}\cdot\mathbf{k} = \|\mathbf{x}\|\cos\theta$ gives the (signed) length of the component of $\mathbf{x}$ lying parallel to $\mathbf{x}$.

EXERCISE 148. The functions $f(x\pm ct)$ have the form $F(y(x,t))$. Use the chain rule.

EXERCISE 149. Work in Cartesian coordinates.

EXERCISE 150. Pressure has dimension [force/area].

EXERCISE 151. Use the chain rule.

EXERCISE 152. Substitute from Exercise 151.

EXERCISE 153. The sphere travels with constant velocity when net body force equals drag.

EXERCISE 154. Assuming that $t_{\text{done}} = CK^{\alpha_1}R^{\alpha_2}u_0^{\alpha_3}u_{\text{done}}^{\alpha_4}$, we find that t_{done} is proportional to R^2. But mass $= 4\pi R^3\rho/3$.

EXERCISE 155. Observe that $\operatorname{tr}\mathsf{T}_\kappa = (3\lambda + 2\mu)\operatorname{tr}\mathsf{E}$.

EXERCISE 156. Substitute for λ and μ using E and ν.

EXERCISE 157. Work in Cartesian coordinates.

EXERCISE 158. The substitution is straightforward.

EXERCISE 159. Use the spectral decomposition of A.

EXERCISE 160. From Equation (5.4.10), Φ must satisfy $\Delta\Phi = \operatorname{Div}\mathbf{u}$; thus Φ is a solution to the Poisson equation, discussed at the end of Section 5.1. Since $\mathbf{u} - \operatorname{Grad}\Phi$ is solenoidal, it is the curl of a vector field, as discussed in Section 4.1.

EXERCISE 161. Use the identities $\text{Div}(\text{Curl}\,\boldsymbol{\Psi}) = \mathbf{0}$ and $\text{Curl}(\text{Grad}\,\Phi) = \mathbf{0}$.

CHAPTER 6

EXERCISE 162. By Exercise 67, $\mathsf{Q}'\mathsf{Q}^\top\mathsf{Q} = -\mathsf{Q}(\mathsf{Q}')^\top\mathsf{Q}$.

EXERCISE 163. Consider an observer transformation involving R.

EXERCISE 164. Examine the mass of a typical part $\mathcal{P}$ of the body in the two reference configurations. Use the change-of-variables theorem.

EXERCISE 166. Recall $\mathsf{C} = \mathsf{U}^2$.

EXERCISE 165. Closure follows from the identity $(\det \mathsf{AB}) = \det \mathsf{A} \det \mathsf{B}$.

EXERCISE 167. Because $\mathsf{J} \in \Gamma_\kappa(X)$,

$$\begin{aligned} G_\lambda(\mathsf{C}_\lambda, \ldots) &= G_\kappa(\mathsf{C}_\kappa, \ldots) \\ &= G_\kappa(\mathsf{J}^\top\mathsf{C}_\kappa\mathsf{J}, \ldots) \\ &= G_\lambda(\mathsf{K}^{-\top}\mathsf{J}^\top\mathsf{C}_\kappa\mathsf{J}\mathsf{K}^{-1}, \ldots) \\ &= G_\lambda(\mathsf{K}^{-\top}\mathsf{J}^\top\mathsf{K}^\top\mathsf{C}_\lambda\mathsf{K}\mathsf{J}\mathsf{K}^{-1}, \ldots) \\ &= G_\lambda((\mathsf{K}\mathsf{J}\mathsf{K}^{-1})^\top\mathsf{C}_\lambda(\mathsf{K}\mathsf{J}\mathsf{K}^{-1}), \ldots). \end{aligned}$$

These steps make repeated use of the statements (1) and (2).

EXERCISE 168. (1) $\|\mathsf{Q}\mathbf{a}\| = \sqrt{(\mathsf{Q}\mathbf{a}) \cdot (\mathsf{Q}\mathbf{a})} = \sqrt{(\mathsf{Q}^\top\mathsf{Q}\mathbf{a}) \cdot \mathbf{a}} = \sqrt{\mathbf{a} \cdot \mathbf{a}}$.
(2) $\det(\mathsf{QAQ}^\top) = \det \mathsf{Q} \det \mathsf{A} \det \mathsf{Q}^\top = \det(\mathsf{QQ}^\top)\det \mathsf{A} = \det \mathsf{A}$.
(3) Show $\det(\mathsf{Q}\,a\mathsf{I}\,\mathsf{Q}^\top) = \det(a\mathsf{I})$.
(4) The induction step is $(\mathsf{QAQ}^\top)^n = (\mathsf{QAQ}^\top)^{n-1}\mathsf{QAQ} = \mathsf{QA}^{n-1}\mathsf{Q}^\top\mathsf{QAQ}^\top = \mathsf{QA}^n\mathsf{Q}^\top$.

EXERCISE 169. Use $\mathsf{Q} = -\mathsf{I}$ in the condition $\varphi(\mathsf{QAQ}^\top) = \mathsf{Q}\varphi(\mathsf{A})$.

EXERCISE 170. For any $\mathsf{Q} \in \mathrm{O}(\mathbb{E})$, the eigenvalues of $\mathsf{QAQ}^\top$ are identical to those of A, so $\varphi(\mathsf{QAQ}^\top) = f(\lambda_1, \lambda_2, \lambda_3) = \varphi(\mathsf{A})$.

EXERCISE 171. Work with the basis $\{\mathbf{e}_1, \mathbf{e}_2, \mathbf{e}_3\}$ to establish closure. Find an element of $\mathrm{FC}_{21;31}(\mathbb{E})$ that is not orthogonal.

EXERCISE 172. Multiply $\mathbf{e}_1$ by a generic element of $\mathrm{FC}_{21;31}(\mathbb{E})$.

EXERCISE 173. Focus on the zero structures of the matrices with respect to $\{\mathbf{e}_1, \mathbf{e}_2, \mathbf{e}_3\}$.

EXERCISE 174. Recall that $\det(c\mathsf{A}) = c^3 \det \mathsf{A}$.

EXERCISE 175. Use the mass balance.

EXERCISE 176. The inequality holds only if $a_0 \geqslant 0$, $a_1 = 0$, $a_2 \geqslant 0$.

EXERCISE 177. Since $\operatorname{tr}\mathsf{E} = E_{11} + E_{22} + E_{33}$,

$$\frac{\partial}{\partial E_{ij}} \operatorname{tr}\mathsf{E} = \frac{\partial}{\partial E_{ij}}(E_{11} + E_{22} + E_{33}) = \mathsf{I},$$

$$\frac{\partial}{\partial E_{ij}} (\operatorname{tr}\mathsf{E})^2 = 2(\operatorname{tr}\mathsf{E})\frac{\partial}{\partial E_{ij}} \operatorname{tr}\mathsf{E} = 2\,(\operatorname{tr}\mathsf{E})\mathsf{I},$$

$$\frac{\partial}{\partial E_{ij}} (\operatorname{tr}\mathsf{E}^2) = \frac{\partial}{\partial E_{ij}}(E_{kl}E_{kl}) = 2\mathsf{E}.$$

CHAPTER 7

EXERCISE 178. Use the definition of $\mathbf{v}$.

EXERCISE 179. Use the definition of the diffusion velocity.

EXERCISE 180. Use the definitions of ρ and Ψ.

EXERCISE 181. Apply the constituent mass balance.

EXERCISE 182. Use the definition of c_α.

EXERCISE 183. Use the definitions of ρ and $\mathbf{v}$ for mixtures.

EXERCISE 184. The results follow directly from the definitions of the quantities involved.

EXERCISE 185. Use the definition (1.2.24) of the divergence of a tensor-valued function.

EXERCISE 186. Use the constituent mass balance together with the identity $\operatorname{div}\mathbf{v}_\alpha = \mathsf{I} : \mathsf{D}_\alpha$.

EXERCISE 187. Use the definitions of ρ and $\mathbf{v}$ for mixtures.

EXERCISE 188. The contraction of a symmetric tensor with a skew tensor vanishes.

EXERCISE 189. Apply $\mathbf{v}_\alpha \cdot$ to the constituent momentum balance.

EXERCISE 190. $\hat{\mathbf{v}}_F - \hat{\mathbf{v}}_S = \mathsf{Q}D^F\boldsymbol{\chi}_{F,\kappa}/Dt + \mathsf{Q}'\boldsymbol{\chi}_{F,\kappa} + \mathbf{c}' - \mathsf{Q}D^S\boldsymbol{\chi}_{S,\kappa}/Dt - \mathsf{Q}'\boldsymbol{\chi}_{S,\kappa} - \mathbf{c}'$. At time t, $\boldsymbol{\chi}_{F,\kappa}(\boldsymbol{\chi}_{F,\kappa}^{-1}(\mathbf{x}, t)) = \mathbf{x} = \boldsymbol{\chi}_{S,\kappa}(\boldsymbol{\chi}_{S,\kappa}^{-1}(\mathbf{x}, t))$. It follows that $\hat{\mathbf{v}}_F - \hat{\mathbf{v}}_S = \mathsf{Q}(\mathbf{v}_F - \mathbf{v}_S)$.

EXERCISE 191. Expand $(\partial/\partial t)(\phi_F\gamma_F) + \operatorname{div}(\phi_F\gamma_F\mathbf{v}_F) = 0$; observe $\operatorname{div}\mathbf{v}_F = \mathsf{I} : \mathsf{D}_F$.

EXERCISE 192. Use arguments similar to those given in Section 6.5.

EXERCISE 193. The entropy inequality holds for any velocity field $\mathbf{v}$.

EXERCISE 194. The Stokes drag has the form $C\mu R\mathbf{v}_\infty$.

EXERCISE 195. The momentum exchange involves a term that accounts for variations in volume fraction.

EXERCISE 196. Use vector identities to show that $\operatorname{curl}\mathbf{E} = -\operatorname{grad}\gamma^{-1} \times \operatorname{grad} p$.

EXERCISE 197. The outward unit normal vector at the wellbore is $\mathbf{n} = -\text{grad}\, f/\|\text{grad}\, f\|$, where $f(\mathbf{x}) = 0$ is any equation for the well's circular boundary.

EXERCISE 198. Use the results of Exercise 130.

EXERCISE 199. The boundary condition at $\|\mathbf{x}\| = R$ yields $C_2 = C_1 \log R + \Phi_R$. Using Exercise 197, show that the boundary condition at $\|\mathbf{x}\| = r_w$ yields $C_1 = Q\mu/(2\pi\gamma bK)$.

EXERCISE 200. $\boldsymbol{\nu}_A = -(\rho_B/\rho_A)\boldsymbol{\nu}_B$, and $\boldsymbol{\nu}_B = -(\rho_A/\rho_B)\boldsymbol{\nu}_A$. Substitute, and multiply the first equation by ρ_B and the second by ρ_A.

EXERCISE 201. Tactically substitute $c_\alpha = \rho_\alpha/\overline{\rho}$.

EXERCISE 202. Solve the ordinary differential equation that holds along characteristic curves to get $c(x,t) = c_0(x - vt)\exp(-qt)$.

APPENDIX B

EXERCISE 203. Split $\mathcal{S}$ into two surfaces, each of which has a boundary.

EXERCISE 204. Apply the divergence theorem to $\text{grad}\, f$, $f\, \text{grad}\, g$, and $f\, \text{grad}\, g - g\, \text{grad}\, f$.

REFERENCES

1. J.E. Adkins, "Non-linear diffusion I. Diffusion and flow of mixtures of fluids," *Philosophical Transactions of the Royal Society of London. Series A, Mathematical and Physical Sciences 255:*1064, 1963, 607–633.
2. R. Aris, *Vectors, Tensors, and the Basic Equations of Fluid Mechanics*, Dover, New York, 1962.
3. R.J. Atkin and R.E. Craine, "Continuum theories of mixtures: basic theory and historical development," *Quarterly Journal of Mechanics and Applied Mathematics 29*, 209–244, 1976.
4. G.I. Barenblatt, *Scaling, Self-Similarity, and Intermediate Asymptotics*, Cambridge University Press, New York, 1996.
5. A. Bedford and D.S. Drumheller, *Elastic Wave Propagation*, John Wiley & Sons, Inc., Chichester, U.K., 1994.
6. R.M. Bowen, "Theory of mixtures," in *Continuum Physics* vol. III, ed. by A.C. Eringen, Academic Press, New York, 1–127, 1976.
7. R.M. Bowen, "Incompressible porous media models by use of the theory of mixtures," *International Journal of Engineering Science 18*, 1129–1148, 1980.
8. R.M. Bowen, "Diffusion models implied by the theory of mixtures," Appendix 5A in C.A. Truesdell, *Rational Thermodynamics*, Springer-Verlag, New York, 237–263, 1984.

Continuum Mechanics: The Birthplace of Mathematical Models, First Edition. M.B. Allen.

9. R.M. Bowen, *Introduction to Continuum Mechanics for Engineers* (revised edition), Dover, New York, NY, 2010.

10. R.M. Bowen, *Porous Elasticity: Lectures on the Elasticity of Porous Materials as an Application of the Theory of Mixtures*, 2010, retrieved July 2012 from http://hdl.handle.net/1969.1/91297.

11. J. Casey, "On the derivation of jump conditions in continuum mechanics," *International Journal of Structural Changes in Solids 3*:2, 61–84, 2011.

12. P. Chadwick, *Continuum Mechanics: Concise Theory and Problems*, Dover, New York, 1976.

13. A.J. Chorin and J.E. Marsden, *A Mathematical Introduction to Fluid Mechanics*, 3rd ed., Springer, New York, 2000.

14. B.D. Coleman and W. Noll, "The thermodynamics of elastic materials with heat conduction and viscosity," *Archive for Rational Mechanics and Analysis 13*, 168–178, 1963.

15. J.H. Cushman, L.S. Bennethum, and B.X. Xu, "A primer on upscaling tools for porous media," *Advances in Water Resources 25*, 1043–1067, 2002.

16. H. Darcy, *Les fontaines publiques de la ville de Dijon,* Dalmont, Paris, 1856.

17. A.C. Eringen, ed., *Continuum Physics: Volume II—Continuum Mechanics of Single-Substance Bodies,* Academic Press, New York, 1975.

18. A.C. Eringen, *Mechanics of Continua*, Krieger, Huntington, NY, 1980.

19. A.C. Eringen and J.D. Ingram, "A continuum theory of chemically reacting media—I," *International Journal of Engineering Science 3*, 197–212, 1965.

20. M.E. Gurtin, *An Introduction to Continuum Mechanics,* Academic Press, San Diego, 2003.

21. S.M. Hassanizadeh and W.G. Gray, "General conservation equations for multiphase systems: 3. Constitutive theory for porous media," *Advances in Water Resources 3*, 25–40, 1980.

22. S.M. Hassanizadeh and W.G. Gray, "Mechanics and thermodynamics of multiphase flow in porous media including interphase boundaries," *Advances in Water Resources 13*, 169–186, 1990.

23. S. Heinz, *Mathematical Modeling*, Springer-Verlag, Berlin, 2011.

24. I. Herrera and G.F. Pinder, *Mathematical Modeling in Science and Engineering: An Axiomatic Approach*, John Wiley & Sons, Inc. Hoboken, NJ, 2012.

25. M.H. Holmes, *Introduction to the Foundations of Applied Mathematics*, Springer, Dordrecht, Netherlands, 2009.

26. M.K. Hubbert, "Darcy's law and the field equations of the flow of underground fluids," *Transactions of the AIME 207*, 222–239, 1956.

27. K. Hutter and K. Jöhnk, *Continuum Methods of Physical Modeling: Continuum Mechanics, Dimensional Analysis, Turbulence*, Springer-Verlag, Berlin, 2004.

28. J.D. Ingram and A.C. Eringen, "A continuum theory of chemically reacting media—II: constitutive equations of reacting fluid mixtures," *International Journal of Engineering Science 5*, 289–322, 1967.

29. W. Jaunzemis, *Continuum Mechanics*, Macmillan, New York, 1967.

30. F. John, *Partial Differential Equations*, 4th ed., Springer-Verlag, New York, 1991.

31. O.D. Kellog, *Foundations of Potential Theory*, Dover, New York, 1953.

32. L.D. Landau and E.M Lifschitz, *Fluid Mechanics* (volume 6 of *A Course of Theoretical Physics*), 2nd ed., Butterworth-Heinemann, Oxford, 1987.

33. P. Lax, *Linear Algebra and its Applications*, 2nd ed., John Wiley & Sons, Inc., Hoboken, NJ, 2007.

34. I.-S. Liu and I. Müller, "Thermodynamics of mixtures of fluids," Appendix 5B in C.A. Truesdell, *Rational Thermodynamics*, Springer-Verlag, New York, 1984, 264–285.

35. I-S. Liu, *Continuum Mechanics*, Springer, Berlin, 2002.

36. J.D. Logan, *Applied Mathematics*, 4th ed., John Wiley & Sons, Inc., Hoboken, NJ, 2013.

37. L.E. Malvern, *Introduction to the Mechanics of a Continuous Medium*, Prentice-Hall, Englewood Cliffs, NJ, 1969.

38. J.E. Marsden and T.J.R. Hughes, *Mathematical Foundations of Elasticity*, Prentice-Hall, Englewood Cliffs, NJ, 1982.

39. J.E. Marsden and A.J. Tromba, *Vector Calculus*, 6th ed., Freeman, New York, 2011.

40. M. Muskat, R.D. Wyckof, H.G. Botset, and M.W. Meres, "Flow of gas-liquid mixtures through sands," *Transactions of AIME 123*, 1937, 69-96.

41. W. Noll, "A mathematical theory of the mechanical behavior of continuous media," *Archive for Rational Mechanics and Analysis 2*, 197–226, 1958.

42. J.T. Oden, *An Introduction to Mathematical Modeling: A Course in Mechanics*, John Wiley & Sons, Inc., Hoboken, NJ, 2011.

43. P.A.C. Raats, "Applications of the theory of mixtures in soil physics," Appendix 5D in C.A. Truesdell, *Rational Thermodynamics*, Springer-Verlag, New York, 326–343, 1984.

44. J.G. Simmonds, *A Brief on Tensor Analysis,* 2nd ed., Springer, New York, 1994.

45. A.J.M. Spencer, *Continuum Mechanics*, Longman Group, Essex, England, 1980.

46. L.A. Segel, *Mathematics Applied to Continuum Mechanics*, Macmillan, New York, 1977.

47. G. Thiem, *Hydrologische methoden*, J.M. Gebhardt, Leipzig, 1906.

48. D.K. Todd, *Groundwater Hydrology,* 2nd ed., John Wiley & Sons, Inc., Hoboken, NJ, 1980.

49. C.A. Truesdell and R. Toupin, "The classical field theories," in *Handbuch der Physik* vol. 3, ed. by S. Flügge, Springer, Berlin, 1960.

50. C.A. Truesdell, *Essays in the History of Mechanics*, Springer-Verlag, New York, 1968.

51. C.A. Truesdell, *Rational Thermodynamics*, Springer-Verlag, New York, 1984.

52. C.A. Truesdell, *The Elements of Continuum Mechanics*, Springer-Verlag, New York, 1985.

53. C.A. Truesdell, *A First Course in Rational Continuum Mechanics,* 2nd ed., Academic Press, San Diego, 1991.

54. R.E. Williamson, R.H. Crowell, and H.F. Trotter, *Calculus of Vector Functions,* 3rd ed., Prentice-Hall, Englewood Cliffs, NJ, 1972.

55. K. Wilmanski, *Continuum Thermodynamics: Part I: Foundations*, World Scientific, Singapore, 2008.

INDEX

Continuum Mechanics: The Birthplace of Mathematical Models, First Edition. M.B. Allen.